AF357839

Internal Environment and Thermal Performance of Buildings

Internal Environment and Thermal Performance of Buildings

Guest Editors

Katarzyna Ratajczak
Łukasz Amanowicz

Basel • Beijing • Wuhan • Barcelona • Belgrade • Novi Sad • Cluj • Manchester

Guest Editors

Katarzyna Ratajczak
Environmental Engineering
and Energy
Poznan University of Technology
Poznan
Poland

Łukasz Amanowicz
Environmental Engineering
and Energy
Poznan University of Technology
Poznan
Poland

Editorial Office
MDPI AG
Grosspeteranlage 5
4052 Basel, Switzerland

This is a reprint of the Special Issue, published open access by the journal *Energies* (ISSN 1996-1073), freely accessible at: www.mdpi.com/journal/energies/special_issues/internal_environment.

For citation purposes, cite each article independently as indicated on the article page online and using the guide below:

Lastname, A.A.; Lastname, B.B. Article Title. *Journal Name* **Year**, *Volume Number*, Page Range.

ISBN 978-3-7258-2666-7 (Hbk)
ISBN 978-3-7258-2665-0 (PDF)
https://doi.org/10.3390/books978-3-7258-2665-0

Contents

About the Editors

Katarzyna Ratajczak

Katarzyna Ratajczak is an associate professor at Poznań University of Technology, Poland in the Institute of Environmental Engineering, Faculty of Environmental Engineering and Energy. She deals with energy efficiency of HVAC systems in swimming pools and other types of buildings and energy performance of buildings in various certification systems. She is interested in indoor air quality and its impact on people's comfort and functioning and the possibilities of its improvement, especially in existing buildings by using decentralized systems. She uses different research techniques such as laboratory experiments, in situ measurements and simulations. For her research, which was the subject of her doctoral dissertation, she received scholarships funded by the European Union, the Poznań City Award for outstanding dissertations, and the National Swimming Pool Foundation Fellowship (USA) twice.

Łukasz Amanowicz

Łukasz Amanowicz is an associate professor at Poznań University of Technology, Poland, in the Institute of Environmental Engineering, Faculty of Environmental Engineering and Energy. He is a member of PZITS—Polish Union of Sanitary Engineers and Technicians. He is a supervisor of the Environmental Engineering Scientific Circle at PUT. He deals with energy efficiency in construction, energy efficiency of HVAC systems and their cooperation with renewable energy sources. He is focused on the issues of energy efficiency of ventilation systems, including decentralized systems, as well as multi-pipe earth-to-air heat exchangers and energy certification of buildings. In his scientific work, he uses experimental research techniques, in situ measurements, and CFD simulations.

He is interested in the use of artificial intelligence and machine learning algorithms for forecasting purposes, e.g., energy demand or behavior and the habits of building users. He is active in publishing in international journals. He is also the co-author of the REHVA report "Replacement of Gas Boilers with Heat Pumps, District Heating, and Hybrid Solutions: Report on the Shift Away from Natural Gas in Buildings" and a member of the REHVA task force working on a guidebook about "Application and maintenance of heat pumps in existing buildings for HVAC professionals". He supports the international academic community by systematically reviewing articles for prestigious journals such as *Energy Conversion and Management, Renewable Energy, Energies, Energy and Buildings, Thermal Science and Engineering Progress, Sustainability*, and many others. He is also involved as a guest editor in special issues.

Preface

The thermal performance of buildings is at the center of global interest. There are many related topics that significantly influence the energy performance of a building: building structure/envelope, heating, ventilation and air conditioning (HVAC) systems, heat and cold sources, controls, users, and much more.

Lowering the energy demand in a building should not be performed at the expense of worsening the quality of the building's internal environment. There is a need to combine analyses related to ensuring adequate air quality in buildings while paying attention to the energy performance. It is also important to pay attention to the quality of the outside air. In some regions with high pollution and cold climates, the supply of outdoor air cannot ensure the required indoor air quality.

We believe that each analysis carried out within this Special Issue contributes to the development of knowledge in the field of buildings thermal performance with respect to the internal air quality. The Special Issue includes review papers on the latest research results and trends in the following areas:

- Reducing energy consumption by ventilation systems;

- Thermal comfort in the context of ventilation and ensuring appropriate living conditions in various types of building.

Among the research articles, you will find works on the following topics:

- Economic evaluation of thermal modernization of production space;

- Subjective perception of overall comfort, indoor air quality, and humidity—questionnaire survey;

- Radiator adjustment in multi-family buildings;

- Air quality in bedrooms equipped with different types of ventilation systems;

- The concept of a small modular house that achieves low energy consumption with an acceptable level of thermal comfort and good indoor air quality;

- Statistical analysis of energy efficiency indicators for multifamily buildings;

- Long-term studies on thermal comfort in lightweight passive houses;

- Assessments of ANN algorithms for the concentration prediction of indoor air pollutants concentration in child daycare centers.

We wish you a fruitful reading!

Katarzyna Ratajczak and Łukasz Amanowicz
Guest Editors

Review

Recent Advancements in Ventilation Systems Used to Decrease Energy Consumption in Buildings—Literature Review

Łukasz Amanowicz [1,*], Katarzyna Ratajczak [1] and Edyta Dudkiewicz [2]

[1] Institute of Environmental Engineering and Building Installations, Poznan University of Technology, ul. Berdychowo 4, 61-131 Poznan, Poland
[2] Faculty of Environmental Engineering, Wrocław University of Science and Technology, Wybrzeże Wyspiańskiego 27, 50-370 Wroclaw, Poland
* Correspondence: lukasz.amanowicz@put.poznan.pl

Abstract: The need for healthy indoor conditions, the energy crisis, and environmental concerns make building ventilation systems very important today. The elements of ventilation systems to reduce energy intensity are constantly the subject of much scientific research. The most recent articles published in the last three years are analyzed in this paper. Publications focused on the topic of reducing energy consumption in ventilation systems were selected and divided into five key research areas: (1) the aspect of the airtightness of buildings and its importance for the energy consumption, (2) the methods and effects of implementing the concept of demand-controlled ventilation in buildings with different functions, (3) the possibilities of the technical application of decentralized ventilation systems, (4) the use of earth-to-air heat exchangers, (5) the efficiency of exchangers in exhaust air heat-recovery systems. The multitude of innovative technologies and rapid technological advances are reflected in articles that appear constantly and prompt a constant updating of knowledge. This review constitutes a relevant contribution to recognizing current advancements in ventilation systems and may be helpful to many scientists in the field.

Keywords: ventilation; energy efficiency; airtightness; DCV; thermal performance of buildings; heat recovery; earth-to-air heat exchangers; decentralized ventilation; solar chimney

Citation: Amanowicz, Ł.; Ratajczak, K.; Dudkiewicz, E. Recent Advancements in Ventilation Systems Used to Decrease Energy Consumption in Buildings— Literature Review. *Energies* **2023**, *16*, 1853. https://doi.org/10.3390/en16041853

Academic Editor: Xi Chen

Received: 28 January 2023
Revised: 10 February 2023
Accepted: 10 February 2023
Published: 13 February 2023

1. Introduction

1.1. Ventilation System Requirements

The thermal and humidity conditions in the room must meet the user's expectations and meet legal and technological requirements. Technical equipment systems, in particular HVAC (heating, ventilation, and air conditioning), are responsible for shaping the microclimate in buildings. To maintain appropriate thermal conditions, ventilation and heating systems are used. If, in addition, the humidity of the air in the room is also regulated, this is called a full air-conditioning system.

Although the design of ventilation in energy-efficient buildings is a key issue, searching the Scopus database for the phrase "designing of ventilation systems" leads to finding many papers from not only the last 3, but even the last 5 years, that present useful tips for designing ventilation systems in such buildings. Therefore, information from articles in local (Polish) journals [1,2] was used to outline the general requirements. Their results present requirements that seem to be universal and, in principle, coincide with the content presented in the article [3], which describes ventilation systems that meet the requirements of "ASHRAE Standard 62.2-2004":

- the building envelope should be airtight to achieve energy efficiency in the building,
- ventilation should be controlled: demand-controlled ventilation (DCV) systems should be used,
- the selection of the ventilation airflow should be based on hygienic or technological reasons,

- the heat from the exhaust air should be possible to be recovered,
- Renewable energy sources (RES), such as, e.g., earth-to-air heat exchangers, heat-pumps, etc. are recommended to be used,
- decentralized systems are recommended.

The general division of the systems used to shape the proper parameters of the air in the room is shown in Figure 1 [4].

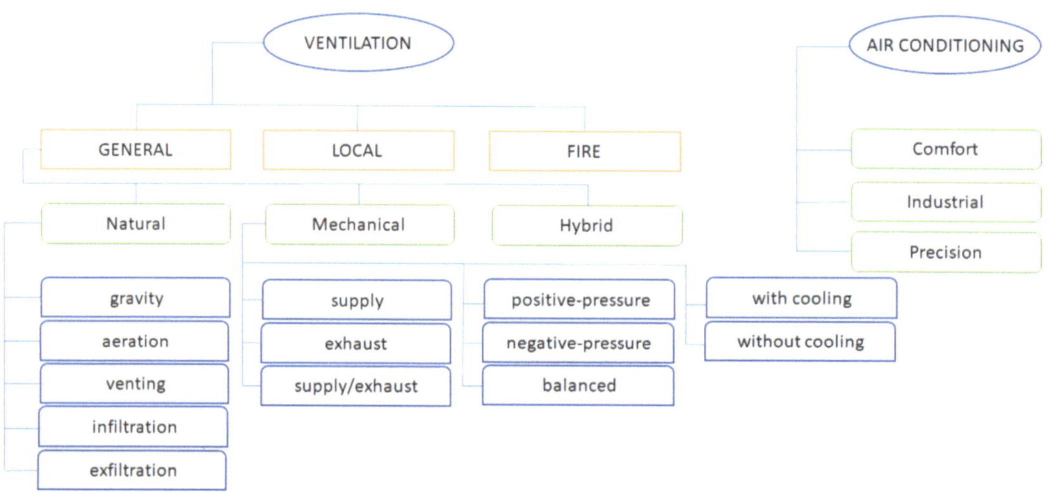

Figure 1. General division of ventilation and air-conditioning systems [4].

The main purpose of ventilation is to exchange air and ensure proper indoor air quality [5]. An appropriate ventilation intensity allows the removal of pathogens from the indoor air and at the same time increase the air quality. The use of appropriate filtration and possibly air sterilization contributes to high indoor air quality [6]. A higher intensity of ventilation also reduces the probability of transmission of airborne infections [7–9]. Although much research and analysis on ventilation has already been carried out, the multitude of innovative technologies and rapid progress in civilization is reflected in articles that appear constantly and encourage a constant updating of knowledge. This is due to the new challenges for ventilation systems, especially in energy-efficient and intelligent buildings [10].

The energy demand for ventilation results from the need to heat the supply air stream and the stream that infiltrates from the outside through leaks in the building envelope. It has been proven that the heat losses resulting from these air flows depend in particular on the type of ventilation system and the airtightness of its partitions [11,12]. In existing buildings, which generate more than 40% of the global energy demand, modernization is necessary, including the introduction of modern solutions that aim to reduce energy demand while ensuring the quality of the internal environment [5,13,14]. A reply to the frequently increasing share of heat demand for ventilation in the total heat demand of a building is the utilisation of advancements in ventilation systems such as demand-controlled ventilation as a ventilation-control strategy, elements of decentralized ventilation, earth-to-air heat exchangers, and heat recovery from exhaust air.

1.2. The Aim of the Paper

The aim of this article is to present the results of a review of the latest research from the last three years on ventilation systems that help reduce energy consumption in buildings. This objective is achieved with a literature review of five key issues concerning the airtightness of buildings, demand-controlled ventilation strategies, decentralized ventilation, ground heat exchangers, and heat recovery from the air removed from rooms.

1.3. Literature Review—Materials and Methods

As part of the literature review, scientific articles that were included in the Scopus and Web of Science databases were analyzed, assuming as search criteria: research article, full text, and conference papers published in 2020–2022. There were also publications from other years that had to be mentioned as key in a given topic. Selected keywords, appropriate for individual chapters, allowed us to select 98 articles for the narrative review. Based on the best knowledge of the authors, the most effective technologies with the highest potential were selected. To make it easier for the reader to recognize the issues discussed in individual chapters, the publications are summarized in five tables. They also provide the number of citations for these publications and the keywords provided by their authors.

2. Airtightness of the Building's Envelope as a Basic Requirement for Decreasing Energy Consumption

The energy efficiency of a building could be improved by improving its airtightness, measurement of which by different methods can have different errors [15]. In energy-efficient buildings with good thermal insulation, airtightness, expressed in air changes per hour by the n_{50} factor, is particularly important for annual energy demand. An increase in airtightness results in significant energy savings [16]. They are greater the better the thermal insulation of the building. Minimum thermal insulation requirements for buildings' partitions force the use of solutions with heat transfer coefficients appropriate for the country, and those in Poland are much stricter than the requirements in other European countries [17]. The reduction in energy consumption in well-insulated buildings is due to a lower infiltration airflow, which, coming through leaks as uncontrolled air, nullifies assumptions about energy efficiency. The percentage differences are greater the lower the energy demand for other building purposes. Therefore, energy-efficient buildings should be airtight, as can be demonstrated using 52 buildings in cities in the United States using a simplified calculator available online [16]. The apparent effect of increased airtightness on the building's thermal performance supports the decision-making process involved in renovating a building envelope with an emphasis on taking care of its airtightness.

Analysis [2] confirms that the energy demand of a building depends, among other factors, on the airtightness of the building. In the case of a building with a low value of the average heat transfer coefficient U (Variant II: U = 0.2 W/(m^2·K)), airtightness affects the energy demand to a greater extent compared to variant I, i.e., a building with a much worse thermal insulation parameter of partitions (U = 0.4 W/(m^2·K)). This is presented in Figures 2 and 3, which show, respectively, calculated heat losses through transmission and ventilation, assuming different classes of airtightness (Figure 2), and savings of usable energy for heating and natural ventilation due to better tightness—lower value of the airtightness coefficient n_{50} (Figure 3).

The airtightness of the building (expressed in L/s/m^2 at a pressure difference between the interior of the building and the environment of 75 Pa) was tested in six commercial buildings in Canada [18]. The authors diagnosed the location of the leakages using an infrared camera. Calculations have shown that the additional energy consumption due to detected leaks ranges from 47 to 64 kWh/m^2/year in these buildings. The diagnosed places of leakage include wall-to-roof connections, window and door assembly processing, and technical passages of pipes and channels through various elements of the building envelope.

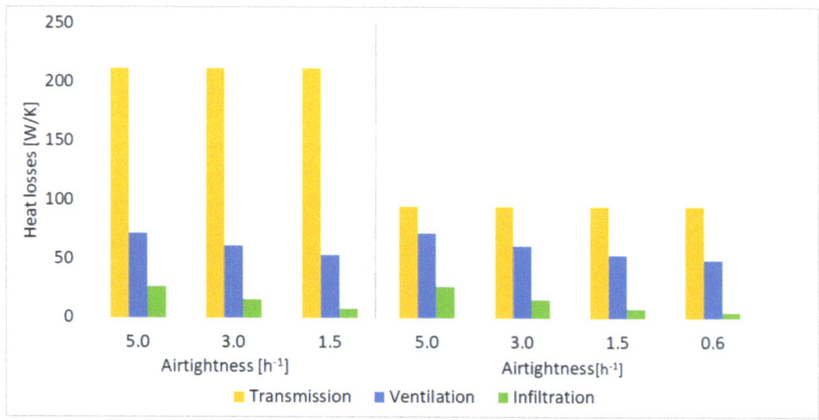

Figure 2. Transmission, ventilation, and infiltration heat losses for exemplary single-family building in the climate of Poland, assuming different airtightness, prepared based on data from [2].

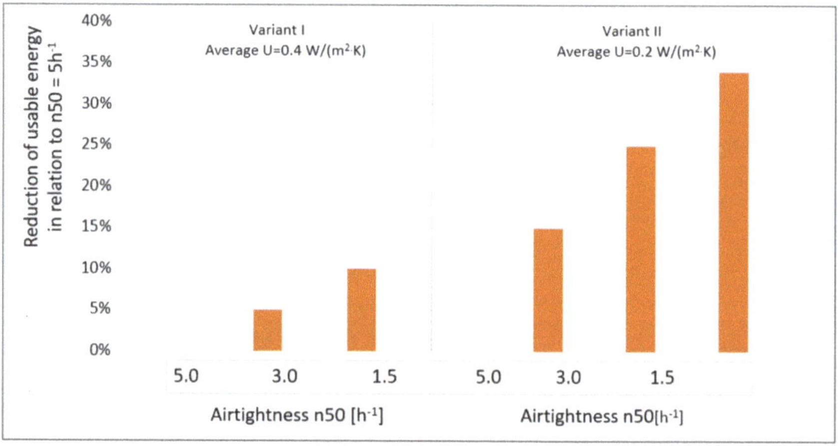

Figure 3. Reduction in usable energy for heating and ventilation in relation to the case $n_{50} = 5.0$, prepared based on data from [2].

In another paper [19], the significant importance of air leakage and infiltration on the energy consumption of an airport in China was demonstrated. The authors found that air infiltration is almost the most relevant factor influencing space heating and is equal to 18% to 71% of the total heat loss. In addition, they showed that by reducing leaks and using underfloor heating in airport terminals, the annual demand for heating can be reduced by as much as 84%.

The theme of "air leakage" was the leading topic during one of the sections of the 12th Nordic Symposium on Building Physics in 2020. Among the many interesting papers (E3S Web of Conferences Volume 172), attention is drawn to [20], in which the variability of airtightness was analyzed in the case of 41 energy-saving buildings in the period of use from 0.5 to 12 years. Interestingly, the air permeability increased in 29 buildings and decreased in four buildings. On average, at a pressure difference of 50 Pa during the tests, the leakage of the buildings increased by 38%, while at the same time causing an increase in the specific infiltration by 0.15 $m^3/(h \cdot m^2)$.

A similar analysis was presented by the authors of the publication [21]. They described the results of research on changes in the n_{50} coefficient in the short (1–3 years) and long

(3–10) period of time. Leakage rates increased by 18% and 20% respectively. The results of the research allowed the authors to conclude on the influence of the type of building structure on the deterioration of airtightness over time ("the number of levels, the type of roof and the type of building material and air-barrier").

In another report [22] on the example of simulations carried out for single- and two-family residential buildings, it was shown that air leakages significantly affect the value of the peak demand for thermal power for heating purposes. Underestimating it may result in an incorrectly sized and functioning heating system. What should be emphasized is that the authors showed that due to the lack of airtightness of buildings, the peak demand for thermal power for heating purposes was recorded with the strongest wind. It increases infiltration and causes an additional load for the building's heating system, which is more than in case of the lowest outdoor air temperatures. This is confirmed by the results of the numerical simulations presented in [23]. For different locations of the same building in Poland characterized by the coefficient $n_{50} = 0.63\ \mathrm{h}^{-1}$, the values of infiltration heat losses differed significantly due to different wind distribution, as shown in Figure 4.

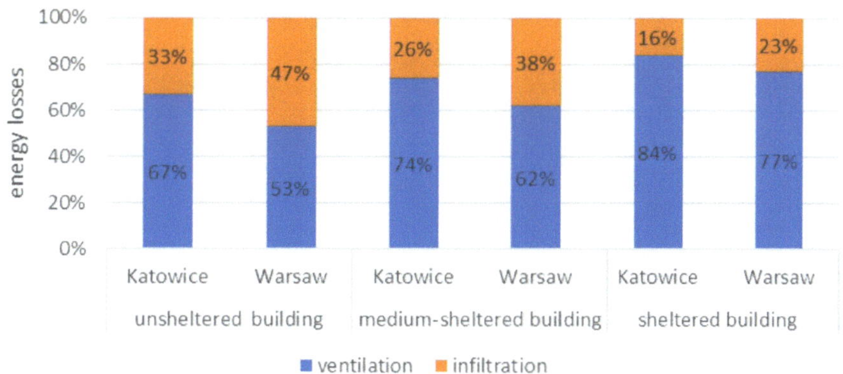

Figure 4. Percentage share of ventilation and infiltration heat losses in total heat losses for ventilation of a building with $n_{50} = 0.63$ located in two cities of Poland for different sheltering conditions [23].

The authors of [23] highlight the importance of location and related climate as well as the building's cover on the intensity of infiltration and energy consumption. Simulations show that the difference in annual energy consumption for a building in the same location, but with different airtightness standards, can be as much as 90%. Additionally, taking into account the different degrees of building sheltering, these differences can reach up to 200%. The authors emphasized that this is the reason why the legal requirements concerning, e.g., airtightness should be differentiated depending on the building's exposure to undesirable factors resulting from the climatic zone (wind, temperature) and/or the building's surroundings in a given location. In support of their appeal, the authors also cited other scientific works.

The authors of another conference paper [24] on airtightness measurements in middle-size stores and distribution centres showed that the proper infiltration flow at a pressure difference of 50 Pa is $1.04\ \mathrm{m}^3/\mathrm{m}^2/\mathrm{h}$ and $1.35\ \mathrm{m}^3/\mathrm{m}^2/\mathrm{h}$, respectively. In their conclusions, they proposed an increase in legal requirements and establishment of new standards for these types of buildings. By increasing airtightness, they could significantly improve their energy performance.

The airtightness of more than 400 single- and multi-family dwellings located in different regions of Spain and built in different years was investigated in a paper [25]. The energy impact of measured leakages on heating demand was in the range 2.43–19.07 $\mathrm{kWh}/\mathrm{m}^2/\mathrm{year}$, showing a great challenge and potential for energy efficiency of buildings not only in Spain. However, the authors did not note the significant importance of airtightness for the energy demand for cooling purposes.

The results of dynamic simulations in the TRNSYS program are presented in the article [26] also confirm the lack of influence of airtightness on energy demand for cooling purposes. The authors simulated a dwelling equipped with a mechanical ventilation system for different locations in Europe (Spain, France, Italy, Germany, UK). In northern and colder locations, the energy impact of infiltration was +13% in energy demand for heating and cooling, in Mediterranean areas from +4 to +7% and in southern locations only +3%. Although the authors highlighted that the passivhaus standard requires the same value of n_{50} regardless of location, their research shows that it should be differentiated. It should be higher in locations with frequent low temperatures ($0.6\ h^{-1}$) and lower where outside air temperatures are high (even $1\text{--}2\ h^{-1}$).

In [27] the importance of airtightness on the energy consumption of social buildings built in the years 1950–1979 in Spain (different variants of the Mediterranean climate) was analysed. The impulse for the research for the authors was, among others, the quote of almost the highest value in Europe of the number of mortalities associated with winter weather. The authors link it with the low quality of social housing in Spain, which was in force before the changes in the legal requirements regarding the quality of the building envelope. Their research results showed that in regions with more severe climates, air leaks are responsible for an additional consumption of more than $10\ kWh/m^2/year$. It was calculated that an increase in airtightness by $0.1\ h^{-1}$ would reduce annual heat demand by 5% in warmer zones and by 7.2% in the colder regions of Spain. Their results are presented in Figure 5.

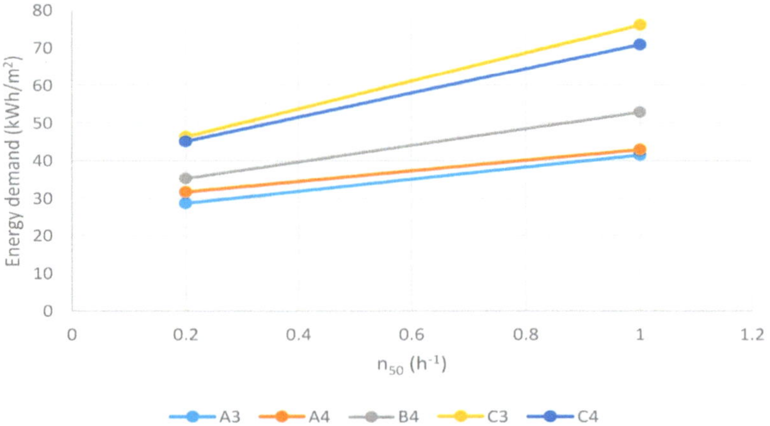

Figure 5. Dependence of the value of the annual infiltration heat loss index for an example building with different degrees of airtightness (n_{50} from 0.2 to 1.0) located in different climates of Spain "most representative of Andalusia, ranging from very mild (zone A) to cold (zone C) winters and warm (zone 3) to hot (zone 4) summers" [27].

Another interesting work is [28], which presents the range of variability of the n_{50} coefficient value for flats of similar area, but located in different parts of the building and on different floors of a multi-family building. As shown in Figure 6, the values differed by up to 25% between apartments with extreme values of area, because, as the authors conclude: "local air leakages or minor construction defects of larger flats statistically had less influence on the general airtightness". The flats located in the end parts of the building had up to 20% higher n_{50} than those in the middle, which "can be explained by a longer length of structural joints in the end units" [28] (Figure 6).

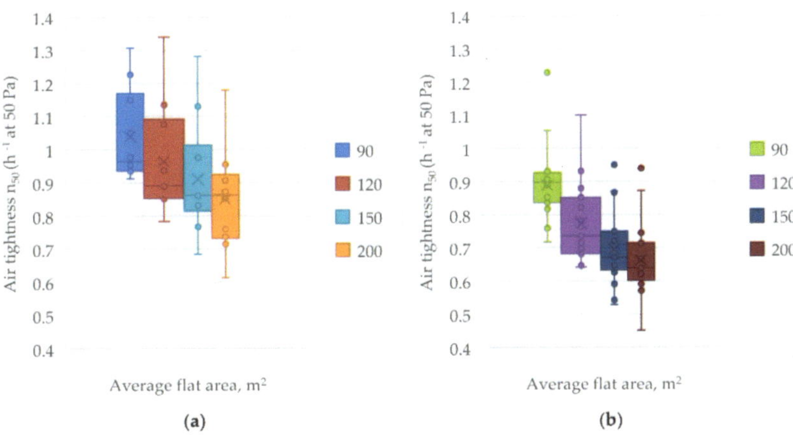

Figure 6. Airtightness of buildings: (**a**)—end units (flats), (**b**)—inside units [28].

The authors of [12] presented interesting results of tightness tests of two single-family buildings in the post-conference paper. The tests were carried out in two variants: (1) for the entire building and (2) separately for the residential part and the garage. The results showed that the airtightness of the residential part (n_{50} = 2.65 h^{-1} or 1.47 h^{-1}) is several times higher than that of the garage (n_{50} = 8.91 h^{-1} or 8.11 h^{-1}, respectively). Assuming the average value of n_{50} in the calculations of energy for heating resulted in a 12% higher energy demand than the value resulting from the assumption of a more realistic (measured) n_{50} for both zones separately. It is interesting that the result depends on the direction of airflow; if infiltration air flows from the garage to the occupant area, the difference in energy usage can be lower or even negligible. This confirms the need for dynamic simulations of air flows between building zones.

Such simulations are presented in [29]. The article describes an accurate method for the dynamic simulation of infiltration that can be used to assess its effect on building energy demand more accurately than in comparison to steady-state methods with an accuracy of 2.5% in comparison to field tests.

Airtightness measurement methods are a separate issue. An interesting overview of the reasons for the uncertainty in conducting leak tests is described in [30]. In the paper, various sources of error are presented and the research perspectives are described for better understanding the problem. The methods to measure the airtightness of buildings, including two nontypical—in transient conditions and acoustic—are presented in [31]. The methods of predicting airtightness performance are worth mentioning; they are being developed all over the world and can be useful for conducting energy simulations of buildings. One of those approaches is described in [32], where a statistical predictive model was developed based on data from real facilities in Spain. The model takes into account variables such as "the age of the building, typology, building state, construction system, and dimensions".

Summarising the review of research on the importance of the airtightness of the building envelope on thermal performance, it can be said that the topic is important, which is confirmed by the number of publications and citations listed in Table 1. Undoubtedly, energy-efficient buildings with various functions should be characterized by high airtightness; otherwise, its absence causes an increase in the demand for thermal power for heating and ventilation purposes (from 2.4 to even 64 kWh/m^2/year), while it practically does not affect the demand for cooling. It should be noted that the location of the building and the resulting climatic conditions are important, as well as the degree of its shielding, which increases the energy demand. The most important is the type of building partition construction and the introduction of more restrictive requirements in various countries is indicated. Also important is the flow of air between zones inside the building, the use of dynamic methods to predict heat losses as a result of leaks, and paying attention to the methods of measuring airtightness.

Table 1. Reviewed papers on airtightness.

Authors, Year	Title	Journal	Citations	Keywords
Amanowicz Ł., Ratajczak K. 2021, [2]	Practical aspects of designing energy-saving ventilation systems (in Polish)	Rynek Instalacyjny	2	Ventilation, energy-efficient buildings, energy performance, PE indicator, technical conditions
Verbeke S., Audenaert A. 2020, [20]	A prospective Study on the Evolution of Airtightness in 41 low energy Dwellings	E3S Web of Conferences	2	-
Simson R. 2020, [22]	The impact of infiltration on heating systems dimensioning in Estonian climate	E3S Web of Conferences	4	-
Miszczuk A., Heim D. 2020, [23]	Parametric Study of Air Infiltration in Residential Buildings—The Effect of Local Conditions on Energy Demand	Energies	6	Airtightness, climate data, building exposure, airflow network, performance simulation
Nitijevskis A. et al., 2020, [24]	Overview of large building testing in Baltic countries	E3S Web of Conferences	0	-
Heim D. et al., 2020, [29]	Modelling building infiltration using the airflow network model approach calibrated by air-tightness test results and leak detection	Building Services Engineering Research & Technology	2	Airtightness, building energy conservation, building simulation
Zheng X. et al., 2020, [31]	A practical review of alternatives to the steady pressurisation method for determining building airtightness	Renewable and Sustainable Energy Reviews	19	Building airtightness, steady pressurisation, blower door, unsteady technique, pulse technique, acoustic method
Liu X. et al., 2021, [19]	Energy saving potential for space heating in Chinese airport terminals: The impact of air infiltration	Energy	21	Airport terminal, space heating, field investigation, air infiltration, indoor thermal environment, energy saving
Moujalled B. et al., 2021, [21]	Mid-term and long-term changes in building airtightness: A field study on low-energy houses	Energy and Buildings	5	Airtightness durability, field measurements, building envelope, low-energy house
Poza-Casado I. et al., 2021, [25]	Airtightness and energy impact of air infiltration in residential buildings in Spain	International Journal of Ventilation	6	Air leakage, blower door, energy impact, residential buildings, fan pressurisation test
Paukštys V. 2021, [28]	Airtightness and Heat Energy Loss of Mid-Size Terraced Houses Built of Different Construction Materials	Energies	1	Airtightness, blower door, heat energy loss, thermographic photo research, building energy performance
Banister C. et al., 2022, [18]	Energy and emissions effects of airtightness for six non-residential buildings in Canada with comparison to contemporary limits and assumptions	Journal of Building Engineering	0	Airtightness, air leakage, infiltration, energy effects, building codes, building emissions
Mélois A. et al., 2022, [30]	Uncertainty in building fan pressurization tests: Review and gaps in research	Journal of Building Engineering	1	-
Poza-Casado I. et al., 2022, [32]	An envelope airtightness predictive model for residential buildings in Spain	Building and Environment	1	Predictive model, airtightness, blower door, dwellings, database, statistical analysis

The authors of this article are of the opinion that the design of the ventilation system in an energy-efficient building should be associated with ensuring the high airtightness of the building. In the case of air leakage, the effectiveness of even a very advanced energy-saving ventilation system will be destroyed by the influence of air infiltration through leaks in the building envelope, and heat losses will be noticeable as a result of increased operating costs. According to the authors' experience, the impact of leakage will be negligible when the value of n_{50} is lower by about one order of magnitude than the value of the base flow rate for ventilation. Also of interest are the research results cited, which show that airtightness changes over the life of a building, which is currently not taken into account in multi-year energy analyses.

3. DCV as a Ventilation Control Strategy

To obtain the highest energy efficiency and low energy consumption, ventilation systems should be adjusted to the actual needs of the occupants. This type of ventilation is called DCV (demand-controlled ventilation). These systems should be highly energy-efficient, as during periods of reduced air demand, lower costs are incurred in pumping air through the duct network, while at the same time the cost of thermal treatment is significantly reduced. Maintaining a constant nominal (design) air flow causes unnecessary costs to be incurred for heating/cooling the air, but also for pumping it through the system (overcoming flow resistance). Fans are commonly powered by electricity, and this, in many countries, still comes from burning nonrenewable fossil fuel resources. Matching the performance of the ventilation system to current needs is therefore justified by concerns in terms of energy, economics, and the environment. The DCV concept is mostly implemented by:

- the use of indoor sensors for CO_2, occupancy, humidity, etc.
- use of variable air volume (VAV) controllers for central systems,
- dividing a building into zones (zoning) with similar usage characteristics with separate ventilation units responsible for air quality in a given zone, such as in several rooms.

The authors of [33] reviewed the latest literature in the field of DCV systems controlled on the basis of carbon dioxide (CO_2) measurements. The three main types of CO_2 control strategies in DCV systems are presented—rule-based (i.e., controlling the CO_2 set-point), model-based (i.e., predictive control), and learning-based—along with their pros and cons. The meaning of sensor performance, sensor placement, and their errors on the effectiveness of the CO_2-based DCV system is extensively discussed. Moreover, many simulation and field test case studies are described and compared.

A review of articles on the impact of demand-controlled ventilation on energy use led to the article [34]. Although published in 2018, it not only shows potential energy savings (ranging from 64% to 84%) compared to the constant air volume (CAV) system, but also presents an innovative approach to the problem of maintaining constant pressure in the main duct. The authors indicate that reducing the pressure setting together with the decreasing demand for air can bring additional energy savings for driving fans at the level of 10% to even 50% in the case of heavily used and lightly used rooms, respectively. To complete this task, throttle position sensors in VAV controllers were used. The savings are greater the more diverse the room use profiles are, especially in the case of rooms that are rarely or not fully used (e.g., lecture halls, where the variability of users may range from 1 to 120 people).

The energy effects of using the DCV system were also analyzed in [35–37]. The simulations carried out in [35] show that in multi-family buildings the DCV system allows for higher indoor air quality, that is, a lower concentration of CO_2 and appropriate humidity in relation to the CAV system. Furthermore, it allows one to save 22% energy for heating when comparing both systems equipped with heat recovery or even 86% when comparing both systems without the possibility of heat recovery from the exhaust air. In [35] a wide variety of humidity loads and occupants in different dwellings was evaluated. These are the elements that are confirmed by users of DCV systems. The authors of the article suggest the use of a two-way control, DCV control according to the moisture load and volatile organic compounds, to better match the real needs of each apartment. In [36] it was shown that a DCV system used in an office building allowed for approximately 30% energy savings. In [37] energy savings of about 25% for ventilation needs due to the use of DCV were shown in school buildings located in the hot climate of Saudi Arabia.

An interesting analysis showing that DCV systems are not always more energy-efficient than CAV is presented in [38]. The influence of longitudinal heat conduction (LHC) on the efficiency of heat recovery using heat-wheel-type heat exchangers was analyzed. It turns out that the effect of lowering the efficiency of heat recovery is significantly greater in the case of low air flow, i.e., when the DCV system reduces its efficiency to match the reduced needs of the user. In some cases, it may turn out that the energy loss resulting from the LHC is greater than the energy savings due to the reduced airflow. However, it should be emphasized that the situation may occur in specific weather conditions (i.e., for several hours a year). Nevertheless, it is also worth taking into account the possibility of this problem when analyzing the potential causes of the so-called "performance gap"—i.e., the difference between the computational and real characteristics of a building. This issue is described, among others, in [39,40].

Experimental studies [41] of indoor air quality were carried out in school classrooms in Australia supplied with fresh air using DCV and traditional constant-flow systems. The concentration of CO_2 and VOCs (volatile organic compounds) were measured. The maximum concentration of CO_2 in the classroom without DCV was equal to 2981 ppm, while in the classroom with DCV it was 1335 ppm. The VOC measurements also showed a better quality of air in the DCV-supplied classroom. An additional interesting element of the research was the survey of students, which showed that not only CO_2 concentration but a combination of these two parameters at once affects the feeling of fatigue and distraction during classes.

The challenges posed by the operation of DCV systems are discussed in the example of eight public buildings in Finland in [42]. The following were measured: air streams, temperature, and CO_2 concentration in individual zones of the building. It was discovered that only in one of the eight buildings did the system work under the designed parameters. Although the users did not complain of poor air quality or uncomfortable temperatures, it turned out that air flows were not as intended in the design. This sheds light on problems with DCV systems that go undetected in many buildings until they affect occupant comfort. However, the purpose of DCV is also to save energy. If these systems do not work according to the designed strategy, they will be more energy-consuming in reality than they could be theoretically.

When CO_2 sensors are used to control the efficiency of the ventilation system, there is always a doubt as to where they should be installed. In [43], the results of the CFD analysis simulating airflow and CO_2 distribution in the lecture hall are presented to select a representative location of the sensors. The results of the analysis showed an uneven distribution of CO_2 in the room and led to the conclusion that in the analyzed room, the places near the ceiling represent the best average CO_2 concentration.

Not only is the location of sensors in the room important, but for air quality management strategy, it is also important because of the sensitivity to influences and recognizing when, how often, and for how long these influences are active [44]. It is also important to indicate the type of pollutant whose concentration is to be measured as a signal for DCV regulation. In [45] the methodology is presented to select the measured parameter in order to obtain a good indoor air quality. The authors noted that the current state of the external air (e.g., $PM_{2.5}$) also affects the indoor air quality. Based on measurements of five parameters, the concentration of CO_2, concentration of particulate matter $PM_{2.5}$, temperature, relative humidity, and formaldehyde in the office, gym, and kitchen, they conclude that the absolute humidity and temperature are correlated, and formaldehyde is correlated to temperature and CO_2; however, CO_2 and temperature did not capture most of the peaks in $PM_{2.5}$.

In Oslo, Norway, experimental research was carried out on the thermal comfort of young people in rooms with the DCV system [46]. The results showed that depending on the age range of theroom occupants, DCV systems should use different CO_2 set-points. The assessment of air quality, especially in the case of younger children, is highly dependent on their age. The authors of another article [47] also came to similar conclusions. They showed that the feeling of warmth is affected by the concentration of CO_2, but ensuring proper ventilation is also necessary in terms of energy efficiency.

Three articles [48–50] refer to the failure of CO_2 sensors used in DCV systems. In [49], based on failure models, a method is presented to assess the impact of sensor failures, while in [50] the automatic 4S3F fault-detection system for DCV systems was presented, based on four generic types of symptoms of faults (4S) and three generic types of fault (3F). In [48] a model for simulating sensor failures was presented, in which the fault injection framework was used to simulate potential failure scenarios (Figure 7). This model can be used to realistically simulate the energy consequences of CO_2 sensor failures in DCV systems.

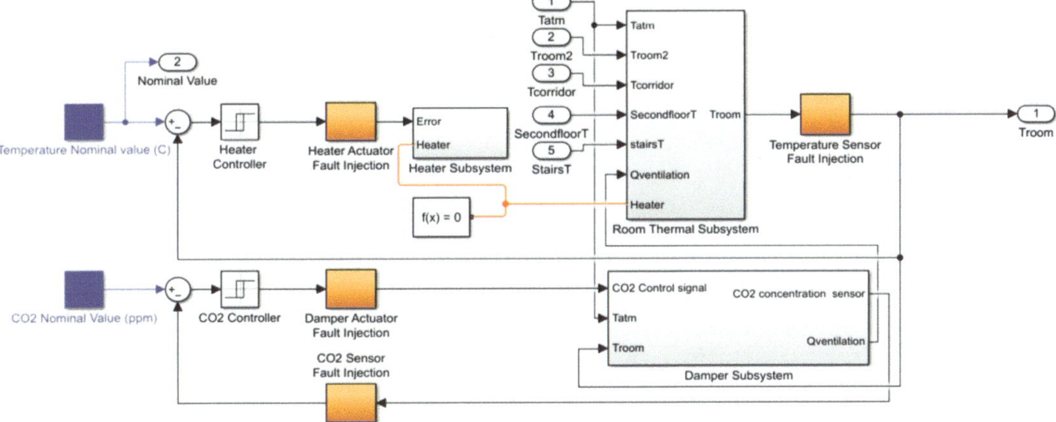

Figure 7. Model of HVAC system with DCV ventilation including fault injector blocks [48].

In summary, it can be seen that adjusting the efficiency of the ventilation stream to the current needs in the room is a result of energy and environmental benefits. Due to the air quality management strategy, the sensitivity of air to impacting factors resulting from human activity and the recognition by the ventilation system of when, how often, and for how long these factors are active are important. Additionally, the correct selection of sensors due to the correct measurement and their placement in the room are key issues for properly operating energy-saving DCV ventilation, which is confirmed by 18 selected recent publications collected in Table 2.

Table 2. Reviewed papers in DCV.

Authors, Year	Title	Journal	Citations	Keywords
Hamid A.A. et al., 2020, [35]	The impact of a DCV-system on the IAQ, energy use, and moisture safety in apartments—a case study	International Journal of Ventilation	1	-
Lu X. et al., 2020, [49]	A novel simulation-based framework for sensor error impact analysis in smart building systems: A case study for a demand-controlled ventilation system	Applied Energy	27	Demand-controlled ventilation (DCV), error impact analysis, sensors, simulation, sensitivity analysis
Taal A. et al., 2020, [50]	Fault detection and diagnosis for indoor air quality in DCV systems: Application of 4S3F method and effects of DBN probabilities	Building and Environment	18	4S3F framework, fault detection and diagnosis (FDD), demand-controlled ventilation (DCV), diagnostic Bayesian networks (DBN), building management systems (BMS), energy performance, indoor air quality (IAQ)
Abuimara T. et al., 2021,[36]	Exploring the adequacy of mechanical ventilation for acceptable indoor air quality in office buildings	Science and Technology for the Built Environment	1	-
Bandurski K. et al., 2021, [39]	Difference Between Calculated and Measured Energy Consumption for Heating in Multi-Family Buildings in Poland (in Polish)	Ciepłownictwo, Ogrzewnictwo, Wentylacja	1	Performance gap, multi-family residential building, multi-unit residential building, energy certification, building energy performance, occupant behavior, building energy modelling, monthly method, energy-efficient buildings
Haddad S. et al., 2021, [41]	On the potential of demand-controlled ventilation system to enhance indoor air quality and thermal condition in Australian school classrooms	Energy and Buildings	25	Indoor environmental quality, air quality, thermal comfort, school buildings, ventilation
Shin H. et al., 2021, [47]	A study on changes in occupants' thermal sensation owing to CO_2 concentration using PMV and TSV	Building and Environment	8	Indoor environmental quality, indoor air quality, CO_2 concentration, predicted mean vote (PMV), thermal sensation vote (TSV), discrepancy between TSV and PMV (DV)
Lu X. 2022, [33]	Advances in research and applications of CO2-based demand-controlled ventilation in commercial buildings: A critical review of control strategies and performance evaluation	Building and Environment	2	CO_2 control strategies, demand-controlled ventilation, energy efficiency, performance evaluation, CO_2 sensor
Alaidroos A. et al., 2022, [37]	Evaluation of the performance of demand control ventilation system for school buildings located in the hot climate of Saudi Arabia	Building Simulation	3	Mechanical ventilation, Demand-controlled ventilation (DCV), indoor air quality (IAQ), CO_2 concentration, energy efficiency, school buildings
Liu P. 2022, [38]	Heat recovery ventilation design limitations due to LHC for different ventilation strategies in ZEB	Building and Environment	2	Energy-efficient ventilation, rotary heat exchanger, zero-emission buildings, longitudinal heat conduction
Ratajczak K. et al., 2022, [40]	Differences Between Calculated and Measured Energy Use For Heating and Domestic Hot Water Preparation on the Example of Single-Family Buildings (in Polish)	Ciepłownictwo, Ogrzewnictwo, Wentylacja	0	Performance gap, single-family buildings, energy certification, energy efficiency calculation, final energy, occupant behaviour, building energy modelling, monthly method, energy-efficient buildings, domestic hot water

Table 2. *Cont.*

Authors, Year	Title	Journal	Citations	Keywords
Zhao W. et al., 2022, [42]	Operational Challenges of Modern Demand-Control Ventilation Systems: A Field Study	Buildings	0	demand-controlled ventilation, performance of ventilation system, public buildings, field study
Mou J. et al., 2022, [43]	Computational fluid dynamics modelling of airflow and carbon dioxide distribution inside a seminar room for sensor placement	Measurement: Sensors	1	Carbon dioxide (CO_2) distribution, computational fluid dynamics (CFD), sensor placement, demand-controlled ventilation (DCV)
Szczurek A. et al., 2022, [44]	The Detection of Activities Occurring Inside Quick Service Restaurants That Influence Air Quality	Sensors	1	Indoor air quality, aensing, pattern recognition
Kiamanesh B. et al., 2022, [48]	Realistic Simulation of Sensor/Actuator Faults for a Dependability Evaluation of Demand-Controlled Ventilation and Heating Systems	Energies	1	HVAC systems, fault injection, fault scenario generation, fault model, finite-state machine, stateflow
Alonso M.A. et al., 2022, [45]	A methodology for the selection of pollutants for ensuring good indoor air quality using the de-trended cross-correlation function	Building and Environment	6	IAQ, DCV, CO_2, air temperature, RH, $PM_{2.5}$, formaldehyde

4. Decentralized Ventilation Systems

Maintaining the appropriate parameters and pumping air from the intake to the rooms and on to the exhaust requires the supply of a large amount of energy. Removing the costs associated with air transport is made possible with decentralized ventilation. Decentralized ventilation is a system that uses several smaller ventilation units that are dispersed throughout the facility. In other words, decentralization involves the use of multiple small-capacity units that can be applied to individual rooms [51–54], and, as a result, form a system or systems in which the division of rooms into zones served by separate units has been made [55]. In addition, unlike central systems, most often decentralized, wall-mounted, distributed systems do not take up as much space. This is important because the space taken up by the system is considered an argument that keeps investors and users from considering the introduction of a mechanical ventilation system. When used in many different types of buildings, such as residential, swimming pools, schools, and offices, it enables tangible energy and financial benefits. This is also confirmed by previous studies [56]. Heat recovery, balance of flow, and low specific fan power enabled energy savings in each of the analyzed 20 decentralized systems and 60 central systems in residential buildings [56]. Compared to the variant of natural ventilation without heat recovery, simulations showed that a decentralized ventilation system provided annual primary energy savings of 4.75 Wh/m^3, and the central system 2.93 Wh/m^3.

A significant part of the energy in a building is used for air heating and its circulation, so it is worth considering the possibility of reducing the energy consumption in the building through the use of decentralized ventilation.

The authors of [57] studied the energy effects as a result of the use of a ventilation unit with a capacity of 300 m^3/h with a heat-recovery exchanger for an office building. Using TRNSYS software, the cooperation of ventilation with the heating and cooling system was assessed for buildings located in different climates in Sweden, Germany, and Italy. The solution made it possible to obtain lower electricity consumption thanks to the introduction of the DCV strategy.

The authors of [58] compared three ventilation systems: central mixing, displacement with the air-handling unit located in the basement, and decentralized with wall units. The results showed that the decentralized system ensures lower power consumption while ensuring adequate air quality. This issue was the subject of another study [52] describing the use of reversible fans in the rooms of an apartment in a multi-family building. The analyzed system is shown in Figure 8.

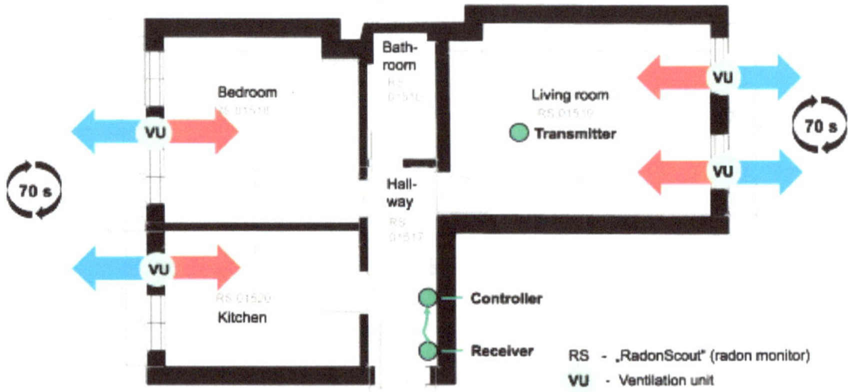

Figure 8. Decentralized ventilation system for an apartment analyzed in [52].

The authors of [52] conducted research on the assessment of the impact of open and closed doors in an apartment on the radon concentration. In rooms without mechanical ventilation, the radon concentration reached 7000 Bq/m^3 and increased when the door was open. The installation of four ventilation units, operating at full capacity in the heat recovery, made it possible to reduce the radon concentration to 300–800 Bq/m^3, that is, to the value recommended by the World Health Organization.

Subsequent research by these authors [51] showed more favorable air conditions in summer and allowed them to conclude that the operation of decentralized ventilation units may depend more on weather conditions than in the case of traditional central systems. The dependence between the operation of decentralized ventilation units and weather conditions was also studied by the authors of [59]. They found that there is a strong relationship between the wind and the efficiency of air supply to a home equipped with two-room ventilation units. The negative influence of the wind can be reduced using dampers controlled as a function of relative humidity. In other studies [60], the influence of wind on the operation of units with axial fans was assessed to be significant. The solution to the problem, according to the authors' suggestions, might be to change the type of fans to radial ones.

Another group of studies related to decentralized ventilation shows the specific use of ventilation units to improve indoor air quality. They focus on the fact that decentralized ventilation in the form of wall devices does not require a large amount of space for its installation. In [53], a strategy for ventilating rooms used as a nursery was proposed by installing wall units with reversible fans equipped with a ceramic exchanger for heat recovery. This has brought the benefits of ensuring a CO_2 concentration of 1000 ppm. The savings resulting from changing the ventilation method from ventilation by opening windows to forced ventilation were also calculated, which can be up to as much as 75% of the annual costs of heating the supplied air. Previous analyses [54] have also shown that a ventilation solution with reversible fans in the nursery will not increase the cost of electricity to supply the ventilation. Good air quality was also obtained through the use of façade devices in studies conducted in office buildings [61].

Due to the need to make openings in the outer partition, it is not always possible to use reversible fans. An example would be a historic school building where there was a problem of high CO_2 concentration during lessons [62]. The authors proposed masking channels in the window opening (Figure 9), which made it possible to maintain not only the CO_2

concentration at the assumed level of 1200 ppm, but also the proper radon concentration. In addition, the use of a heat-recovery exchanger allowed the authors to obtain the desired air temperature in winter, which was impossible when ventilating classrooms during breaks.

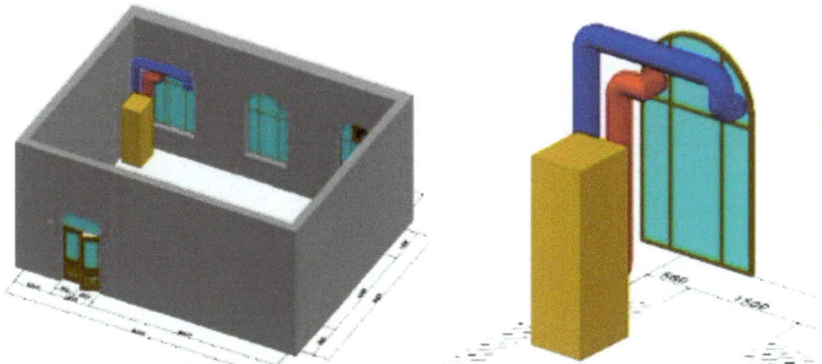

Figure 9. Model of HVAC system with DCV ventilation including fault injector analyzed in [62].

In the case of decentralized wall (façade) units, where the intake and exhaust are located in the same place and the air is removed and taken in through the same channel, there may be doubts as to the hygienic safety of such a solution. Laboratory tests described in [63] have shown that these units are safe from a hygienic point of view and there is no risk of mixing the intake and exhaust streams. In the same research, the authors carried out calculations showing that thanks to the use of a decentralized ventilation system in thermally modernized buildings, it is possible to reduce primary energy by up to 70%.

The authors of [64] tested a new algorithm for controlling decentralized units located in an apartment. The algorithm takes into account the preferences of users in terms of room ventilation, CO_2 concentration, and thermal comfort. Based on the research, energy savings of 20% were obtained while meeting the expectations of users. It should be noted that sometimes taking into account all user suggestions can increase energy consumption, so a compromise should be found between the needs of users and a reasonable level of energy consumption [65].

In the case of the assessment of decentralized ventilation units, the concentration of particulate matter PM is an important parameter. The comparative assessment of the central and decentralized systems [58] showed the correct air parameters for both systems in which air filters were used (MERV 8, MERV 10, MERV 14, MERV 16). Although the central system provided better air purification efficiency, the decentralized system was able to ensure adequate and sufficient air purity.

The decentralization of the ventilation system to divide the space into smaller ones with different needs was considered in the example of a swimming pool facility [55]. Figure 10 shows the two air distribution systems: the traditional central system that prepares the air for the entire swimming pool hall and the decentralized system that takes into account the division of the facility into zones with different needs: pool basins, external partitions, and the auditorium.

Traditional centralized ventilation system

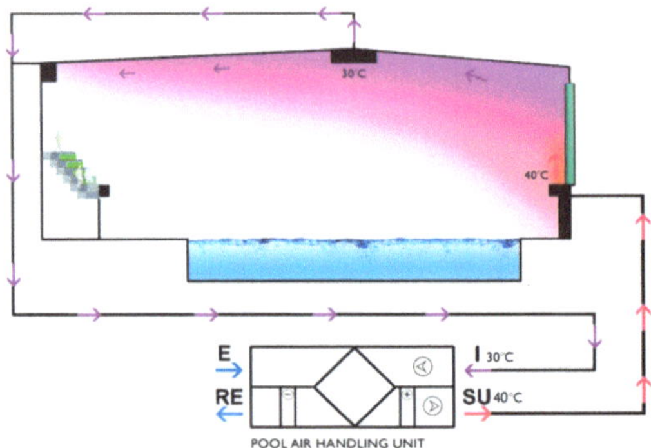

Decentralized ventilation system as an energy consumption decreasing strategy

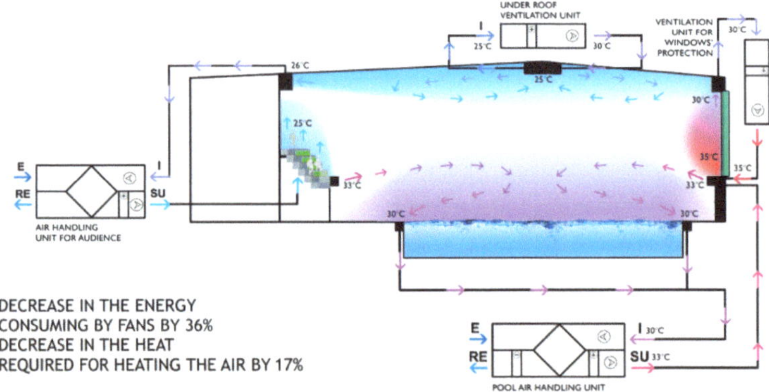

Figure 10. Traditional and decentralized ventilation systems for swimming pool analyzed in [55].

The demand for energy to heat the ventilation air was lower by 30–36% (depending on the method of air heating), and the electricity consumption needed to supply air in the decentralized system was lower by 42%. In addition, it is possible to maintain appropriate conditions in various zones of the facility as required.

Another topic of research is the assessment of the thermal comfort of users of rooms equipped with ventilation units [66]. The authors found that in such rooms people feel discomfort, which is associated with the supply of low-temperature air, because the façade devices are not equipped with air heaters. The author has come to similar conclusions before [67] while testing the created external façade unit (Figure 11), thanks to the application that good air quality was obtained in the tested room.

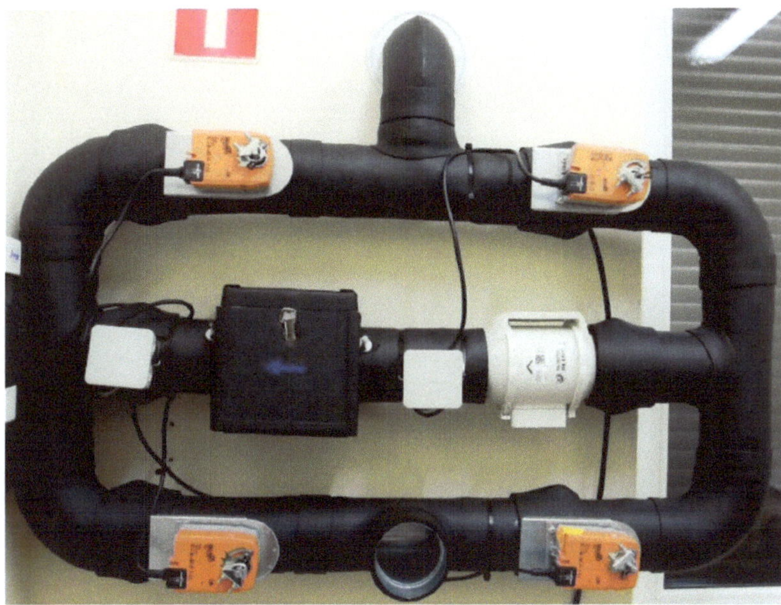

Figure 11. Façade ventilation unit analyzed in [67].

Recently, solutions have also been sought to improve the work of products on the market. The authors of [68] evaluated the length of the work cycle throughout the year. For the commonly used work cycle lasting 70 s, they obtained a heat-recovery efficiency of 67%, and increasing the time to 120 s, the efficiency was as much as 82%. Furthermore, the authors pointed out, similarly to [51,52], that taking into account the seasonal change in the parameters of the outside air will improve the operation of the system. Another element of wall (façade) systems that is used as elements of decentralized ventilation systems is a heat-recovery exchanger. The authors of [69] tested the efficiency of heat recovery in devices available on the market (Figure 12).

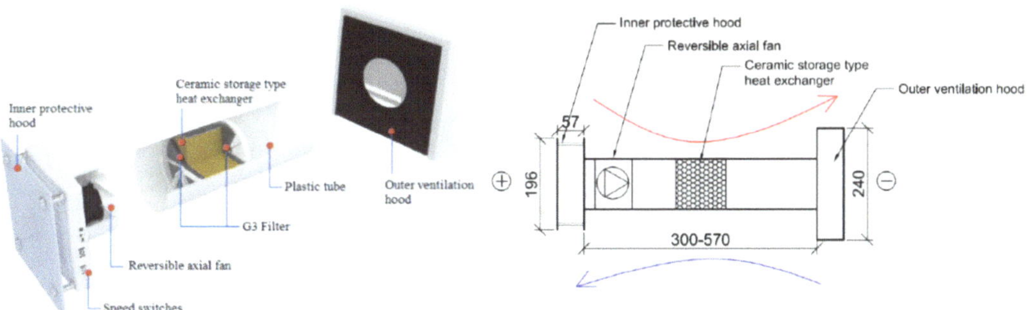

Figure 12. Decentralized ventilation unit with reversible fan and ceramic heat exchanger [69].

As a result of their research, they found that the efficiency of heat recovery provided by manufacturers may differ significantly from what will occur when working in real conditions. This may affect the results of the building's energy performance calculations. In the case of the tested unit, the manufacturer gave an efficiency of 85%, while the tests obtained a range of 20–50%, which was related to the pressure difference. High efficiency

(73%) occurred at a pressure difference of 0 Pa, and a high efficiency of 85% only in the initial period of the fans' work cycle.

A novelty in wall-mounted units may be the replacement of the ceramic heat exchanger with phase-changed material (PCM) filling [70], creating a latent heat-recovery ventilation system (HRV). The combination of ventilation systems and PCM as a solution enabling energy savings is the use of PCM façades as the first stage of heat recovery for decentralized ventilation units [71]. Due to the fact that outside air is introduced into the space behind the glass façade, the air temperature is 3 °C higher in winter and 5 °C lower in summer. The proposed solution can be used in new and existing buildings.

Research on decentralized ventilation systems is carried out in various scopes: efficiency; reduction in the demand for heat, cold, or electricity; the technical possibilities of use in buildings; and the effects measurable with parameters of indoor air, e.g., CO_2, radon, or dust. Modern PCM and façade solutions are being developed. All the articles collected in Table 3 present analyses that indicate that the decentralization of the ventilation system has many benefits in terms of its application, improvement of air quality, and reduction in energy demand, and encourage further research. At the same time, it was shown that there are premises for further research on decentralized systems, e.g., in terms of providing greater comfort to users or further improving the operation of small units that could be used in modernized buildings.

Table 3. Reviewed papers on decentralized ventilation systems.

Authors, Year	Title	Journal	Citations	Keywords
Ratajczak K. et al., 2020, [55]	Energy consumption decreasing strategy for indoor swimming pools—Decentralized Ventilation system with a heat pump	Energy and Buildings	15	-
Bonato, P. 2020, [57]	Modelling and simulation-based analysis of a façade-integrated decentralized ventilation unit	Journal of Building Engineering	14	-
Carbonare N. et al., 2020, [60]	Simulation and Measurement of Energetic Performance in Decentralized Regenerative Ventilation Systems	Energies	0	Decentralized ventilation, heat recovery, honeycomb heat exchanger, computational fluid dynamics, Modelica
Zender-Świercz E. 2020, [61]	Improvement of indoor air quality by way of using decentralised ventilation	Journal of Building Engineering	15	Decentralised façade, ventilation systems
Ratajczak K. et al., 2020, [63]	Assessment of the air streams mixing in wall-type heat recovery units for ventilation of existing and refurbishing buildings toward low energy buildings	Energy and Buildings	20	Integrated wall-type air intake-outtake unit, air streams mixing assessment, indoor air quality, heat recovery ventilation unit, energy efficiency in buildings, thermomodernization
Zemitis J. Bogdanovics R. 2020, [69]	Heat-recovery efficiency of local decentralized ventilation devices	Magazine of Civil Engineering	0	Decentralized ventilation, heat recovery, efficiency, pressure difference
Dehnert J. 2021, [52]	Radon protection in apartments using a ventilation system wireless-controlled by radon activity concentration	Journal of Radiological Protection	2	-
Ratajczak K. Basińska M. 2021, [54]	The well-being of children in nurseries does not have to be expensive: The real costs of maintaining low carbon dioxide concentrations in nurseries	Energies	4	Decentralized ventilation, façade ventilation
Fu N. 2021, [58]	Comparative Modelling Analysis of Air Pollutants, $PM_{2.5}$ and Energy Efficiency Using Three Ventilation Strategies in a High-Rise Building: A Case Study in Suzhou, China	Sustainability	2	Decentralized ventilation system, centralized ventilation, indoor air quality, high-rise building, infiltration, air filter efficiency

Table 3. *Cont.*

Authors, Year	Title	Journal	Citations	Keywords
Filis V. et al., 2021, [59]	The impact of wind pressure and stack effect on the performance of room ventilation units with heat recovery	Energy and Buildings	6	Decentralized ventilation, room ventilation units, façade ventilation, heat recovery, rotary heat exchanger, stack effect, centrifugal fans, axial fans, humidity-controlled damper
Carbonare N. 2021, [64]	Design and implementation of an occupant-centered self-learning controller for decentralized residential ventilation systems	Building and Environment	3	Mechanical ventilation, occupant-centered control, adaptive control, building simulation, residential buildings, occupant behavior
Dudkiewicz E. et al., 2021, [65]	Users' Sensations in the Context of Energy Efficiency Maintenance in Public Utility Buildings	Energies	1	CO_2 concentration, lecture room, students' preferences, sultriness, thermal comfort; willingness to work
Zednre-Świercz E. 2021, [67]	Assessment of Indoor Air Parameters in Building Equipped with Decentralised Façade Ventilation Device	Energies	4	Air quality, CFD simulation, decentralised façade ventilation systems
Pekdogan T. 2021, [68]	Experimental investigation of a decentralized heat recovery ventilation system	Journal of Building Engineering	6	Indoor air quality, ventilation, decentralized heat recovery, sensible energy storage
Pekdogan T. 2021, [70]	Experimental investigation on heat transfer and air flow behaviour of latent heat storage unit in a I integrated ventilation system	Journal of Energy Storage	2	-
Altendorf D. 2022, [51]	Decentralised ventilation efficiency for indoor radon reduction considering different environmental parameters	Isotopes in Environmental and Health Studies	0	Decentralised ventilation
Ratajczak K. 2022, [53]	Ventilation Strategy for Proper IAQ in Existing Nurseries Buildings—Lesson Learned from the Research during COVID-19 Pandemic	Aerosols and Air Quality Research	3	Ventilation strategy, nurseries, indoor air quality, COVID-19 pandemic, decentralized ventilation
Zender-Świercz E. et al., 2022, [66]	Assessment of Thermal Comfort in Rooms Equipped with a Decentralised Façade Ventilation Unit	Energies	0	Thermal comfort, decentralised façade ventilation units, storage heat exchangers for heat recovery
Shahrzad S. 2022, [71]	Parametric optimization of multifunctional integrated climate-responsive opaque and ventilated façades using CFD simulations	Applied Thermal Engineering	7	Decentralized ventilation

5. Preheating/Cooling of Outdoor Air in Earth-to-Air Heat Exchanger (EAHE)

Fresh ventilation air can be preheated (in winter) or cooled (in summer), achieving a partial air-conditioning effect through the use of earth-to-air heat exchangers (EAHEs), a kind of ground heat exchanger. EAHEs can take the form of pipes buried in the ground, be made as gravel-type, i.e., with an accumulation layer of supporting stones, or be plate-type, i.e., with plates of various shapes buried in the ground under which air flows. In recent years, multi-pipe earth-to-air heat exchangers have attracted the interest of researchers. These are heat exchangers built of multiple parallel pipes, which are used, for example, for facilities with high air ventilation demand. They allow a reduction in flow resistance and the use of smaller-diameter pipes, which are easier to place under the ground surface. A schema of single-pipe and multi-pipe EAHE is presented in Figure 13.

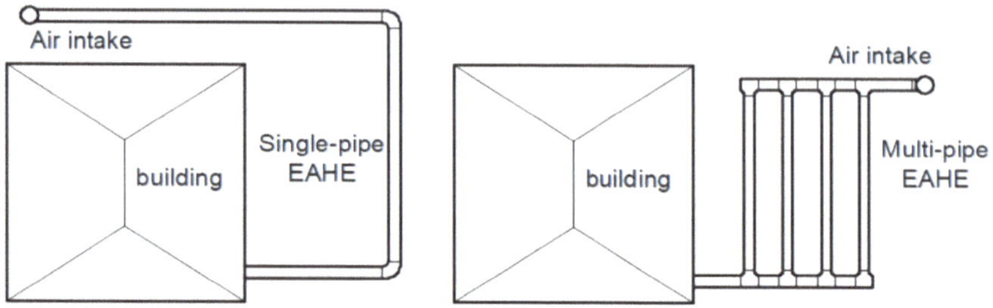

Figure 13. Schema of single-pipe and multi-pipe EAHE [72].

Another idea, the ISOMAX system, allows the use of solar energy stored in a ground buffer or in a foundation plate [73–76]. The system is based on a co-axial pipe in a pipe counter recuperator for ventilation purposes and was investigated in Slovakia [77,78]. The system is still being developed and allows for an improvement in the energy efficiency of heating and cooling sources [79].

Earth-to-air heat exchangers make it possible to preheat air in the ground in winter and thus protect heat-recovery systems from frost. This is a more efficient solution than bypass [80] and is based on the high thermal inertia of the ground, the temperature of which at a certain depth, depending on the type of soil and climatic zone, is always higher than 0 °C. In summer conditions, on the other hand, the air flowing through the heat exchanger can be cooled and also dehumidified by condensation on the cool surface of the exchanger. In energy terms, the profit of EAHE operation is the use of renewable energy from the ground, while the cost is the electricity required to overcome the resistance of the airflow through the exchanger and ducts. To ensure efficient heat exchange in the exchanger, devices are used to increase its intensity, and at the same time the flow resistance. These solutions always require a comparative analysis to judge which variant will be a compromise or optimal, e.g., in terms of energy or finance. Also important is the issue of unequal distribution of airflows in EAHE pipes, which has a bearing on energy requirements for heating and cooling. This issue is illustrated from the company's own CFD simulation results in Figure 14, where the distribution of velocities in individual pipes is shown for multi-pipe EAHE where blue colour means the lowest velocity and yellow to red colour, the highest values.

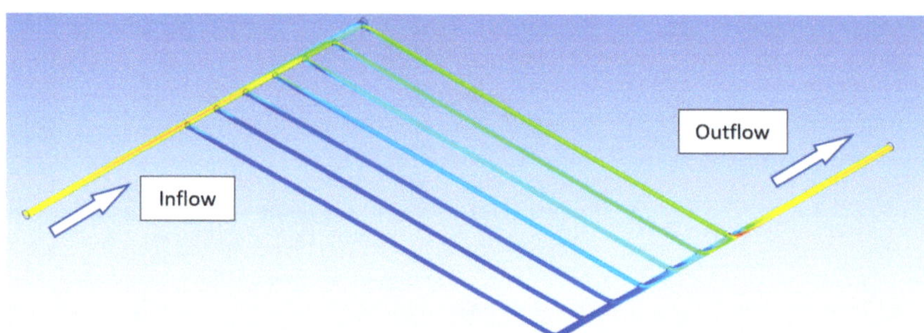

Figure 14. Distribution of air velocity in the axis of individual pipes of a multi-pipe EAHE: blue color—the lowest velocity, yellow to red—the highest; results of own simulation.

Earth-to-air heat exchangers are used around the world and in virtually every climate. The vast majority of the research work is on tubular heat exchangers, although gravel-type EAHEs have also recently been published [81]. The transfer of the recovered heat to cold and dry outside air results in a significant reduction in primary energy requirements for other purposes [82].

The number of research papers on earth-to-air heat exchangers is huge. Among them, there are well-known works dealing with mathematical modelling issues without any proof of the practical application of these devices in building ventilation systems. In this review, we focus on selected works that address the energy aspects of the exchanger's operation in the context of its use in ventilation systems.

The most recent review article on the subject of EAHE is [83], which summarizes the results of experimental studies, numerical simulations, and case-study analyses from around the world. The chapters discuss the methods used to model the performance, conduct experimental studies, and then describe parametric studies, hybrid systems with other renewable energy sources, and economic assessments. This is a very comprehensive work that leads to the conclusion that EAHEs can be successfully applied worldwide, bringing benefits in reducing energy demand and decarbonizing the building sector, as well as lowering the operating costs of HVAC systems. In an overview article [84] the authors analyzed the results of various publications that show the cooperation of EAHE with other systems: air conditioners, air-source heat pumps, heat-recovery units, evaporative cooling or air humidity control devices, photovoltaic (PV) systems, solar air heaters, solar chimneys (SC), wind towers, phase-change materials, roofs, and floors. This is a valuable overview that confirms the advantages of using EAHEs in various applications.

For proper operation of EAHE heat exchangers, an intensification of heat transfer is required, which is achieved using flow disturbances that increase the airflow turbulence. They usually increase pressure losses, but there are also solutions that intensify the heat transfer without increasing the flow resistance. An example is the research results presented in [85]. The study analyzed the impact of using different configurations and shapes of high-heat-transfer-coefficient fins mounted on EAHE ducts. An improvement in the heat exchanger's thermal efficiency by 33% over the baseline variant of the duct with a rectangular cross-section without fins was obtained. Another example of a way to intensify exchanger performance is to use backfilling material with a high value of the conduction coefficient, as analyzed in [86]. Based on the study, it was concluded that the thermal conductivity of backfilling materials should be no higher than 2.5 W/(m·K) to ensure the improved thermal performance of the analyzed system.

In recent years, solar chimneys (SC) coupled with EAHEs (SC + EAHE = SCEAHE) have become very popular in international journals. The use of a solar chimney creates a sufficient pressure differential to induce air flow through the exchanger without the need for electricity. The schematic diagram of an SCEAHE is presented in Figure 15.

In [87] the influence of various parameters of the solar chimney and EAHE was analyzed using numerical simulations. It has been shown that the diameter of the EAHE pipes and their length are much more crucial than the height of the chimney and the length of the solar collector. An analysis of the use of solar energy as a driving force showed that during the day airflows of 260–280 m^3/h were achieved, while at night only 50–100 m^3/h were achieved. Thanks to the solar chimney, EAHE obtains renewable energy from the ground in a passive way and has allowed the indoor air temperature to be reduced by 4.4 °C in summer and increased by 6.4 °C in winter. In another article [88] experimental studies of the solar chimney natural ventilation system operating with EAHE were carried out. The flow rate was 209 m^3/h during the day and 139 m^3/h at night. The indoor air temperature for the transitional season was 19.7–22.7 °C with the outdoor air temperature in the range of 12.5–25.0 °C. Experimental studies of natural ventilation in a system with an earth-to-air heat exchanger and a solar chimney are also presented in [89]. In summer, the air flow was at the level of 56.5 m^3/h at night and 291.5 m^3/h during the day, while in winter it was 90.9 and 388.8 m^3/h, respectively. The results confirm the phenomenon of a decrease in ventilation intensity in the absence of solar radiation at night. At the same time, it was shown that the ventilation air streams were higher in winter than in summer—thermal efficiency in winter and summer was 61% and 86%, respectively. Further experimental studies for SCEAHE with diameters of 0.3 m and 0.2 m are described in the article [90]. The paper notes that the "daily natural ventilation cycle" consists of the following stages: "thermal mass driven ventilation, coupling ventilation, solar chimney driven ventilation, coupling ventilation, and thermal mass driven ventilation". The maximum air flows achieved for the exchanger diameters are 252 m^3/h and 166 m^3/h during the day and 50–70 m^3/h and 45–50 m^3/h at night, respectively. The peak cooling capacity of the EAHE with SC was calculated to be 1179 W with a 0.3 m pipe diameter and 629 W with a 0.2 m pipe diameter. This is explained by the higher air flow, as the maximum temperature drop in both cases was similar: at approx. 12.5–13 K, similar capacities, i.e., 252 m^3/h during the day and 50–70 m^3/h at night, were obtained as a result of experimental tests presented in [91]. In [92] numerical simulations were presented to determine the cooling potential of SCEAHE, in which SC is filled with a phase-change material (PCM). With this modification, the airflow during the night was increased by 50%, while the internal air temperature was 0.8 K lower than the SCEAHE system without PCM. The maximum airflow during the day in the system with PCM was 17.8% lower than without PCM and amounted to 209.5 m^3/h. The use of phase-change material resulted in a reduction in the difference between the extreme values of airflow observed during the day and at night. In addition, as expected, the percentage increase in the stream at night was greater than the decrease in the stream during the day, which eliminated too-low air flow during the night due to the lack of solar radiation. The results of the discussed tests of the natural ventilation system with SCEAHE are summarized in Table 4.

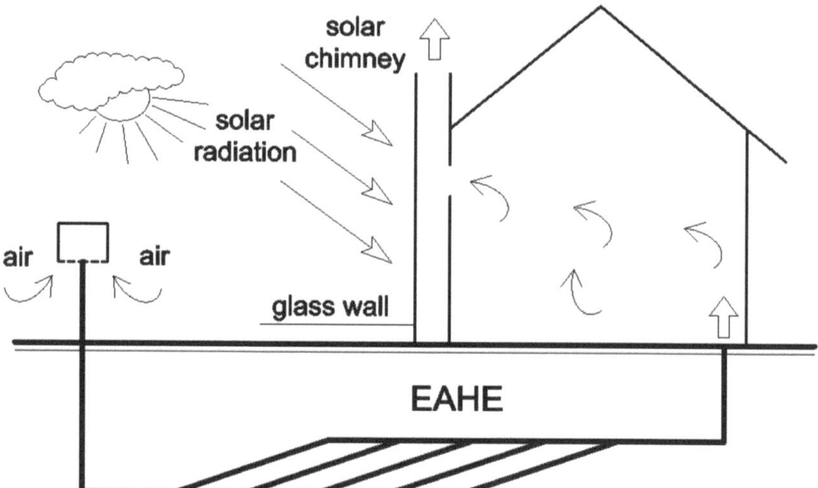

Figure 15. Schematic diagram of solar chimney (SC) coupled with earth-to-air heat exchanger (EAHE) = SCEAHE.

Table 4. A summary of SCEAHE research results.

Paper	Day-Time Airflow	Night-Time Airflow	Thermal Effect
[87]	260–280 m^3/h	50–100 m^3/h	Reduction in indoor air temperature by 4.4 °C in summer and increase by 6.4 °C in winter
[88]	209 m^3/h	139 m^3/h	Indoor air temperature 19.7–22.7 °C with outdoor air temperature 12.5–25.0 °C
[89]	291.5 m^3/h (summer) 388.8 m^3/h (winter)	56.5 m^3/h (summer), 90.9 m^3/h (winter)	The efficiency of obtaining energy from the ground was 86% in summer and 61% in winter
[90]	252 m^3/h (pipes diameter 0.3 m) 166 m^3/h (pipes diameter 0.2 m)	50–70 m^3/h (pipes diameter 0.3 m) 45–50 m^3/h (pipes diameter 0.2 m)	Reduction in the temperature of the supply air to the room by 12.5–13 K, translating into a cooling power of 1179 W for 0.3 m pipe diameter and 629 W for 0.2 m pipe diameter
[91]	252 m^3/h	50–70 m^3/h	Reduction in room supply air temperature by 12.5 K, maximum total cooling capacity (latent + sensible): 1398 W
[92]	209.5 m^3/h with PCM 255 m^3/h without PCM	95 m^3/h with PCM 50 m^3/h without PCM	Air temperatures at the outlet of EAHE with and without PCM of 24.8–26.5 °C and 24.4–27.2 °C, respectively; more stable indoor thermal comfort with PCM

A sensitivity analysis of the SCEAHE performance to changes in selected parameters affecting flow through the system is presented in [93]. It can be read that "increasing the solar collector length or the chimney height can both increase the system performance. However, the effect of chimney height is not as significant as that of solar collector length. The higher the solar intensity, the higher the buoyancy force, the airflow rate, the outlet air temperature, and the cooling capacity. The cooling capacity is increased by 101.4% by increasing the solar intensity from 100 W/m^2 to 600 W/m^2". While researching the optimal geometry of EAHE it was found that a pipe of 0.6 m diameter and 60 m long provides the highest performance. Further increasing the length of the pipe reduces the airflow rate.

The application of SCEAHE for ventilation of a swimming pool located in a hot, arid climate is presented in [94] using the CFD simulation tool. The results demonstrated that on the most critical design day, most date of temperature, velocity, relative humidity and CHCl3 mass fraction lay in the standard ranges. The system met the expectations of ventilation systems and is energy-efficient because of the use of passive technique.

In a warm, dry climate EAHEs can harvest a desirable coolness from the ground. If it is humid, they can additionally dry the air thanks to the condensation phenomenon on the pipe walls. Experimental studies of cooling capacity in hot and humid climates are presented in [95], where increasing the depth and decreasing the diameter decrease the air temperature/moisture content. The results of the study showed that increasing the depth of the EAHE location is conducive to achieving greater cooling and dehumidification of ventilation air (which is desirable in the climate studied). However, the limit value that would be economically viable was not analyzed, and the cost of making a deeper excavation was not considered.

In [96] an experimental study of the effect of EAHE on indoor humidity was conducted in buildings located in arid regions. The study involved a 66 m long, 0.11 m diameter pipe exchanger located in Algeria 1.5 m below the ground surface. An increase in relative humidity (RH) of 19% during the period requiring humidification of the air and a decrease in RH of 27% during the period requiring a decrease in humidity were observed. It was concluded that "the EAHE technique has excellent potential for the enhancement of building hygrometry in arid regions". Another experimental set-up was built in Algeria to study the transient thermal performance of EAHE [97]. Investigations revealed a temperature drop of ventilation air at a level of 19 K. The decrease in the air cooling capacity was 0.85 K after 95 h of continuous operation of the exchanger with a constant maximum air flow velocity of 3.5 m/s.

The application of EAHE in subtropical climates (warm and humid) is presented in [98]: "EAHE system reduced the temperature in a single room building by a maximum of 2.18 °C, saving 415.92 kWh energy (on average 59.91 kWh) during a 3-months summer".

The application of EAHE in the temperate climate of Poland (Central Europe) has been analyzed in recent years, including in [99], where experimental studies showed that EAHE "reduces the demand for energy in the ventilation system by around 45%, and it can reduce the energy load of the entire building by around 20%".

In [100] calculations of the hourly energy demand for a building cooperating with EAHE were carried out. The EN ISO 13790 standard was used to model the building (describing the so-called "5R1C" model) and the EN 16798-5-1 standard to calculate the outlet air temperature after EAHE. The results of calculating the temperature change due to air flow through a ground exchanger installed in Poland are shown in Figure 16. It is shown that the use of a "bypass and switching between the EAHE and ambient air as the source of ventilation for the building resulted in annual energy savings of 123 kWh".

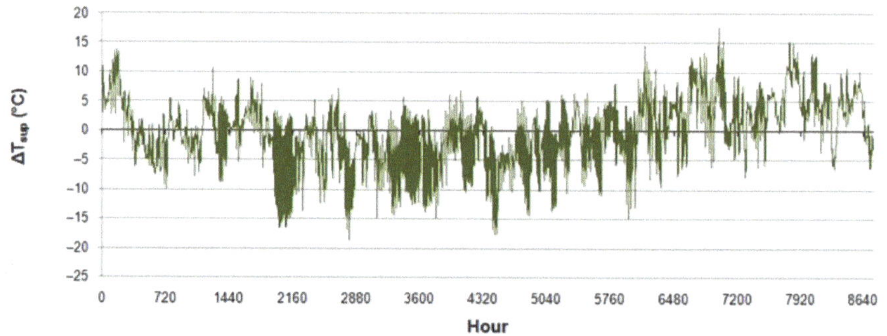

Figure 16. The change in the temperature of the air flowing through the exchanger over a year (0 h = 1 January, 8760 h = 31 December) [100].

In [101], which is a continuation of the previous paper, the authors analyzed the effect of changes in air density in Poland's climate on simulated EAHE performance. It was noted that the comparison of density variation with temperature for the typical meteorological year (at constant pressure) and the method of EN 16798-5-1 causes an hourly difference of unit gain of up to 4.3 W/m^2 and 2.0 W/m^2 for heating and cooling, respectively. In the case of International Weather for Energy Calculation files the differences were from 5.5 W/m^2 to 1.1 W/m^2. Differences in the values of the volumetric heat capacity of air, used in the next step of the EAHE thermal performance calculations, resulting from different calculation methods are shown in Figure 17.

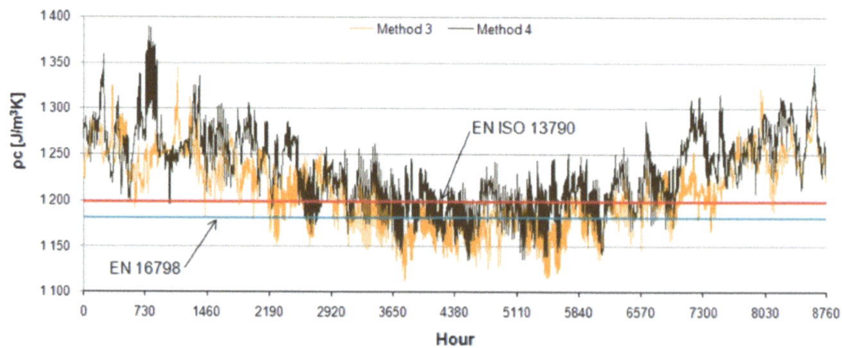

Figure 17. Volumetric specific heat of air according to various methods [101].

In [72] calculations were performed of the amount of energy saved by the operation of single-pipe and multi-pipe earth-to-air heat exchangers in the climate of Central Europe. Based on the study of their structure, it was proven that the appropriate selection of single-pipe EAHE is able to provide an equivalent energy yield to the multi-pipe structure with similar or lower pressure losses. The yearly energy usage for driving the fans for multi-pipe structures of length L = 55.4 m and single-pipe structures of equivalent length in the context of heating capacity with different pipe diameters (d = DN200, DN250 or DN315, V = 600 m^3/h) is presented in Figure 18. The paper presents a methodology for the simple and quick comparison of different EAHE structures and geometric and environmental parameters affecting the energy yield. The method takes into account the correction factor resulting from the analyses presented in [102].

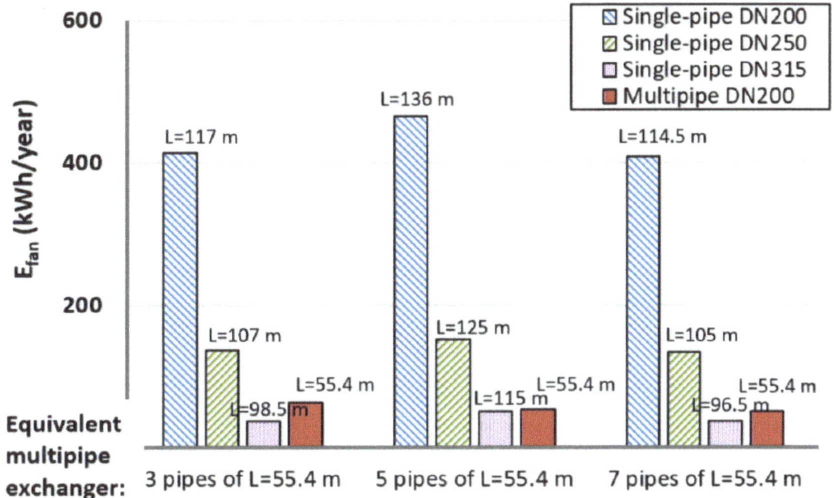

Figure 18. The yearly energy usage for driving the fans in the case of multi-pipe and single-pipe structures [72].

In [102] the relationship between the airflow distribution in the EAHE tubes and the heating and cooling yield is presented. It is shown that as a result of an uneven distribution of airflows, the energy gained can be up to 28% less than that determined for ideal conditions, i.e., assuming an even distribution of air in the exchanger tubes. Airflows in individual branches of multi-pipe EAHEs were determined in experimental studies and published in [103]. The given approximations make it possible to calculate the instantaneous values of the airflow in each branch of the multi-tube exchanger under varying ventilation system loads (different total airflow through the exchanger).

The topic of the effect of exchanger geometry on the uniformity of air distribution and on the reduction in energy consumption was continued by the research team in experimental studies [104] and numerical simulations [105]. These were also used to determine the effect of geometric parameters on pressure loss and heat capacity [106]. The results of the study confirmed the conclusions of [104], in which lower pressure losses, better uniformity of air distribution in the exchanger branches, and higher efficiency of exchangers with a U-type structure than exchangers with a Z-type structure were demonstrated. These studies were also confirmed in [107], where a case study of the use of multi-pipe EAHE for greenhouse ventilation was presented.

A not-very-popular type of EAHE is the gravel-type heat exchanger. The possibility of its use in the ventilation system of a single-family house in various configurations is described in [81] and demonstrated in Figure 19. Due to the direct contact of the ventilation air with the ground layer and the accumulation mass (gravel), it is heated and humidified in winter, and cooled and dried in summer. The estimated simple payback time (SPBT) was 3.65 years.

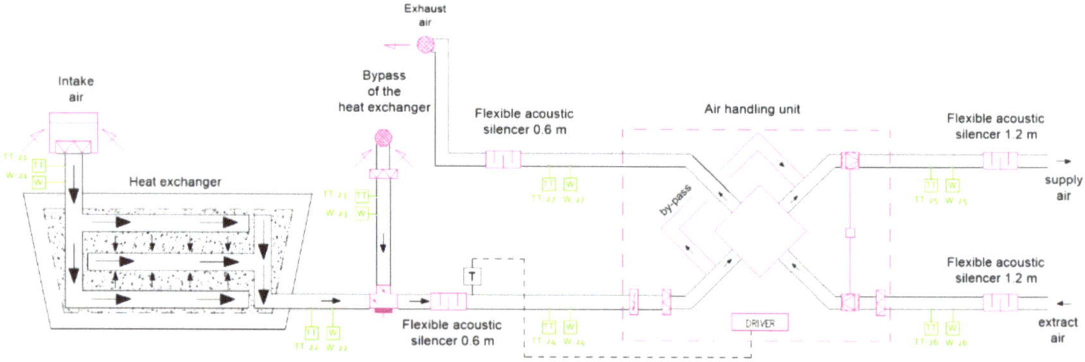

Figure 19. Diagram of integration of gravel-type EAHE into the mechanical ventilation system of a single-family building [81].

Currently, various efficiency-modeling methods are being used and experimental research is being conducted to develop an exchanger that works in the most efficient system with other renewable energy sources. Efforts are being made to reduce pressure losses and achieve the highest possible efficiency per day and per year. The interest in the topic of EAHEs is huge, which confirms many scientific publications and their citations (Table 5). Numerous studies by many authors confirm that different configurations of EAHEs can be used all over the world in different climatic conditions. Their installation allows a reduction in the energy demand for ventilation and decarbonization in the building sector.

Table 5. Reviewed papers on earth-to-air heat exchangers.

Authors, Year	Title	Journal	Citations	Keywords
Kalus D. et al., 2022 [79]	Experience in Researching and Designing an Innovative Way of Operating Combined Building–Energy Systems Using Renewable Energy Sources	Applied Sciences	0	Combined building–energy systems, RES, energy roof, ground-source heat storage; peak heat source, ground-heat recovery, air–heat exchanger, cooling circuits
Wei H, et al., 2020, [95]	Field experiments on the cooling capability of earth-to-air heat exchangers in hot and humid climate	Applied Energy	35	Earth-to-air heat exchanger, renewable energy, cooling capacity, configuration parameter, hot and humid climate
Sakhri N. et al., 2020, [96]	Experimental investigation of the performance of earth-to-air heat exchangers in arid environments	Journal of Arid Environments	27	Earth-to-air heat exchanger, soil temperature, arid region, hygrometry regime, natural ventilation
Amanowicz Ł. Wojtkowiak J. 2020, [102]	Thermal performance of multi-pipe earth-to-air heat exchangers considering the non-uniform distribution of air between parallel pipes	Geothermics	16	Multi-pipe earth-to-air heat exchanger, airflow distribution, thermal performance, tnergy efficiency, multi-tube
Amanowicz Ł. Wojtkowiak J. 2020, [103]	Approximated flow characteristics of multi-pipe earth-to-air heat exchangers for thermal analysis under variable airflow conditions	Renewable Energy	21	Earth-to-air heat exchanger, multi-pipe, flow characteristics, airflow distribution, thermal analysis
Soares N. et al., 2021, [84]	Advances in standalone and hybrid earth-air heat exchanger (EAHE) systems for buildings: A review	Energy and Buildings	12	Earth-air heat exchanger (EAHE) hybrid system, geothermal energy, design parameters, operation conditions, thermal performance

Table 5. *Cont.*

Authors, Year	Title	Journal	Citations	Keywords
Li Y. et al., 2021, [91]	An experimental investigation on the passive ventilation and cooling performance of an integrated solar chimney and earth–air heat exchanger	Renewable Energy	20	Earth–air heat exchanger, solar chimney, natural ventilation, cooling capacity, full-scale experimental study
Long T. et al., 2021, [93]	Numerical investigation of the working mechanisms of solar chimney coupled with earth-to-air heat exchanger (SCEAHE)	Solar Energy	8	Solar chimney, earth-to-air heat exchanger, passive cooling, natural ventilation, numerical simulation
Pouranian F. et al., 2021, [94]	Performance assessment of solar chimney coupled with earth-to-air heat exchanger: A passive alternative for an indoor swimming pool ventilation in hot-arid climate	Applied Energy	7	Indoor swimming pool, renewable energies, solar chimney, passive ventilation, thermal comfort, computational fluid dynamics
Skotnicka-Siepsiak A. 2021, [99]	An Evaluation of the Performance of a Ground-to-Air Heat Exchanger in Different Ventilation Scenarios in a Single-Family Home in a Climate Characterized by Cold Winters and Hot Summers	Energies	4	Ground-to-air heat exchanger, sustainable indoor ventilation system, energy-efficient ventilation
Amanowicz Ł. Wojtkowiak J. 2021, [72]	Comparison of Single- and Multipipe Earth-to-Air Heat Exchangers in Terms of Energy Gains and Electricity Consumption: A Case Study for the Temperate Climate of Central Europe	Energies	8	Earth-to-air heat exchangers, pressure losses, multipipe, renewable energy, geothermal energy, building energy performance, ventilation
Qi D. et al., 2021, [106]	Comparative analysis of earth to air heat exchanger configurations based on uniformity and thermal performance	Applied Thermal Engineering	15	Multi-pipe earth to air heat exchanger (MEAHE), uniformity, thermal performance, pressure drop
Mihalakakou G. et al., 2022, [83]	Applications of earth-to-air heat exchangers: A holistic review	Renewable and Sustainable Energy Reviews	11	Earth-to-air heat exchangers, studies of EAHE Systems, experimental studies, hybrid EAHE Systems, economic assessment
Radomski B. et al., 2022, [81]	The Direct-Contact Gravel, Ground, Air Heat Exchanger—Application in Single-Family Residential Passive Buildings	Energies	0	Air direct-contact; gravel; ground heat exchanger; heating and cooling support; passive buildings
Ramalho J. et al., 2022, [85]	Accessing the thermal performance of Earth–air heat exchangers surrounded by galvanized structures	Sustainable Energy Technologies and Assessments	0	Earth–air heat exchangers, (EAHE), fins, galvanized structures, thermal conductivity, thermal performance, numerical simulations
Gao X. et al., 2022, [86]	Thermal potential improvement of an earth-air heat exchanger (EAHE) by employing backfilling for deep underground emergency ventilation	Energy	3	Geothermal, EAHE system backfilling, energy-efficiency ventilation, deep underground buildings
Long T. et al., 2022, [87]	Numerical simulation of diurnal and annual performance of coupled solar chimney with earth-to-air heat exchanger system	Applied Thermal Engineering	2	Solar chimney, earth-to-air heat exchanger, diurnal and annual performance, dynamic simulation
Long T. et al., 2022, [88]	Natural ventilation performance of solar chimney with and without earth-air heat exchanger during transition seasons	Energy	7	Earth–air heat exchanger, solar chimney, buoyancy force, natural ventilation, indoor thermal environment
Bai Y. et al., 2022, [89]	Experimental investigation of natural ventilation characteristics of a solar chimney coupled with earth-air heat exchanger (SCEAHE) system in summer and winter	Renewable Energy	2	Solar chimney, earth-air heat exchanger, natural ventilation, summer and winter, experimental study

Table 5. *Cont.*

Authors, Year	Title	Journal	Citations	Keywords
Long T. et al., 2022, [90]	Investigation on the cooling performance of a buoyancy driven earth-air heat exchanger system and the impact on indoor thermal environment	Applied Thermal Engineering	3	Earth-air heat exchanger, solar chimney, thermal mass, buoyancy force, indoor thermal environment
Long T. et al., 2022, [92]	Benefits of integrating phase-change material with solar chimney and earth-to-air heat exchanger system for passive ventilation and cooling in summer	Journal of Energy Storage	3	Solar chimney, earth-to-air heat exchanger, phase-change material, passive ventilation passive cooling
Belloufi Y. et al., 2022, [97]	Transient assessment of an earth air heat exchanger in warm climatic conditions	Geothermics	2	Earth air heat exchanger, transient thermal performance, continuous operation, derating factor, summer cooling
Ahmed SF. et al., 2022, [98]	Thermal performance of building-integrated horizontal earth-air heat exchanger in a subtropical hot humid climate	Geothermics	2	Building energy consumption, ground heat exchanger, thermal modelling, thermal performance, renewable energy, cnergy savings
Michalak P. 2022, [100]	Hourly Simulation of an Earth-to-Air Heat Exchanger in a Low-Energy Residential Building	Energies	4	Earth-to-air heat exchanger; EAHE, EAHX, outlet air temperature, ground temperature, EN 16798-5, EN ISO 13790, 5R1C model, bypass
Michalak P. 2022, [101]	Impact of Air Density Variation on a Simulated Earth-to-Air Heat Exchanger's Performance	Energies	1	Earth-to-air heat exchanger, air density, specific heat of air, barometric formula, EAHE, outlet temperature, ground temperature, EN ISO 13790, 5R1C model, hourly simulation
Qi D. et al., 2022, [107]	Structural optimization of multi-pipe earth to air heat exchanger in greenhouse	Geothermics	5	Multi-pipe parallel earth-to-air heat exchanger, air distribution, heat exchanger performance, greenhouse

6. Heat Recovery from Exhaust Air

The drive toward near-zero-energy buildings is driving interest in a variety of energy harvesting methods. For ventilation systems, great potential exists in the use of heat from exhaust air. The use of heat-recovery exchangers to recover heat from moist exhaust air at low outdoor temperatures makes it possible to recover both sensible heat and latent heat. The recovered heat can be used to heat the supply air to the rooms, but also for heating or hot water preparation, as long as heat pumps are used for the purpose of providing sufficiently high temperatures. Striving for the highest possible heat-recovery efficiency is most often associated with a simultaneous increase in flow resistance through heat-recovery exchangers. It is therefore necessary to find a compromise. Due to the possibility of water vapor freezing, which can condense on the surface of heat-recovery exchangers, there is a need to defrost them, which consumes additional energy. New exchanger designs are still being sought and control strategies are being developed for air-handling units that will minimize the energy consumed in defrosting.

A very popular solution for heat recovery is the use of a plate heat exchanger, the energy effects of which are analyzed, for example, in [108] (Figure 20). The problem of water vapour condensation in such a heat exchanger was investigated experimentally in [96]. An increase in the heat capacity of the exchanger was observed, with a simultaneous increase in the flow resistance on the side where condensation occurred.

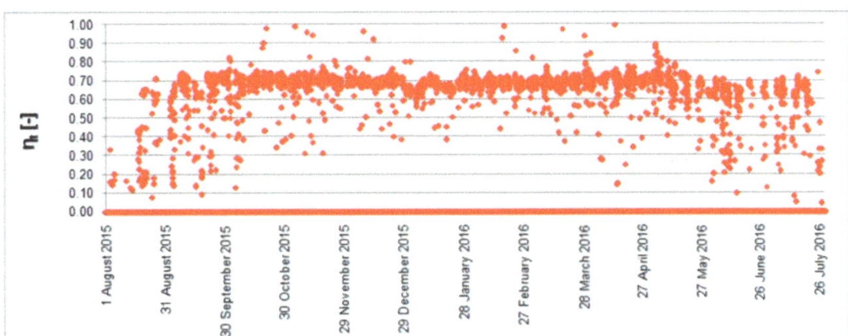

Figure 20. Efficiency of heat recovery between 1 August and 31 July [108].

However, in air-handling units, typical plate heat exchangers are sometimes replaced by membrane heat exchangers, in which the membrane used allows not only heat transfer by penetration through their wall, but also diffusion of water vapor (moisture recovery) based on the concentration difference between air streams separated by the membrane. In [109] the authors reviewed membrane heat exchangers, their design and operating efficiency under various conditions, and the mathematical models used to simulate their operating efficiency. In [110] one can read about assisting the heat-transfer process in a membrane heat exchanger with ultrasound.

Another way of recovering heat from the exhaust air is the use of an accumulating rotary heat-wheel heat exchanger, for which the authors of [111] note that its efficiency is reduced by longitudinal heat conduction. They proposed a wheel made of plastic offering high effectiveness of about 90% at acceptable pressure losses and low cost of material. A similar concept in principle of an exchanger for heat and moisture recovery was proposed in [112]. The hollow fiber membrane heat exchanger was developed and its latent and sensible heat-recovery effectiveness was examined (also experimentally for validation of simulation mode) resulting in an effectiveness of 80% and 62%, respectively, for latent and sensible heat. A similar problem was also observed and analyzed in [46], where it was shown that the average annual temperature recovery efficiency can be 10% lower if the effect of longitudinal heat conduction is taken into account.

A very interesting and comprehensive way to recover heat from exhaust air is the system presented in [113]. The technology is based on the use of an air-handling unit with a high-efficiency heat-recovery exchanger, a cooling system for PV panels and preheating of fresh air using ventilation air. The added effect is an increase in the efficiency of the PV panels and greater electricity production. Similarly, a complex system that is used to recover heat and at the same time dehumidify the air in a building is described in [114]. Exhaust air is directed to a module on the roof of the building, where heat and water vapor exchange takes place. Depending on the mode of operation (day/night), the exhaust air is superheated and/or humidified, and then directed to the heat exchanger in the air-handling unit. There it causes a significant increase in the temperature of the fresh air supplied to the building, also due to the enthalpy of the water vapor in it. In turn, in summer conditions, the system allows one to reduce the moisture load, and thus reduce the energy consumed by the air conditioner. On the contrary, a similar system, but using double glazing of the building, is described in [115]. The removed air is routed between the glazing, causing a reduction in the intensity of heat transfer through the partition, thus reducing heat gain in summer and heat loss in winter, generating the lower energy demand shown in Figure 21 depending on the recovery systems. The systems also have a positive effect on thermal comfort as a result of the elimination of the asymmetric thermal field. The authors note the problem of condensation on the glazing and the formation of fog that obscures the view through the window.

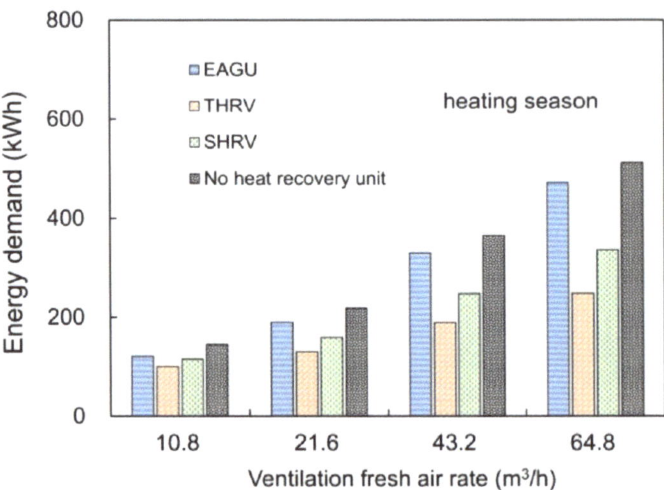

Figure 21. Yearly energy demand for ventilation using different heat-recovery systems: EAGU = exhaust air glass unit, THRV = conventional total heat recovery ventilator, SHRV = sensible heat recovery ventilator [115].

Ventilation systems with a so-called "warm air intake," where fresh air is drawn through a glazed element that acquires radiant heat from the sun, are described in the paper [116]. "Sunspaces" were investigated in the climate of Spain, where reductions in energy savings ranging from 2.51 $kWh/m^2/year$ in areas of the country with milder climates to as much as 39.54 $kWh/m^2/year$ in areas with colder climates were obtained. In contrast, [117] tested a system for heat recovery and fresh air supply integrated into a window frame in which heat pipes were installed. The CFD simulations showed a heat-recovery efficiency of 65–95% depending on the size of the airflow.

A comprehensive review of various types of technologies for heat recovery from exhaust air was published in [118]. The authors pay special attention to the problem of frosting of appliances in cold climates. This is a phenomenon associated with condensation of water vapor and its subsequent freezing on cold components. The solution to this problem still requires various strategies of protection against frosting or defrosting, which, however, are associated with additional energy consumption, a temporary reduction in air quality in the building, maintenance of the continuity of air supply, and a lack of stable pressure distribution in the building. The economic energy aspect, which is not always beneficial, of heat recovery in various ventilation systems is also presented by the author of a review publication [119].

In [120,121] a heat-recovery system integrated with a pump-driven loop heat pipe and a heat pump is presented. The proposed combination of compressor and pump circuits, as well as the principle of the heat pipe, allows the operation mode to be adjusted to current conditions, ensuring energy savings. The system requires four properly interconnected heat exchangers and is particularly effective in cold climates.

In [122] a typical indoor air heat-recovery system based on a compressor heat pump installed inside a ventilation unit is presented. Innovative in this solution is the dual-cylinder compressor, and complementing the study of the heat-recovery system itself are simulations of the annual operation of this device in the ventilation system of a building.

The results of a study, rarely used in the ventilation of a thermoelectric heat pump, are presented in [123]. The system, although less efficient than the compressor heat pump in the proposed solution, can be competitive for buildings with low heating power requirements of <10 W/m^2. A wide and smooth performance control range and high control accuracy are presented as advantages of this solution. The results of system optimization were presented by the same team of authors in the article [124].

The authors of [125] investigated the feasibility of using a thermoelectric heat pump mounted on a window frame as aheat-recovery element from exhaust air. Pilot studies were conducted in the UK. The conclusions emphasized that the system would be most environmentally friendly if it worked in conjunction with PV panels.

Heat recovered from exhaust air can also be used for heating purposes in a multi-family building. This solution, taking into account the location of the building as well as the different operating parameters of the heating system, was analyzed in [11] and is presented in Figure 22. The highest average coefficient of performance (COP) of the system was found in Koszalin (mild climate on the Baltic Sea in Poland) for the lowest operating parameters of the heating system (35/28 °C). The article highlights that CO_2 emissions for heating the analyzed buildings can be further reduced if the electricity to power the heat pump comes from PV panels. The authors' team also performed an analysis of the efficiency of air heat recovery to heat domestic hot water [126]. For the case analyzed and the Polish energy production conditions, the final energy demand for hot water preparation was 35.1% lower compared to the use of a condensing gas boiler only.

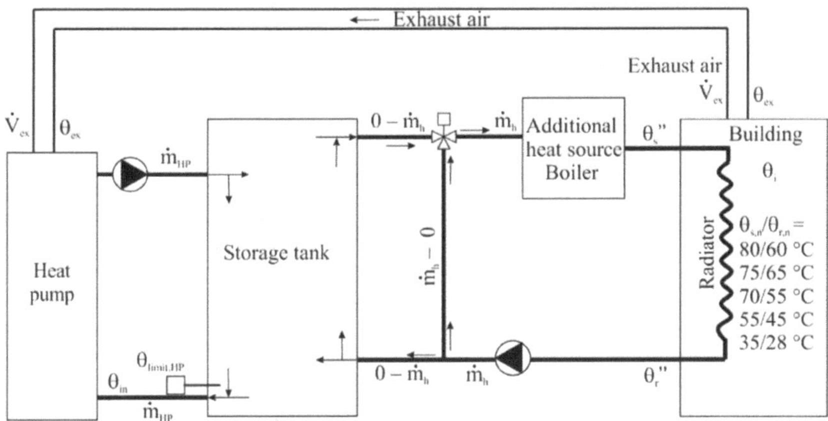

Figure 22. Schema of the system of heat recovery from exhausted air, presented in [11].

A system of two recovery exchangers as an idea to increase heat-recovery efficiency is analyzed in [127]. In summer, exhaust air is directed to the secondary heat exchanger to heat the air behind the cooler and then goes to the primary unit to pre-cool the fresh air. Compared to the baseline solution, with a single heat-recovery exchanger, the solution reduced the energy consumption of the cooler by more than 20% and that of the heater by 80%.

The test results in [128] especially should move those responsible for maintaining mechanical ventilation systems. The study showed that the pressure loss of a dirty exchanger was 12% higher and the thermal efficiency 8% lower compared to a clean exchanger. Attention is paid to the need for cleanliness and periodic inspection of system components.

In recent years, there have been many publications on innovations in the field of heat-recovery technology from exhaust air. The 23 most interesting articles, in the opinion of the authors, have been selected to show trends and developments (Table 6). It is possible to create a heat-recovery system from exhausted air owing to numerous configurations of heat exchangers, i.e., plate, membrane, heat wheels, and using heat pumps. Composite systems with several exchangers are being developed, allowing multi-stage heat recovery. To ensure the high efficiency of heat exchangers, it is necessary to protect them from freezing and keep them clean.

Table 6. Reviewed papers on heat recovery from exhausted air.

Authors Year	Title	Journal	Citations	Keywords
Cepiński W. et al., 2020, [126]	Waste heat recovery by electric heat pump from exhausted ventilating air for domestic hot water in multi-family residential buildings	Annual Set The Environment Protection	2	Air-to-water electric heat pump, waste heat recovery, heat recovery in ventilation, domestic hot water, TRNSYS
Zender–Świercz E. 2021, [119]	A Review of Heat Recovery in Ventilation	Energies	15	Heat recovery, ventilation systems, recovery efficiency, energy-consumption
Michalak, P. 2021, [108]	Annual Energy Performance of an Air Handling Unit with a Cross-Flow Heat Exchanger	Energies	3	Air-handling unit, cross-flow heat exchanger, heat recovery, temperature effectiveness, temperature efficiency, EN 308, ventilation loss
Kowalski P. et al., 2021, [11]	Waste Heat Recovery by Air-to-Water Heat Pump from Exhausted Ventilating Air for Heating of Multi-Family Residential Buildings	Energies	1	TRNSYS, air-to-water electric heat pump, AWHP, hybrid heat source, waste heat, building energy performance, primary energy, exhaust ventilation, energy efficiency
Bai H.Y. 2022, [118]	A review of heat recovery technologies and their frost control for residential building ventilation in cold climate regions	Renewable and Sustainable Energy Reviews	9	Heat recovery, cold climate, indoor air quality, frosting control, energy-efficient ventilation
Li J. et al., 2022, [109]	A review of air-to-air membrane energy recovery technology for building ventilation	Energy and Buildings	3	Energy recovery, air-to-air membrane, moisture transfer effectiveness, ventilation
Gurubalan A., et al., 2022, [110]	Performance Improvement of Membrane Energy Exchanger Using Ultrasound for Heating, Ventilation, and Air Conditioning Application	Journal of Thermal Science and Engineering Applications	0	Ultrasound, energy exchanger, membrane dehumidifier, membrane humidifier, membrane energy recovery ventilator, energy efficiency, energy systems, heat and mass transfer, thermal systems
Liu P. et al., 2022, [111]	Development and optimization of highly efficient heat recoveries for low carbon residential buildings	Energy and Buildings	4	Energy-efficient ventilation heat recovery, heat wheel, low-carbon buildings
Cho H.J., 2022, [112]	Development of empirical models to predict latent heat exchange performance for hollow fiber membrane-based ventilation system	Applied Thermal Engineering	0	Air-to-air hollow fiber, membrane, energy recovery ventilation, latent heat exchanger, experimental analysis, empirical models
Tang Y et al., 2022, [113]	Performance prediction of a novel double-glazing PV curtain wall system combined with an air handling unit using exhaust cooling and heat recovery technology	Energy Conversion and Management	3	Solar energy, AHU, double-skin PV façade, exhaust air, heat recovery
Chen Y, et al., 2022, [114]	Energy saving potential of passive dehumidification system combined with energy recovery ventilation using renewable energy	Energy and Buildings,	1	Passive dehumidification, moisture adsorption, moisture desorption, energy recovery ventilation, renewable energy, latent heat-load reduction
Guo J. et al., 2022, [115]	Utilization of Window System as Exhaust Air Heat Recovery Device and Its Energy Performance Evaluation: A Comparative Study	Energies	1	Exhaust air heat recovery, exhaust air glass unit, low-energy building, window system; comparative study
Gainza-Barrencua J, et al., 2022, [116]	Use of sunspaces to obtain energy savings by preheating the intake air of the ventilation system: Analysis of its main characteristics in the different Spanish climate zones	Journal of Building Engineering	1	Sunspace, solar heating, mechanical ventilation, heat recovery, heat storage

Table 6. *Cont.*

Authors Year	Title	Journal	Citations	Keywords
Barreto G. 2022, [117]	An innovative window heat recovery (WHR) system with heat pipe technology: Analytical, CFD, experimental analysis and building retrofit performance	Energy Reports	3	Building ventilation, heat recovery, window heat recovery, heat pipe, energy performance, thermal comfort
Liu S. et al., 2022, [120]	Experimental study of ventilation system with heat recovery integrated by pump-driven loop heat pipe and heat pump	Journal of Building Engineering	3	Ventilation heat recovery, integrated heat pump, pump-driven loop heat pipe, switch temperature
Shuailing L. et al., 2022, [121]	Performance of a mechanically-driven loop heat pipe heat recovery system	Applied Thermal Engineering	3	Heat recovery, ventilation, heat pipe, refrigerant pump, booster
Jia X. et al., 2022, [122]	The applicability and energy consumption of a parallel-loop exhaust air heat pump for environment control in ultra-low energy building	Applied Thermal Engineering	2	Exhaust air heat pump, heat recovery, ultra-low energy building, energy consumption, COP
Diaz de Garayo S. et al., 2022, [123]	Annual energy performance of a thermoelectric heat pump combined with a heat recovery unit to HVAC one passive house dwelling	Applied Thermal Engineering	5	Thermoelectricity, heat pump, passive house, HVAC, heat recovery
Diaz de Garayo S. et al., 2022, [124]	Optimal combination of an air-to-air thermoelectric heat pump with a heat recovery system to HVAC a passive house dwelling	Applied Energy	6	Hermoelectricity, Heat pump, Heat-recovery unit, Passive house, HVAC
Xu Q. et al., 2022, [125]	Ecopump: a novel thermoelectric heat pump/heat recovery ventilator system for domestic building applications	International Journal of Low-Carbon	0	-
Goldanlou A.S. et al., 2020 [127]	Energy usage reduction in an air handling unit by incorporating two heat recovery units	Journal of Building Engineering,	72	Air-handling unit, energy demand, exergy, building
Abdul Hamid A. et al., 2020, [128]	Determining the impact of air-side cleaning for heat exchangers in ventilation systems	Building Services Engineering Research and Technology	4	-
Abadi I.R. et al., 2022, [129]	Experimental investigation of condensation in energy recovery ventilators	Energy and Buildings	5	Energy-recovery ventilator, total heat exchanger, energy exchanger, heat and moisture exchanger, condensation, membrane

7. Conclusions

The literature review was performed to leverage the current knowledge base to present the advances in ventilation systems that reduce energy consumption. This issue is still a challenge, as evidenced by the many studies conducted and analyses performed during the last three years. As many as 98 papers were subjectively chosen based on the high value of compelling evidence within. The results of the publications reviewing advancement in ventilation systems may be useful to many scientists—young and experienced—as in the case of other similar reviews, e.g., [130]. They can adopt these as an initial step to write articles, as well as to provide a guide pointing out cutting-edge technologies and strategies in ventilation systems. Although this review does not define a stance, there is a relevant contribution to recognizing the issues of the requirements for ventilation systems, the airtightness of the building's envelope, the demand-controlled ventilation strategy, decentralized ventilation systems, earth-to-air heat exchangers, and the recovery of exhaust air heat. Summaries of these aspects are presented in the particular paragraphs.

- This literature review reinforces the belief that:

- airtightness of the building's envelope is as a basic requirement for efficiency of buildings; in many countries, regulations need to be introduced or revised to suit the current global energy situation,
- lower demand for ventilation airflow and the associated lower amount of energy to drive fans are the main advantages of the ventilation control strategy known as DCV,
- decentralized systems, which do not require the use of long ducts, are an interesting alternative used to save energy, as it is known that the use of a central ventilation system requires more power consumption,
- due to the multitude of solutions and operating conditions, there is no simple answer to the question: which type of ground exchanger is better in terms of energy—multi-pipe or single-pipe? In order to answer this question, a detailed analysis of a given case should be carried out, taking into account the financial and/or energy aspect, or a SWOT analysis (strengths and weaknesses, opportunities and threats),
- the use of mechanical ventilation with heat recovery is actually a necessity regardless of the purpose of the building; such solutions are advantageous for enabling the maintenance of adequate indoor air quality, while improving the thermal performance of buildings.

We hope that, together with the discoveries of other scientists, this paper will be helpful for further developments in ventilation systems because the application of the advancements of technology in ventilation systems is an effective way to reduce the energy consumption in buildings currently and in the future.

Author Contributions: Conceptualization, Ł.A; methodology, Ł.A; writing—original draft preparation, ŁA, E.D. and K.R.; writing—review and editing, Ł.A., E.D. and K.R.; supervision, Ł.A.; project administration, Ł.A and K.R.; funding acquisition, Ł.A. and K.R. All authors have read and agreed to the published version of the manuscript.

Funding: This research was funded by the Polish Ministry of Science and Higher Education, grant number 504 101/0713/SBAD/0958.

Data Availability Statement: Not applicable.

Conflicts of Interest: The authors declare no conflict of interest.

Abbreviations

CAV	constant air volume
CO_2	carbon dioxide
DCV	demand-controlled ventilation
EAHE	earth-to-air heat exchanger
HRV	heat-recovery ventilation
HVAC	heating, ventilation, and air conditioning
LHC	longitudinal heat conduction
MERV	minimum efficiency reporting value
n_{50}	airtightness coefficient
PCM	phase-changed materials
PM	particulate matter
PV	photovoltaic
RES	renewable energy sources
RH	relative humidity
SC	solar chimney
U	heat transfer coefficient
VAV	variable air volume
VOCs	volatile organic compounds

References

1. Amanowicz, Ł.; Szczechowiak, E. Zasady projektowania systemów wentylacji budynków energooszczędnych. *Ciepłownictwo Ogrzew. Went.* **2017**, *48*, 72–78. [CrossRef]
2. Amanowicz, Ł.; Ratajczak, K. Praktyczne aspekty projektowania energooszczędnych systemów wentylacyjnych. *Rynek Instal.* **2021**, *6*, 46–52.
3. Russell, M.; Sherman, M.; Rudd, A. Review of Residential Ventilation Technologies. *HVACR Res.* **2007**, *13*, 325–348. [CrossRef]
4. Dudkiewicz, E.; Szczęśniak, S. Natural and Mechanical Ventilation of Industrial Halls. *Nowocz. Hale* **2017**, *3*, 66–71.
5. Spodyniuk, N.; Voznyak, O.; Sukholova, I.; Dovbush, O.; Kasynets, M.; Datsko, O. Diagnosis of Damage to the Ventilation System. *Diagnostyka* **2021**, *22*, 91–99. [CrossRef]
6. Cepiński, W.; Szałański, P. Air Filtration and Sterilization in Ventilation Systems According to the New Paradigm. *J. Ecol. Eng.* **2022**, *23*, 25–34. [CrossRef]
7. Cepiński, W.; Szałański, P.; Misiński, J. Redukcja rozprzestrzeniania koronawirusa SARS-CoV-2 i choroby COVID-19 poprzez instalacje wentylacyjne i klimatyzacyjne. *Instal* **2020**, *6*, 28–36. [CrossRef]
8. Szałański, P.; Cepiński, W.; Misiński, J. Przegląd zaleceń dla instalacji wentylacyjnych i klimatyzacyjnych w związku z zagrożeniem koronawirusem SARS-CoV-2 i chorobą COVID-19. *Instal* **2020**, *5*, 17–21.
9. Szałański, P.; Cepiński, W. Prawdopodobieństwo przenoszenia wirusa SARS-CoV-2 drogą powietrzną w pomieszczeniach wentylowanych. *Instal* **2022**, *2*, 23–29. [CrossRef]
10. Orman, Ł.J.; Majewski, G.; Radek, N.; Pietraszek, J. Analysis of Thermal Comfort in Intelligent and Traditional Buildings. *Energies* **2022**, *15*, 6522. [CrossRef]
11. Kowalski, P.; Szałański, P.; Cepiński, W. Waste Heat Recovery by Air-to-Water Heat Pump from Exhausted Ventilating Air for Heating of Multi-Family Residential Buildings. *Energies* **2021**, *14*, 7985. [CrossRef]
12. Kowalski, P.; Szałański, P. Airtightness Test of Single-Family Building and Calculation Result of the Energy Need for Heating in Polish Conditions. *E3S Web Conf.* **2018**, *44*, 00078. [CrossRef]
13. Cholewa, T.; Siuta-Olcha, A.; Smolarz, A.; Muryjas, P.; Wolszczak, P.; Guz, Ł.; Bocian, M.; Balaras, C.A. An Easy and Widely Applicable Forecast Control for Heating Systems in Existing and New Buildings: First Field Experiences. *J. Clean. Prod.* **2022**, *352*, 131605. [CrossRef]
14. Kordana-Obuch, S.; Starzec, M.; Słyś, D. Evaluation of the Influence of Catchment Parameters on the Required Size of a Stormwater Infiltration Facility. *Water* **2023**, *15*, 191. [CrossRef]
15. Royuela-del-Val, A.; Padilla-Marcos, M.Á.; Meiss, A.; Casaseca-de-la-Higuera, P.; Feijó-Muñoz, J. Air Infiltration Monitoring Using Thermography and Neural Networks. *Energy Build.* **2019**, *191*, 187–199. [CrossRef]
16. Bhandari, M.; Hun, D.; Shrestha, S.; Pallin, S.; Lapsa, M. A Simplified Methodology to Estimate Energy Savings in Commercial Buildings from Improvements in Airtightness. *Energies* **2018**, *11*, 3322. [CrossRef]
17. Dudkiewicz, E.; Fidorów-Kaprawy, N. Analiza energetyczna wydajności stropu grzewczego na przykładzie budynku mieszkalnego w Niemczech, Polsce i Ukrainie. *Mater. Bud.* **2022**, *9*, 84–87. [CrossRef]
18. Banister, C.; Bartko, M.; Berquist, J.; Macdonald, I.; Vuotari, M.; Wills, A. Energy and Emissions Effects of Airtightness for Six Non-Residential Buildings in Canada with Comparison to Contemporary Limits and Assumptions. *J. Build. Eng.* **2022**, *58*, 104977. [CrossRef]
19. Liu, X.; Zhang, T.; Liu, X.; Li, L.; Lin, L.; Jiang, Y. Energy Saving Potential for Space Heating in Chinese Airport Terminals: The Impact of Air Infiltration. *Energy* **2021**, *215*, 119175. [CrossRef]
20. Verbeke, S.; Audenaert, A. A Prospective Study on the Evolution of Airtightness in 41 Low Energy Dwellings. *E3S Web Conf.* **2020**, *172*, 05005. [CrossRef]
21. Moujalled, B.; Leprince, V.; Berthault, S.; Litvak, A.; Hurel, N. Mid-Term and Long-Term Changes in Building Airtightness: A Field Study on Low-Energy Houses. *Energy Build.* **2021**, *250*, 111257. [CrossRef]
22. Simson, R.; Rebane, T.; Kiil, M.; Thalfeldt, M.; Kurnitski, J. The Impact of Infiltration on Heating Systems Dimensioning in Estonian Climate. *E3S Web Conf.* **2020**, *172*, 05004. [CrossRef]
23. Miszczuk, A.; Heim, D. Parametric Study of Air Infiltration in Residential Buildings—The Effect of Local Conditions on Energy Demand. *Energies* **2020**, *14*, 127. [CrossRef]
24. Nitijevskis, A.; Keviss, V. Overview of Large Building Testing in Baltic Countries. *E3S Web Conf.* **2020**, *172*, 05007. [CrossRef]
25. Poza-Casado, I.; Meiss, A.; Padilla-Marcos, M.Á.; Feijó-Muñoz, J. Airtightness and Energy Impact of Air Infiltration in Residential Buildings in Spain. *Int. J. Vent.* **2021**, *20*, 258–264. [CrossRef]
26. Guillén-Lambea, S.; Rodríguez-Soria, B.; Marín, J.M. Air Infiltrations and Energy Demand for Residential Low Energy Buildings in Warm Climates. *Renew. Sustain. Energy Rev.* **2019**, *116*, 109469. [CrossRef]
27. Domínguez-Amarillo, S.; Fernández-Agüera, J.; Campano, M.Á.; Acosta, I. Effect of Airtightness on Thermal Loads in Legacy Low-Income Housing. *Energies* **2019**, *12*, 1677. [CrossRef]
28. Paukštys, V.; Cinelis, G.; Mockienė, J.; Daukšys, M. Airtightness and Heat Energy Loss of Mid-Size Terraced Houses Built of Different Construction Materials. *Energies* **2021**, *14*, 6367. [CrossRef]
29. Heim, D.; Miszczuk, A. Modelling Building Infiltration Using the Airflow Network Model Approach Calibrated by Air-Tightness Test Results and Leak Detection. *Build. Serv. Eng. Res. Technol.* **2020**, *41*, 681–693. [CrossRef]

30. Mélois, A.; Carrié, F.R.; El Mankibi, M.; Moujalled, B. Uncertainty in Building Fan Pressurization Tests: Review and Gaps in Research. *J. Build. Eng.* **2022**, *52*, 104455. [CrossRef]
31. Zheng, X.; Cooper, E.; Gillott, M.; Wood, C. A Practical Review of Alternatives to the Steady Pressurisation Method for Determining Building Airtightness. *Renew. Sustain. Energy Rev.* **2020**, *132*, 110049. [CrossRef]
32. Poza-Casado, I.; Rodríguez-del-Tío, P.; Fernández-Temprano, M.; Padilla-Marcos, M.-Á.; Meiss, A. An Envelope Airtightness Predictive Model for Residential Buildings in Spain. *Build. Environ.* **2022**, *223*, 109435. [CrossRef]
33. Lu, X.; Pang, Z.; Fu, Y.; O'Neill, Z. Advances in Research and Applications of CO2-Based Demand-Controlled Ventilation in Commercial Buildings: A Critical Review of Control Strategies and Performance Evaluation. *Build. Environ.* **2022**, *223*, 109455. [CrossRef]
34. Delwati, M.; Merema, B.; Breesch, H.; Helsen, L.; Sourbron, M. Impact of Demand Controlled Ventilation on System Performance and Energy Use. *Energy Build.* **2018**, *174*, 111–123. [CrossRef]
35. Abdul Hamid, A.; Johansson, D.; Wahlström, Å.; Fransson, V. The Impact of a DCV-System on the IAQ, Energy Use, and Moisture Safety in Apartments—A Case Study. *Int. J. Vent.* **2022**, *21*, 35–52. [CrossRef]
36. Abuimara, T.; Hobson, B.W.; Gunay, B.; O'Brien, W. Exploring the Adequacy of Mechanical Ventilation for Acceptable Indoor Air Quality in Office Buildings. *Sci. Technol. Built Environ.* **2022**, *28*, 275–288. [CrossRef]
37. Alaidroos, A.; Almaimani, A.; Krarti, M.; Dahlan, A.; Maddah, R. Evaluation of the Performance of Demand Control Ventilation System for School Buildings Located in the Hot Climate of Saudi Arabia. *Build. Simul.* **2022**, *15*, 1067–1082. [CrossRef]
38. Liu, P.; Justo Alonso, M.; Mathisen, H.M. Heat Recovery Ventilation Design Limitations Due to LHC for Different Ventilation Strategies in ZEB. *Build. Environ.* **2022**, *224*, 109542. [CrossRef]
39. Bandurski, K. Różnica Między Obliczeniowym i Pomiarowym Wykorzystaniem Energii Do Ogrzewania w Budynkach Wielorodzinnych. *Ciepłownictwo Ogrzew. Went.* **2021**, *1*, 14–18. [CrossRef]
40. Ratajczak, K. Rozbieżności Między Obliczeniowym i Zmierzonym Zużyciem Energii Do Ogrzewania i Przygotowania Ciepłej Wody Użytkowej Na Przykładzie Budynków Jednorodzinnych. *Ciepłownictwo Ogrzew. Went.* **2022**, *1*, 5–11. [CrossRef]
41. Haddad, S.; Synnefa, A.; Ángel Padilla Marcos, M.; Paolini, R.; Delrue, S.; Prasad, D.; Santamouris, M. On the Potential of Demand-Controlled Ventilation System to Enhance Indoor Air Quality and Thermal Condition in Australian School Classrooms. *Energy Build.* **2021**, *238*, 110838. [CrossRef]
42. Zhao, W.; Kilpeläinen, S.; Bask, W.; Lestinen, S.; Kosonen, R. Operational Challenges of Modern Demand-Control Ventilation Systems: A Field Study. *Buildings* **2022**, *12*, 378. [CrossRef]
43. Mou, J.; Cui, S.; Khoo, D.W.Y. Computational Fluid Dynamics Modelling of Airflow and Carbon Dioxide Distribution inside a Seminar Room for Sensor Placement. *Meas. Sens.* **2022**, *23*, 100402. [CrossRef]
44. Szczurek, A.; Azizah, A.; Maciejewska, M. The Detection of Activities Occurring Inside Quick Service Restaurants That Influence Air Quality. *Sensors* **2022**, *22*, 4056. [CrossRef]
45. Justo Alonso, M.; Wolf, S.; Jørgensen, R.B.; Madsen, H.; Mathisen, H.M. A Methodology for the Selection of Pollutants for Ensuring Good Indoor Air Quality Using the De-Trended Cross-Correlation Function. *Build. Environ.* **2022**, *209*, 108668. [CrossRef]
46. Borgen Haugland, M.; Yang, A.; Holøs, S.B.; Thunshelle, K.; Mysen, M. Demand-Controlled Ventilation: Do Different User Groups Require Different CO$_2$-Setpoints? *IOP Conf. Ser. Mater. Sci. Eng.* **2019**, *609*, 042062. [CrossRef]
47. Shin, H.; Kang, M.; Mun, S.-H.; Kwak, Y.; Huh, J.-H. A Study on Changes in Occupants' Thermal Sensation Owing to CO$_2$ Concentration Using PMV and TSV. *Build. Environ.* **2021**, *187*, 107413. [CrossRef]
48. Kiamanesh, B.; Behravan, A.; Obermaisser, R. Realistic Simulation of Sensor/Actuator Faults for a Dependability Evaluation of Demand-Controlled Ventilation and Heating Systems. *Energies* **2022**, *15*, 2878. [CrossRef]
49. Lu, X.; O'Neill, Z.; Li, Y.; Niu, F. A Novel Simulation-Based Framework for Sensor Error Impact Analysis in Smart Building Systems: A Case Study for a Demand-Controlled Ventilation System. *Appl. Energy* **2020**, *263*, 114638. [CrossRef]
50. Taal, A.; Itard, L. Fault Detection and Diagnosis for Indoor Air Quality in DCV Systems: Application of 4S3F Method and Effects of DBN Probabilities. *Build. Environ.* **2020**, *174*, 106632. [CrossRef]
51. Altendorf, D.; Grünewald, H.; Liu, T.-L.; Dehnert, J.; Trabitzsch, R.; Weiß, H. Decentralised Ventilation Efficiency for Indoor Radon Reduction Considering Different Environmental Parameters. *Isot. Environ. Health Stud.* **2022**, *58*, 195–213. [CrossRef]
52. Dehnert, J.; Altendorf, D.; Trabitzsch, R.; Grünewald, H.; Geisenhainer, R.; Oeser, V.; Streil, T.; Weber, L.; Schönherr, B.; Thomas, J.; et al. Radon Protection in Apartments Using a Ventilation System Wireless-Controlled by Radon Activity Concentration. *J. Radiol. Prot.* **2021**, *41*, S109. [CrossRef]
53. Ratajczak, K. Ventilation Strategy for Proper IAQ in Existing Nurseries Buildings—Lesson Learned from the Research during COVID-19 Pandemic. *Aerosol Air Qual. Res.* **2022**, *22*, 210337. [CrossRef]
54. Ratajczak, K.; Basińska, M. The Well-Being of Children in Nurseries Does Not Have to Be Expensive: The Real Costs of Maintaining Low Carbon Dioxide Concentrations in Nurseries. *Energies* **2021**, *14*, 2035. [CrossRef]
55. Ratajczak, K.; Szczechowiak, E. Energy Consumption Decreasing Strategy for Indoor Swimming Pools—Decentralized Ventilation System with a Heat Pump. *Energy Build.* **2020**, *206*, 109574. [CrossRef]
56. Merzkirch, A.; Maas, S.; Scholzen, F.; Waldmann, D. Primary Energy Used in Centralised and Decentralised Ventilation Systems Measured in Field Tests in Residential Buildings. *Int. J. Vent.* **2019**, *18*, 19–27. [CrossRef]
57. Bonato, P.; D'Antoni, M.; Fedrizzi, R. Modelling and Simulation-Based Analysis of a Façade-Integrated Decentralized Ventilation Unit. *J. Build. Eng.* **2020**, *29*, 101183. [CrossRef]

58. Fu, N.; Kim, M.K.; Chen, B.; Sharples, S. Comparative Modelling Analysis of Air Pollutants, $PM_{2.5}$ and Energy Efficiency Using Three Ventilation Strategies in a High-Rise Building: A Case Study in Suzhou, China. *Sustainability* **2021**, *13*, 8453. [CrossRef]
59. Filis, V.; Kolarik, J.; Smith, K.M. The Impact of Wind Pressure and Stack Effect on the Performance of Room Ventilation Units with Heat Recovery. *Energy Build.* **2021**, *234*, 110689. [CrossRef]
60. Carbonare, N.; Fugmann, H.; Asadov, N.; Pflug, T.; Schnabel, L.; Bongs, C. Simulation and Measurement of Energetic Performance in Decentralized Regenerative Ventilation Systems. *Energies* **2020**, *13*, 6010. [CrossRef]
61. Zender-Świercz, E. Improvement of Indoor Air Quality by Way of Using Decentralised Ventilation. *J. Build. Eng.* **2020**, *32*, 101663. [CrossRef]
62. Catalina, T.; Istrate, M.; Damian, A.; Vartires, A.; Dicu, T.; Cucoş, A. Indoor Air Quality Assessment in a Classroom Using a Heat Recovery Ventilation Unit. *Rom. J. Phys.* **2019**, *64*, 9–10.
63. Ratajczak, K.; Amanowicz, Ł.; Szczechowiak, E. Assessment of the Air Streams Mixing in Wall-Type Heat Recovery Units for Ventilation of Existing and Refurbishing Buildings toward Low Energy Buildings. *Energy Build.* **2020**, *227*, 110427. [CrossRef]
64. Carbonare, N.; Pflug, T.; Bongs, C.; Wagner, A. Design and Implementation of an Occupant-Centered Self-Learning Controller for Decentralized Residential Ventilation Systems. *Build. Environ.* **2021**, *206*, 108380. [CrossRef]
65. Dudkiewicz, E.; Laska, M.; Fidorów-Kaprawy, N. Users' Sensations in the Context of Energy Efficiency Maintenance in Public Utility Buildings. *Energies* **2021**, *14*, 8159. [CrossRef]
66. Zender-Świercz, E.; Telejko, M.; Galiszewska, B.; Starzomska, M. Assessment of Thermal Comfort in Rooms Equipped with a Decentralised Façade Ventilation Unit. *Energies* **2022**, *15*, 7032. [CrossRef]
67. Zender–Świercz, E. Assessment of Indoor Air Parameters in Building Equipped with Decentralised Façade Ventilation Device. *Energies* **2021**, *14*, 1176. [CrossRef]
68. Pekdogan, T.; Tokuç, A.; Ezan, M.A.; Başaran, T. Experimental Investigation of a Decentralized Heat Recovery Ventilation System. *J. Build. Eng.* **2021**, *35*, 102009. [CrossRef]
69. Zemitis, J.; Bogdanovics, R. Heat Recovery Efficiency of Local Decentralized Ventilation Device at Various Pressure Differences. *Mag. Civ. Eng.* **2020**, *2*, 120–128. [CrossRef]
70. Pekdogan, T.; Tokuç, A.; Ezan, M.A.; Başaran, T. Experimental Investigation on Heat Transfer and Air Flow Behavior of Latent Heat Storage Unit in a Facade Integrated Ventilation System. *J. Energy Storage* **2021**, *44*, 103367. [CrossRef]
71. Shahrzad, S.; Umberto, B. Parametric Optimization of Multifunctional Integrated Climate-Responsive Opaque and Ventilated Façades Using CFD Simulations. *Appl. Therm. Eng.* **2022**, *204*, 117923. [CrossRef]
72. Amanowicz, Ł.; Wojtkowiak, J. Comparison of Single- and Multipipe Earth-to-Air Heat Exchangers in Terms of Energy Gains and Electricity Consumption: A Case Study for the Temperate Climate of Central Europe. *Energies* **2021**, *14*, 8217. [CrossRef]
73. Kalús, D.; Gašparík, J.; Janík, P.; Kubica, M.; Šťastný, P. Innovative Building Technology Implemented into Facades with Active Thermal Protection. *Sustainability* **2021**, *13*, 4438. [CrossRef]
74. Radlak, G.; Ulbrich, R. Application of Environmental Energy in Modern Solutions of Passive Buildings Based on ISOMAX System. *Zesz. Nauk. Inż. Śr. Politech. Opol.* **2005**, *310*, 141–149.
75. Radlak, G.; Szmolke, N.; Ulbrich, R. The Experiences Concerning Energy Audits in Low Energy Consuming Buildings. *Zesz. Nauk. Inż. Śr. Politech. Opol.* **2006**, *319*, 205–214.
76. Radlak, G. System pozyskiwania ciepła dla potrzeb ogrzewania i wentylacji obiektów. *Zesz. Nauk. Mech. Politech. Opol.* **2009**, *94*, 77–78.
77. Kalús, D.; Koudelková, D.; Mučková, V.; Sokol, M.; Kurčová, M.; Janík, P. Practical Experience in the Application of Energy Roofs, Ground Heat Storages, and Active Thermal Protection on Experimental Buildings. *Appl. Sci.* **2022**, *12*, 9313. [CrossRef]
78. Kalús, D.; Koudelková, D.; Mučková, V.; Sokol, M.; Kurčová, M. Contribution to the Research and Development of Innovative Building Components with Embedded Energy-Active Elements. *Coatings* **2022**, *12*, 1021. [CrossRef]
79. Kalús, D.; Koudelková, D.; Mučková, V.; Sokol, M.; Kurčová, M. Experience in Researching and Designing an Innovative Way of Operating Combined Building–Energy Systems Using Renewable Energy Sources. *Appl. Sci.* **2022**, *12*, 10214. [CrossRef]
80. Pacak, A.; Jedlikowski, A.; Karpuk, M.; Anisimov, S. Analysis of Power Demand Calculation for Freeze Prevention Methods of Counter-Flow Heat Exchangers Used in Energy Recovery from Exhaust Air. *Int. J. Heat Mass Transf.* **2019**, *133*, 842–860. [CrossRef]
81. Radomski, B.; Kowalski, F.; Mróz, T. The Direct-Contact Gravel, Ground, Air Heat Exchanger—Application in Single-Family Residential Passive Buildings. *Energies* **2022**, *15*, 6110. [CrossRef]
82. Przydróżny, E.; Przydróżna, A.; Szczęśniak, S. Możliwość wykorzystania odzysku ciepła z powietrza wywiewanego z suszarni zbóż. *Instal* **2018**, *11*, 26–30.
83. Mihalakakou, G.; Souliotis, M.; Papadaki, M.; Halkos, G.; Paravantis, J.; Makridis, S.; Papaefthimiou, S. Applications of Earth-to-Air Heat Exchangers: A Holistic Review. *Renew. Sustain. Energy Rev.* **2022**, *155*, 111921. [CrossRef]
84. Soares, N.; Rosa, N.; Monteiro, H.; Costa, J.J. Advances in Standalone and Hybrid Earth-Air Heat Exchanger (EAHE) Systems for Buildings: A Review. *Energy Build.* **2021**, *253*, 111532. [CrossRef]
85. Ramalho, J.V.A.; Fernando, H.J.; Brum, R.S.; Domingues, A.M.B.; Navarro Pastor, N.R.; Burlón Olivera, M.R. Accessing the Thermal Performance of Earth–Air Heat Exchangers Surrounded by Galvanized Structures. *Sustain. Energy Technol. Assess.* **2022**, *54*, 102838. [CrossRef]
86. Gao, X.; Xiao, Y.; Gao, P. Thermal Potential Improvement of an Earth-Air Heat Exchanger (EAHE) by Employing Backfilling for Deep Underground Emergency Ventilation. *Energy* **2022**, *250*, 123783. [CrossRef]

87. Long, T.; Zhao, N.; Li, W.; Wei, S.; Li, Y.; Lu, J.; Huang, S.; Qiao, Z. Numerical Simulation of Diurnal and Annual Performance of Coupled Solar Chimney with Earth-to-Air Heat Exchanger System. *Appl. Therm. Eng.* **2022**, *214*, 118851. [CrossRef]
88. Long, T.; Zhao, N.; Li, W.; Wei, S.; Li, Y.; Lu, J.; Huang, S.; Qiao, Z. Natural Ventilation Performance of Solar Chimney with and without Earth-Air Heat Exchanger during Transition Seasons. *Energy* **2022**, *250*, 123818. [CrossRef]
89. Bai, Y.; Long, T.; Li, W.; Li, Y.; Liu, S.; Wang, Z.; Lu, J.; Huang, S. Experimental Investigation of Natural Ventilation Characteristics of a Solar Chimney Coupled with Earth-Air Heat Exchanger (SCEAHE) System in Summer and Winter. *Renew. Energy* **2022**, *193*, 1001–1018. [CrossRef]
90. Long, T.; Li, Y.; Li, W.; Liu, S.; Lu, J.; Zheng, D.; Ye, K.; Qiao, Z.; Huang, S. Investigation on the Cooling Performance of a Buoyancy Driven Earth-Air Heat Exchanger System and the Impact on Indoor Thermal Environment. *Appl. Therm. Eng.* **2022**, *207*, 118148. [CrossRef]
91. Li, Y.; Long, T.; Bai, X.; Wang, L.; Li, W.; Liu, S.; Lu, J.; Cheng, Y.; Ye, K.; Huang, S. An Experimental Investigation on the Passive Ventilation and Cooling Performance of an Integrated Solar Chimney and Earth–Air Heat Exchanger. *Renew. Energy* **2021**, *175*, 486–500. [CrossRef]
92. Long, T.; Li, W.; Lv, Y.; Li, Y.; Liu, S.; Lu, J.; Huang, S.; Zhang, Y. Benefits of Integrating Phase-Change Material with Solar Chimney and Earth-to-Air Heat Exchanger System for Passive Ventilation and Cooling in Summer. *J. Energy Storage* **2022**, *48*, 104037. [CrossRef]
93. Long, T.; Zheng, D.; Li, W.; Li, Y.; Lu, J.; Xie, L.; Huang, S. Numerical Investigation of the Working Mechanisms of Solar Chimney Coupled with Earth-to-Air Heat Exchanger (SCEAHE). *Sol. Energy* **2021**, *230*, 109–121. [CrossRef]
94. Pouranian, F.; Akbari, H.; Hosseinalipour, S.M. Performance Assessment of Solar Chimney Coupled with Earth-to-Air Heat Exchanger: A Passive Alternative for an Indoor Swimming Pool Ventilation in Hot-Arid Climate. *Appl. Energy* **2021**, *299*, 117201. [CrossRef]
95. Wei, H.; Yang, D.; Wang, J.; Du, J. Field Experiments on the Cooling Capability of Earth-to-Air Heat Exchangers in Hot and Humid Climate. *Appl. Energy* **2020**, *276*, 115493. [CrossRef]
96. Sakhri, N.; Menni, Y.; Ameur, H. Experimental Investigation of the Performance of Earth-to-Air Heat Exchangers in Arid Environments. *J. Arid Environ.* **2020**, *180*, 104215. [CrossRef]
97. Belloufi, Y.; Zerouali, S.; Rouag, A.; Aissaoui, F.; Atmani, R.; Brima, A.; Moummi, N. Transient Assessment of an Earth Air Heat Exchanger in Warm Climatic Conditions. *Geothermics* **2022**, *104*, 102442. [CrossRef]
98. Ahmed, S.F.; Khan, M.M.K.; Amanullah, M.T.O.; Rasul, M.G.; Hassan, N.M.S. Thermal Performance of Building-Integrated Horizontal Earth-Air Heat Exchanger in a Subtropical Hot Humid Climate. *Geothermics* **2022**, *99*, 102313. [CrossRef]
99. Skotnicka-Siepsiak, A. An Evaluation of the Performance of a Ground-to-Air Heat Exchanger in Different Ventilation Scenarios in a Single-Family Home in a Climate Characterized by Cold Winters and Hot Summers. *Energies* **2021**, *15*, 105. [CrossRef]
100. Michalak, P. Hourly Simulation of an Earth-to-Air Heat Exchanger in a Low-Energy Residential Building. *Energies* **2022**, *15*, 1898. [CrossRef]
101. Michalak, P. Impact of Air Density Variation on a Simulated Earth-to-Air Heat Exchanger's Performance. *Energies* **2022**, *15*, 3215. [CrossRef]
102. Amanowicz, Ł.; Wojtkowiak, J. Thermal Performance of Multi-Pipe Earth-to-Air Heat Exchangers Considering the Non-Uniform Distribution of Air between Parallel Pipes. *Geothermics* **2020**, *88*, 101896. [CrossRef]
103. Amanowicz, Ł.; Wojtkowiak, J. Approximated Flow Characteristics of Multi-Pipe Earth-to-Air Heat Exchangers for Thermal Analysis under Variable Airflow Conditions. *Renew. Energy* **2020**, *158*, 585–597. [CrossRef]
104. Amanowicz, Ł. Influence of Geometrical Parameters on the Flow Characteristics of Multi-Pipe Earth-to-Air Heat Exchangers—Experimental and CFD Investigations. *Appl. Energy* **2018**, *226*, 849–861. [CrossRef]
105. Amanowicz, Ł.; Wojtkowiak, J. Validation of CFD Model for Simulation of Multi-Pipe Earth-to-Air Heat Exchangers (EAHEs) Flow Performance. *Therm. Sci. Eng. Prog.* **2018**, *5*, 44–49. [CrossRef]
106. Qi, D.; Li, A.; Li, S.; Zhao, C. Comparative Analysis of Earth to Air Heat Exchanger Configurations Based on Uniformity and Thermal Performance. *Appl. Therm. Eng.* **2021**, *183*, 116152. [CrossRef]
107. Qi, D.; Li, S.; Zhao, C.; Xie, W.; Li, A. Structural Optimization of Multi-Pipe Earth to Air Heat Exchanger in Greenhouse. *Geothermics* **2022**, *98*, 102288. [CrossRef]
108. Michalak, P. Annual Energy Performance of an Air Handling Unit with a Cross-Flow Heat Exchanger. *Energies* **2021**, *14*, 1519. [CrossRef]
109. Li, J.; Shao, S.; Wang, Z.; Xie, G.; Wang, Q.; Xu, Z.; Han, L.; Gou, X. A Review of Air-to-Air Membrane Energy Recovery Technology for Building Ventilation. *Energy Build.* **2022**, *265*, 112097. [CrossRef]
110. Gurubalan, A.; Maiya, M.P.; Geoghegan, P.; Simonson, C.J. Performance Improvement of Membrane Energy Exchanger Using Ultrasound for Heating, Ventilation, and Air Conditioning Application. *J. Therm. Sci. Eng. Appl.* **2022**, *14*, 051015. [CrossRef]
111. Liu, P.; Justo Alonso, M.; Mathisen, H.M.; Halfvardsson, A. Development and Optimization of Highly Efficient Heat Recoveries for Low Carbon Residential Buildings. *Energy Build.* **2022**, *268*, 112236. [CrossRef]
112. Cho, H.-J.; Cheon, S.-Y.; Jeong, J.-W. Development of Empirical Models to Predict Latent Heat Exchange Performance for Hollow Fiber Membrane-Based Ventilation System. *Appl. Therm. Eng.* **2022**, *213*, 118686. [CrossRef]

113. Tang, Y.; Ji, J.; Wang, C.; Xie, H.; Ke, W. Performance Prediction of a Novel Double-Glazing PV Curtain Wall System Combined with an Air Handling Unit Using Exhaust Cooling and Heat Recovery Technology. *Energy Convers. Manag.* **2022**, *265*, 115774. [CrossRef]
114. Chen, Y.; Ozaki, A.; Lee, H. Energy Saving Potential of Passive Dehumidification System Combined with Energy Recovery Ventilation Using Renewable Energy. *Energy Build.* **2022**, *268*, 112170. [CrossRef]
115. Guo, J.; Zhang, C. Utilization of Window System as Exhaust Air Heat Recovery Device and Its Energy Performance Evaluation: A Comparative Study. *Energies* **2022**, *15*, 3116. [CrossRef]
116. Gainza-Barrencua, J.; Odriozola-Maritorena, M.; Barrutieta, X.; Gomez-Arriaran, I.; Hernández Minguillón, R. Use of Sunspaces to Obtain Energy Savings by Preheating the Intake Air of the Ventilation System: Analysis of Its Main Characteristics in the Different Spanish Climate Zones. *J. Build. Eng.* **2022**, *62*, 105331. [CrossRef]
117. Barreto, G.; Qu, K.; Wang, Y.; Iten, M.; Riffat, S. An Innovative Window Heat Recovery (WHR) System with Heat Pipe Technology: Analytical, CFD, Experimental Analysis and Building Retrofit Performance. *Energy Rep.* **2022**, *8*, 3289–3305. [CrossRef]
118. Bai, H.Y.; Liu, P.; Justo Alonso, M.; Mathisen, H.M. A Review of Heat Recovery Technologies and Their Frost Control for Residential Building Ventilation in Cold Climate Regions. *Renew. Sustain. Energy Rev.* **2022**, *162*, 112417. [CrossRef]
119. Zender–Świercz, E. A Review of Heat Recovery in Ventilation. *Energies* **2021**, *14*, 1759. [CrossRef]
120. Liu, S.; Ma, G.; Xu, S.; Jia, X.; Wu, G. Experimental Study of Ventilation System with Heat Recovery Integrated by Pump-Driven Loop Heat Pipe and Heat Pump. *J. Build. Eng.* **2022**, *52*, 104404. [CrossRef]
121. Shuailing, L.; Guoyuan, M.; Xiaoya, J.; Shuxue, X.; Guoqiang, W. Performance of a Mechanically-Driven Loop Heat Pipe Heat Recovery System. *Appl. Therm. Eng.* **2022**, *207*, 118066. [CrossRef]
122. Jia, X.; Ma, G.; Zhou, F.; Liu, S.; Wu, G.; Sui, Q. The Applicability and Energy Consumption of a Parallel-Loop Exhaust Air Heat Pump for Environment Control in Ultra-Low Energy Building. *Appl. Therm. Eng.* **2022**, *210*, 118292. [CrossRef]
123. Díaz de Garayo, S.; Martínez, A.; Astrain, D. Annual Energy Performance of a Thermoelectric Heat Pump Combined with a Heat Recovery Unit to HVAC One Passive House Dwelling. *Appl. Therm. Eng.* **2022**, *204*, 117832. [CrossRef]
124. Diaz de Garayo, S.; Martínez, A.; Astrain, D. Optimal Combination of an Air-to-Air Thermoelectric Heat Pump with a Heat Recovery System to HVAC a Passive House Dwelling. *Appl. Energy* **2022**, *309*, 118443. [CrossRef]
125. Xu, Q.; Zhang, S.; Riffat, S. Ecopump: A Novel Thermoelectric Heat Pump/Heat Recovery Ventilator System for Domestic Building Applications. *Int. J. Low-Carbon Technol.* **2022**, *17*, 611–621. [CrossRef]
126. Cepiński, W.; Kowalski, P.; Szałański, P. Waste Heat Recovery by Electric Heat Pump from Exhausted Ventilating Air for Domestic Hot Water in Multi-Family Residential Buildings. *Rocz. Ochr. Śr.* **2020**, *22*, 940–958.
127. Shahsavar Goldanlou, A.; Kalbasi, R.; Afrand, M. Energy Usage Reduction in an Air Handling Unit by Incorporating Two Heat Recovery Units. *J. Build. Eng.* **2020**, *32*, 101545. [CrossRef]
128. Abdul Hamid, A.; Johansson, D.; Lempart, M. Determining the Impact of Air-Side Cleaning for Heat Exchangers in Ventilation Systems. *Build. Serv. Eng. Res. Technol.* **2020**, *41*, 46–59. [CrossRef]
129. Abadi, I.R.; Aminian, B.; Nasr, M.R.; Huizing, R.; Green, S.; Rogak, S. Experimental Investigation of Condensation in Energy Recovery Ventilators. *Energy Build.* **2022**, *256*, 111732. [CrossRef]
130. Kordana-Obuch, S.; Starzec, M.; Wojtoń, M.; Słyś, D. Greywater as a Future Sustainable Energy and Water Source: Bibliometric Mapping of Current Knowledge and Strategies. *Energies* **2023**, *16*, 934. [CrossRef]

Review

Recent Achievements in Research on Thermal Comfort and Ventilation in the Aspect of Providing People with Appropriate Conditions in Different Types of Buildings—Semi-Systematic Review

Katarzyna Ratajczak *, Łukasz Amanowicz, Katarzyna Pałaszyńska, Filip Pawlak and Joanna Sinacka

Faculty of Environmental Engineering and Energy, Poznan University of Technology, 60-965 Poznan, Poland; lukasz.amanowicz@put.poznan.pl (Ł.A.); katarzyna.palaszynska@put.poznan.pl (K.P.); filip.pawlak@put.poznan.pl (F.P.); joanna.sinacka@put.poznan.pl (J.S.)
* Correspondence: katarzyna.m.ratajczak@put.poznan.pl

Abstract: Ventilation systems are mainly responsible for maintaining the quality of indoor air. Together with thermal comfort maintenance systems, they create appropriate conditions for living, working, learning, sleeping, etc., depending on the type of building. This explains the high popularity of research in this area. This paper presents a review of articles published in the years 2020–2023, which are indexed in the Scopus database and found with keywords "ventilation" and "thermal comfort" in conjunction with the type of building or predominant activity. Finally, 88 selected works for five types of buildings were discussed, namely offices, schools, hospitals, bedrooms, and atriums. Data on publications are summarized in the tables, taking into account the publishing year, country of origin of the authors, and keywords. In this way, the latest directions in research were presented, and research groups dealing with this subject were highlighted. For each type of building, synthetic conclusions were presented, summarizing the results of the analyzed research. This review paper would be helpful for scientists and practitioners in the field of ventilation in order to organize knowledge and in a short time be up to date with the latest research showing how ventilation affects the quality of use of buildings by their users.

Keywords: ventilation; thermal comfort; office; school; hospital; residential; atrium; sleep quality

Citation: Ratajczak, K.; Amanowicz, Ł.; Pałaszyńska, K.; Pawlak, F.; Sinacka, J. Recent Achievements in Research on Thermal Comfort and Ventilation in the Aspect of Providing People with Appropriate Conditions in Different Types of Buildings—Semi-Systematic Review. *Energies* **2023**, *16*, 6254. https://doi.org/10.3390/en16176254

Academic Editors: Xi Chen and Antonio Gagliano

Received: 11 July 2023
Revised: 23 August 2023
Accepted: 25 August 2023
Published: 28 August 2023

1. Introduction

Research on indoor air quality, thermal comfort, and ventilation is closely related to the IEQ (indoor environment quality) issue. People in developed countries with cold and temperate climates spend 90% of their lives indoors. Only 10% of their activity is outdoor activity. Therefore, it is extremely important to provide comfortable interior conditions in buildings. All research carried out around the world that concerns issues related to indoor air parameters brings valuable results that can be used in the various activities of engineers worldwide. Conclusions from experimental studies, field studies, surveys, or simulations indicate new directions for the actions of designers: architects, constructors, and installers. Taking into account the results of scientific research in one's professional activity should improve people's quality of life. This is the implementation of the sustainable development goals, in particular, Goal 3: good health and quality of life. The type of ventilation affects air temperature, humidity, and velocity in the room. However, ventilation rates impact carbon dioxide concentration and consequently air quality. The right ventilation strategy for the building allows thermal comfort and avoid a sick building syndrome (SBS). Due to the numerous scientific publications on the subject of thermal comfort and ventilation, it was considered necessary to review the results of the latest research in this area to set directions for further research and collect the most common results in a form that allows them to be used in engineering activities. Results of scientific research that are not disseminated and used in practice lose their importance. It is important, especially at a time when so many scientific articles are published around the world every day, to compile a list of the latest scientific achievements so that a direction for further research can be presented. It is also important to communicate research results to the public in the form of specific guidelines that can be taken into account in practice. An overview of the latest research and improvements in the field of ventilation systems published in 2020–2022 is presented in the review paper [1]. The article focuses on five key aspects related to ventilation systems: airtightness of buildings, demand-controlled ventilation, decentralized systems, earth-to-air heat exchangers, and heat recovery from removed air. Although the article showed a significant impact of the ventilation system on the energy consumption of the building, this topic was discussed in isolation from the issues of indoor air quality and thermal comfort. Our current article aims to fill this research gap. Nevertheless, in the introduction, which is the background of the current review, it is impossible not to pay attention to the issues related to energy demand, which is the cost of maintaining high air quality and/or thermal comfort. Various technologies for heat recovery [2], or methods of intensifying heat exchange in system components [3–5], as well as the use of decentralized ventilation systems [6–9] or earth-to-air heat exchangers [10,11], positively affect the reduction in energy consumption in the energy audit of the building [12]. The energy-saving effect can also be obtained by using heat recovery from domestic hot water [13–16], but it is easier to achieve it in mechanical ventilation systems [17]. It should be emphasized that thermal comfort is also affected by factors other than ventilation, such as the radiation temperature of the surrounding surfaces, which is easily obtained using surface heating [16,18,19], radiators [20] or wall panels with heat pipes [21] and others [22,23], as well as outdoor sources [24,25].

The purpose of this review paper, which concerns issues related to the parameters of the indoor environment, assessed through the prism of thermal comfort and air quality as a function of ventilation for selected types of buildings, is to indicate the direction of scientific research for specific topics, together with the possibility of using the results in engineering practice.

2. Materials and Methods

A semi-systematic review based on the methodology provided by Synder [26] was conducted. This type of review is appropriate for overviewing research and tracking development over a chosen period of time. Our desired contribution to the field is to show the main themes found in the literature related to the specified scientific problem in a broad way. We intended to analyze the content of the articles, show how scientists research a particular topic, and present the traditions of those studies in a given area. We see the need for such a review on the topic of thermal comfort and ventilation in selected buildings because the number of articles published each year is increasing, which makes it difficult to review the latest news in the field even for a short period of time. We decided to review research conducted recently for a community that is familiar with the topic and is interested to find out what is new in the subject without having to spend many hours browsing through dozens of works. Our contribution is an analysis of recent works in order to summarize the most important conclusions and identify areas of current interest for researchers and possible directions for further research (what challenges are indicated by the results of recent works). Such an analysis requires dozens of hours of work and must be carried out by a person with knowledge in a given field. For this reason, our team has chosen the topics in which we conduct research and teaching.

We selected two research questions to which answers were sought:

Research Question 1: What research is conducted on thermal comfort and ventilation in terms of people's functioning in different types of buildings?

Research Question 2: What research methods have been used in recent years?

A literature review based on the proposed methodology was performed. The answer to the first question is a narrative description of the research in the selected articles, presented in Sections 3–7. The answer to question 2 is presented in Section 9 in the form of summary tables and analysis of the data collected in the review.

This review of scientific and conference articles was based on the assumption that the articles should be related to a specific topic and to be published in the years 2020–2023 (May). Publications were selected by choosing specific keywords: thermal comfort and ventilation. In order for an article to be included in the review, the selected keywords had to appear in the title, keyword set, or abstract of the article.

Various types of building uses were also selected and a review was conducted to assess ventilation and thermal comfort studies for offices, schools, bedrooms, hospitals, and atrium buildings. The analysis of the published results was carried out by narrowing down the review to the assessment of air parameters in the room by limiting it to the form of additional keywords related to the activity of people in these selected types of rooms. For offices, additional words: "work" and "productivity" were introduced, thanks to which the articles reviewed were narrowed down to those in which the impact of indoor air quality or thermal comfort parameters on work productivity was analyzed. In the case of schools, the scope was limited to the word "learning" to assess the impact of air parameters on learning and knowledge acquisition. In residential buildings, the influence of air parameters on "sleep quality" was taken into account. Due to the specificity of the operation of hospitals and other rooms, the search was not limited to specific activities of people in these rooms due to the fact that publications on these rooms are not very numerous, and the type of room itself defines the way it is used. Papers that met all the selection criteria but did not strictly concern people's satisfaction or the impact of air parameters on their functioning were rejected by the authors and marked accordingly in the text.

The following methodology was used for this review:

1. Specification of the subject of interest, which is "thermal comfort and ventilation in the aspect of providing people with appropriate conditions in different types of buildings"
2. Selection of the database to analyze. The review is limited to the SCOPUS database.
3. Selection for the types of buildings or rooms to be covered by this review and the time range for the articles. The "office", "school", "hospital", "bedroom", and "atrium" buildings are selected, and the database is limited to those articles, where at least one of the listed building types appears in the title, keyword set, or abstract of every analyzed article. The review is limited to articles published in the years 2020–2023 (May).
4. Selection for the main criteria that articles should meet to fit into the area of interest, which are set as the appearance of the following specific keywords in their titles, keyword sets, or abstracts: "thermal comfort", and "ventilation".
5. Screening of the filtered database and including additional keywords related to the typical expected indoor activity for selected types of rooms:
   a. "Work" and "productivity" for "office" buildings;
   b. "Learning" for "school" buildings;
   c. "Sleep quality".
6. For each type of building selected for this review:
   a. All review articles fitting the selected search criteria are included for review and discussion.
   b. For research and case-study articles, the screening is carried out for eligibility with the topic of interest, and all articles considered to be relevant to the subject of the review are included for review and discussion.

Figure 1 shows the adopted working methodology. Following this methodology, 53 scientific and conference papers were reviewed.

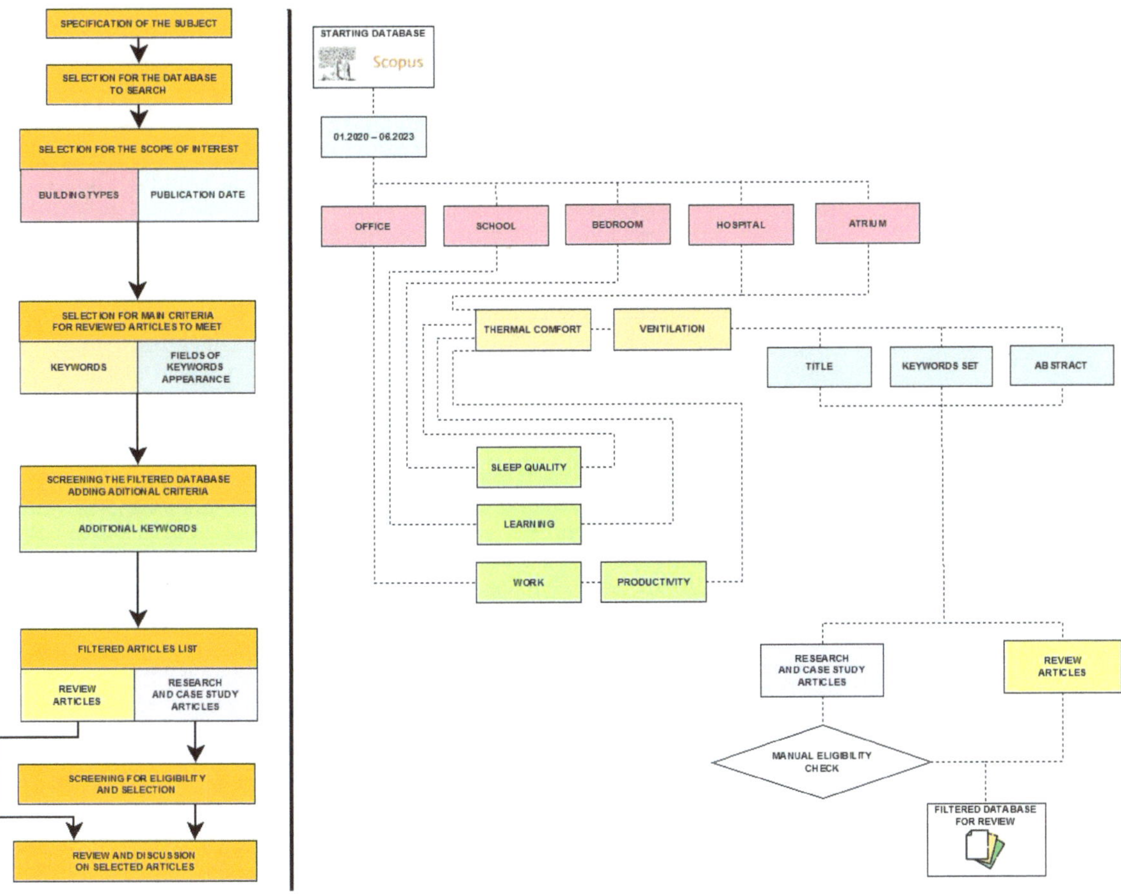

Figure 1. Review methodology.

The article partly concerns the period during which the COVID-19 pandemic lasted. Despite this, the review did not include works that would highlight the importance of the pandemic for this topic. During the COVID-19 pandemic, most public buildings operated in an unusual way—the number of users was limited, time intervals were applied between shifts, remote work was preferred, etc. Knowing that the virus can be transmitted by droplets, ventilation was intensified by increasing the streams of air supplied through the control system (mechanical ventilation) or more often windows were opened/rooms were ventilated (natural ventilation). In this unusual situation, the goal was to avoid the risk of infection without the feeling of thermal comfort or concern for the energy efficiency of the ventilation system. This is reflected in the research conducted in the field of ventilation during the COVID-19 pandemic, which mainly referred to the risk of virus transmission via the ventilation system. The issues of thermal comfort and the impact of the ventilation system on the quality of activity performed in a given room were not taken into account. It is therefore not surprising that none of the articles selected for review in this article were related to the COVID-19 pandemic. The subject of the review carried out as part of this work is thermal comfort and ventilation in the aspect of providing people with appropriate conditions in offices, schools, hospitals, bedrooms, and atriums during the typical use of the building—not in the exceptional situation which was the COVID-19 pandemic.

3. Thermal Comfort and Ventilation in Office Buildings vs. Productivity

Many people doing office work spend most of their time in buildings (usually at least 40 h a week) [27]. Worker well-being and productivity depends on conditions in the workplace. Therefore, it is important to provide indoor environmental quality (IEQ). One of the crucial parameters is the thermal comfort that is maintained by heating, ventilation, and air conditioning (HVAC) systems.

Between 2020 and 2023, according to the Scopus database, with the search criteria selected for this review (title, abstract, or keywords including "thermal" and "comfort" and "ventilation" and "productivity"), no review papers were published; however, eight original studies were found and reviewed in this section.

In the paper [28], Peng B. and S.J. Hsieh noted that thermal comfort measurements are often simplified and do not provide adequate control, and traditional control models are unable to adjust indoor environmental conditions to individual requirements. They proposed a data-driven thermal comfort model and HVAC control system which took into account occupant variability and could improve the thermal comfort of occupants in an open-space office room by 43.4–69.9% with slightly lower energy consumption (1–3.6%). The analysis was conducted for buildings equipped with HVAC systems. The support vector machine (SVM) algorithm (Gaussian kernel function, kernel scale—6.68, box constraint—Inf, epsilon—0.008, standardized, 20% of outlier fraction) and the artificial neural network (ANN) algorithm (parameters—Levenberg–Marquardt algorithm, 10 hidden layers, 20% in the validation set) achieved convergence of $R^2 = 0.99$ with 50 training data points; however, ANN required a shorter training time and had slightly better prediction test performance with 200 training data points ($R^2 = 0.9962$) and mean squared error (MSE = 0.0289). The limitation of data-driven models was the necessity for training, and they performed worse in cold-start scenarios than the PMV model (especially for the first day). For cold start scenarios, high accuracy and rapid computing time, a multistage hybrid model (PMV-SVM-ANN) was developed. The PMV-SVM-ANN model proved to be a statistically significant improvement over the SVM model. Control algorithms and data-driven thermal comfort can be used in commercial office buildings equipped with building management systems and smart thermostats.

In article [29], a transfer-learning-based multilayer perceptron model was presented. It was developed in order to make thermal comfort prediction more precise. The authors used the ASHRAE RP-884, Scales Project and Medium US Office datasets. They took into account only buildings with HVAC systems in 'temperate' climate zones. They developed a transfer-learning-based multilayer perceptron (TL-MLP) model and transfer-learning-based multilayer perceptron model considering climate zone (TL-MLP-C*). They predicted occupants' thermal comfort with insufficient labeled data.

The TL-MLP-C* model has better accuracy of thermal comfort prediction than popular knowledge-driven and data-driven models, (accuracy 54.5%). Combination of age, gender, outdoor environmental features, and the factors from the PMV model gave the best prediction of thermal comfort.

In article [30], measurements of thermal comfort and indoor air quality in an open office within a green-certified building with mechanical ventilation system in the tropics were presented. Levels of thermal comfort, CO_2 concentration, and supply air rates were analyzed, and questionnaires about the indoor environment were conducted among employees. Physical parameters were in acceptable ranges, and 81% of the occupants founded that thermal comfort with high adaptation rates was provided. The neutral temperature was 23.9 °C, which is 1.3 K lower than predicted (25.2 °C). The obtained results can be used to improve thermal comfort by adjusting the air temperature setting according to the real requirements of occupants in offices of a tropical green campus building.

Badeche and Bouchahm [31] studied the impact of large glazing areas on thermal comfort as well as visual comfort. They analyzed office building with natural ventilation and a deep room configuration in Algeria (semi-arid climate). The measurements were conducted according a post occupancy evaluation (POE) method which include IEQ physical measurement, focus group meetings, structured interviews, visual records, occupant survey questionnaires, walkthroughs, and technical measurement of building structure, services and systems. Obtained results indicated that natural ventilation was unable to effectively control air flow and night ventilation absence reduced the release of accumulated heat gains. Extensive solar gains caused overheating. In northeast- and southeast (NE-SE)-oriented rooms, temperature was higher than the temperature of thermal comfort during 96% of working time, during 86% of working time in the northeast (NE)-oriented rooms, and 100% in southeast- (SE) and southwest (SW)-oriented rooms. The mean internal temperature was 29.78 °C, and the maximum temperature was noticed in the southwest orientation and was equal to 38 °C. At the same time, daylight was not sufficient: areas near to glazing were overlit and areas in the rear were ill-lit. The authors suggest using an adaptive solar façade, smart glazing, and moving shading devices.

In article [32], parametric airflow evaluation of skylights, was conducted using software ANSYS Fluent 2019. The research in the admissions office with natural ventilation, located on the ground floor in the eight story educational building in Ecuador, was carried out. Thermal comfort was assessed by use of the predicted mean vote (PMV) and the predicted percentage of dissatisfied (PPD). Eight scenarios that took into account the opening area of skylight, distributing louvers along the skylight, door position (open or close), and impact of furniture on air movement in the room were analyzed. The best internal airflow performance was obtained by use of the 8 louvers 18 inches high by 60 inches wide: individual effective area was equal 0.23 m^2 and a total area was 1.84 m^2. In addition, the solution is more homogeneous, avoids turbulence, and can be more controlled. The results indicated that opening the door increases ventilation rates by 30%. Indoor air temperature equaled 18 °C and air speed in range: 0.25–0.75 m/s represented cool sensation (clothing level = 0.65) and slightly cool sensation (clothing level = 1.0). On the other hand, when the temperature was equal to 25 °C, thermal comfort was met.

Adjusting the opening distance of a tipping window by use of the predicted mean vote (PMV) in paper [33] was studied. Tipping shutter distance equaling 10 cm, 30 cm, and 60 cm was taken into consideration. The office was located on the seventh floor of a university building located directly on a busy street. The obtained PMV value was in the range 2.23–3.27 (warm-hot thermal sensation) and was correlated with wind velocity. In order to achieve a natural range of sensation, the authors recommended to apply a hybrid air conditioning system or mechanical aids.

Borsos, Á. et al. [34] developed a "comfort map" that takes into account individual preferences of indoor environment quality (IEQ). They used a comfort survey to assess thermal comfort, air quality, visual comfort and acoustic comfort as well as health risk factors and negative comfort sensation of each IEQ parameter. The research was conducted in an office building in Hungary and included 216 occupants; 65%, 50%, and 47% of respondents were dissatisfied with the adjustability of noises and sounds, ventilation, and thermal conditions in the work environment, respectively. The authors investigated the relationship between IEQ parameters and the employees' comfort sensation. The developed comfort map shows IEQ parameters: thermal comfort (PMV, PPD), air quality (CO_2 concentration [ppm]), acoustic comfort (RT60 [s], LA [dB]), and visual comfort (daylight factor [%]) for each workstation.

Paper [35] presents the analysis of the working from home experience on the COVID-19 pandemic. As it is a specific case not related to typical office buildings, this paper is not reviewed.

Table 1 presents the list of publications on ventilation, thermal comfort, and productivity in office buildings.

Table 1. List of publications on ventilation, thermal comfort, and productivity in office buildings.

Authors	Year	Title	Journal DOI	Keywords
[28] Peng B., Hsieh S.-J.	2023	Cyber-Enabled Optimization of HVAC System Control in Open Space of Office Building	Sensors 10.3390/s23104857	cyber-physical system; HVAC; thermal comfort model; artificial neural network; support vector machine
[29] Gao N et al.	2022	Transfer learning for thermal comfort prediction in multiple cities	Building and Environment 10.1016/j.buildenv.2021.107725	human–building interaction, thermal comfort, transfer learning, HVAC automation, smart building
[30] Yong, N.H. et al.	2022	Post-occupancy evaluation of thermal comfort and indoor air quality of office spaces in a tropical green campus building	Journal of Facilities Management 10.1108/JFM-12-2020-0092	green campus building; indoor air quality (IAQ); neutral temperature; open-plan office; post occupancy evaluation (POE); thermal comfort
[31] M. Badeche, Bouchahm Y.	2021	A study of Indoor Environment of Large Glazed Office Building in Semi-Arid Climate	Journal of Sustainable Architecture and Civil Engineering 10.5755/j01.sace.29.2.28008	fenestration, indoor environment, post occupancy evaluation, thermal comfort, visual comfort
[32] Ortiz, M.C., Morales, S., Cabrera, V.	2021	Bioclimatic Optimization: Skylight Ground Floor New Building, Udla Park Torre II	Proceedings of International Structural Engineering and Construction 10.14455/ISEC.2021.8(1).AAE-18	bioclimatic design, natural ventilation, effective area, tropical climate, energy efficiency, thermal comfort, building performance, educational building
[33] Lahji K., Puspitasari P.	2023	Thermal comfort analysis by adjusting the tipping window opening distance using PMV	AIP Conference Proceedings 10.1063/5.0120286	not provided
[34] Borsos, Á. et al.	2021	The Comfort Map—A Possible Tool for Increasing Personal Comfort in Office Workplaces	Buildings 10.3390/buildings11060233	exploratory research; indoor environment quality; multidisciplinary approach; workplace health and well-being
[35] McGee B.L. et al.	2023	Work from Home: Lessons Learned and Implications for Post-pandemic Workspaces	Interiority 10.7454/in.v6i1.259	COVID-19, biophilic design, work from home, office design, post-pandemic design

In the analyzed articles, productivity was taken into consideration only as an issue correlated with IEQ parameters. There was no definition and comparison of work productivity in the context of better or faster work. In order to assess thermal comfort questionnaire surveys, experimental measurements and simulations using PMV (Fanger), adaptive and data-driven thermal comfort models were employed. A review of recent articles on ventilation and thermal comfort in the context of work productivity showed two main trends:

- Optimization of HVAC system by use of the machine learning algorithms to predict and control a thermal comfort considering occupants' individual IEQ preferences and climate;
- IEQ parameter assessments in office buildings with natural ventilation indicate that maintaining thermal comfort is difficult. Large glazing areas in office buildings are one of the main reasons. They cause overheating.

4. Thermal Comfort and Ventilation and Its Effects on Learning at Schools

Students spend 6–8 h a day at school [36], over many years. Effective learning requires a friendly environment. Air quality, thermal comfort, but also furniture ergonomics or interior aesthetics will affect the ability to learn and the well-being and health of children and young people learning. In order for children and young people to have a chance for the best possible development and exploration of knowledge, it is our duty to recognize the impact of the environment and minimize its negative impact on pupils and students.

Between 2020 and 2023, according to the Scopus database, with the search criteria selected for this review (title, abstract, or keywords including "thermal" and "comfort" and "ventilation" and "learning" and "school"), 18 original studies were founded and reviewed in this section.

The way in which the space for children while they are learning is organized was the subject of a literature review carried out in [36,37]. Based on an analysis of 68 articles, the authors [36] concluded that a pleasant, warm, and flexible learning environment is a key impact on students' well-being and performance. When designing schools, it is necessary to ensure the presence of charming colors and images, ergonomic furniture, and adequate acoustic and thermal comfort, ventilation and natural lighting. In [37], it was pointed out that contact with nature positively affects creativity and the ability to solve problems. Therefore, in classrooms and other places where students spend any amount of time learning, opening windows should be used to provide natural ventilation, which has been shown to improve both comfort and cognitive function.

Visual and thermal comfort was the subject of research conducted at a primary school in Semarang, Indonesia [38]. The authors showed that in tropical climates, especially in naturally ventilated buildings, relative humidity and temperature (dry air temperature) have an impact on the effective temperature in the room. To increase thermal comfort, the air speed in the rooms should be increased by using fans or providing cross-ventilation by placing openings in the façade that allow wind to penetrate into the building. Another method is to block direct sunlight entering the room to minimize the heat generated by solar radiation. However, it should be remembered that natural lighting should be in the range of 250–750 lux. Natural lighting strategies must be able to limit and control the level of solar radiation, especially from side and overhead lighting, to overcome the problem of heat distribution by providing shading and adjusting the size and placement of windows to produce indirect sunlight.

In Medan, Indonesia [39], primary school students spend 90% of their time in the same class. In situ tests were performed in one of the classrooms in the school building with natural cross-ventilation. It was shown that although there was overheating in the class (air temperature inside was 30.5–34.5 °C), the relative humidity in the class ranged from 57.2% to 74.6%, and the median CO_2 concentration ranged between 602 ppm and 637 ppm, while the median CO_2 concentration in the outdoor air was between 498 ppm and 520 ppm. The authors draw attention to the need for further research to assess the relationship between temperature and relative humidity and CO_2 levels in classes with natural cross-ventilation.

Research conducted in Poland [40] focused on the possibility of using a hybrid mechanical ventilation system and night ventilation in a school located in a passive building in the summer. In order to build a simulation model (simulations in Design Builder), the authors determined the value of students' metabolism, taking into account a smaller body surface area than adults, and at 108 W/person, the CLO value was 0.5. The comfort temperature range of 24 to 27 °C was determined. Simulation and empirical studies have shown that intensive night mechanical ventilation combined with high thermal insulation without the use of a ground heat exchanger or heat pump allows for a significant reduction in thermal discomfort in the building. However, it is not possible to completely protect the building during the hottest period of the year. It was also shown that students and teachers easily adapted clothing to the prevailing conditions. The article also proposes a method for determining the hour of discomfort. This method assigns a kc factor in the range of 0–4 depending on the degree of exceeding the comfort range. The products of the weighting factor and the number of hours for all intervals are summed up.

The authors of [41] performed simulation studies (Design Builder) and in situ measurements that were conducted in Nepal and also addressed the issue of a comfortable temperature for students. Based on the conducted research, they showed that the average comfort temperature was 26.9 °C, and the temperatures were in the range of 25–29 °C. In the results of surveys on students' perception of the indoor thermal environment, students preferred lower temperatures. No significant differences in feelings were noted according to gender.

Also, in Australia [42], a study was conducted among high school students in which it was determined that the comfortable temperature was in the range of 20.4–27 °C, with a neutral temperature of 23.7 °C in summer and 22.6 °C in winter. An equally important issue was air quality. In the subtropical climate of Australia, split air conditioners are often used to lower the temperature in classrooms. Studies have shown that in classrooms without a DCV (demand-controlled ventilation) system, CO_2 concentrations reached up to 2981 ppm in summer and 2418 ppm in winter, and in rooms with DCV ventilation systems, the peak concentration was 1335 ppm. In addition, volatile organic compound (VOCs) analyses were performed from air samples, and an improvement in air quality in the DCV class was demonstrated. The authors highlight the effect of both indoor temperature and CO_2 concentration on students' fatigue. Students demonstrated their ability to adapt to changes in room temperature. The time it takes for students to adapt to a rapid change in average outside air temperature is one week.

Measurement and survey research conducted in Selangor, Malaysia [43] focused on differences in thermal comfort in schools located in urban and rural areas. The optimal temperature range in the room is from 27.1 °C to 29.3 °C and it was exceeded in all analyzed facilities. Measurements showed that in urban areas the temperature increase was rapid, and in rural areas it was constant. The survey results confirmed that students and teachers lost concentration when the room was too warm.

The authors of [44] conducted a review of the literature on indoor environmental quality, air quality, and thermal, acoustic, and visual comfort in Indian schools. India is the seventh most densely populated country in the world, and its area exceeds 3 million km^2. They are divided into five climatic zones. The subject of the analysis was 37 articles. The authors concluded that the available research for India is convoluted and disorganized. They suggest the need to develop a well-established method for assessing classroom environments. This survey review uses the Preferred Reporting Items for Systematic Review and Meta-Analyses (PRISMA) approach as the review methodology.

The prediction of thermal comfort based on machine learning was addressed by the authors [45]. The article draws attention to the fact that most of the machine learning solutions proposed for air-conditioned or ventilated buildings are intended for adults. A month-long study was conducted in five naturally ventilated school buildings in the city of Dehradun, India. The study included 512 primary school students. It was shown that the students were able to more accurately assess their thermal sensations than to show thermal preferences or thermal comfort levels. The ability to generalize thermal comfort models used in machine learning is highly desirable and achievable in neighboring school buildings with natural ventilation. Techniques such as SVMs (support vector machines algorithm) offer high predictive stability but need to be refined in practical applications. There are also new paradigms, such as transfer learning and multitasking learning. For high accuracy, TC models must be trained using spatial or building-specific data. Still, performance can vary in naturally ventilated buildings.

In China [46], numerical studies of CFDs (computational fluid dynamics) were carried out using the RANS model. They were aimed at assessing the structure of the school building in which classrooms are in the middle of the building, and on their sides, there are corridors, and the building is also equipped with ground heat exchangers. The building is located in Northern Shaanxi, where the average outside air temperature from October to

March is below 10 °C. It has been shown that the double-sided construction of the corridor is an energy-saving solution, because, thanks to them, you can achieve the effect of heat preservation and increase the temperature at the entrance. Ground ventilation pipes can increase the room temperature when the room is located downstairs, which can effectively improve the comfort of the room. A significant influence on the temperature in the room has the proportion of width to height of the window. As the ratio of width to the height of the window increases, the area of the window for ventilation will increase, which will accelerate the improvement of indoor air quality. However, if the window area is too large, the effect of thermal insulation of the room will be weakened.

Next, studies based on numerical simulations are described in [47]. It was planned to install an air conditioning (A/C) system in all primary schools in Yuan (Taiwan) within two years. Therefore, the authors performed simulation studies of various air conditioning control strategies for an average 8 h CO_2 concentration in the classroom. The optimal strategy of use is the operation of air conditioning throughout the school day in cooperation with a mechanical ventilation system ensuring the exchange of fresh air at the level of 5 L/s for each person in the room. If air conditioning is not installed and the classroom is cooled and ventilated only by natural ventilation, the average CO_2 concentration in the classroom varies between 583 and 612 ppm, depending on the percentage of students staying in the classroom during the break.

The article [48] presents the results of air quality tests that took place in two classes of a secondary school undergoing thermo-modernization renovation. The school is located in Bucharest, Romania. There were 25 students and one teacher in the study class. During the period of stay in the classrooms, the CO_2 concentration exceeded the maximum allowed value of 1000 ppm for almost the entire period, peaking at 3600 ppm. While maximum concentrations of O_3 during the study period exceeded 200 $\mu g/m^3$. The variability in ozone concentrations in classrooms is caused by ozone emissions from laser printers, but direct solar radiation also plays an important role in the formation of these concentrations. Children are more likely to develop asthma and other respiratory diseases as a result of prolonged exposure to high ozone concentrations. The main conclusion presented by the authors is the need to use ventilation in learning spaces.

The authors [49] performed a dynamic thermal simulation on three levels (housing, classroom, school building) to assess the resources of school buildings and explore the possibilities for improvement. The study was carried out for three locations representative of climate distribution in Europe: Southampton (United Kingdom) as a temperate oceanic climate; Brescia (Italy) as a warm temperate climate; and Thessaloniki (Greece) as a warm Mediterranean climate. Possible renovation strategies have been developed, taking into account both the heat transfer coefficient and thus the heat demand, thermal mass, and the dynamic heat transfer coefficient, i.e., the efficiency of the cooling season. The most effective retrofitting strategies to improve indoor comfort are natural ventilation; good shading; high thermal mass; good insulation; and night ventilation in summer. It was stated that passive strategies must be evaluated in an iterative process to choose the best option, given that the building must be comfortable all year round. In this regard, there is no unique strategy; however, it is important to emphasize that in temperate climates, the most effective involves the control of thermal inertia, taking into account both winter and summer performance. Thermal comfort in educational buildings in summer is possible to achieve in some regions with a temperate climate without the use of mechanical systems. In order to achieve thermal comfort conditions in rooms, it is also important to inform users about monitoring and building management.

The authors of [50] have developed a method to simulate school buildings located in the United Kingdom based on an adaptation of the modeling method originally used for energy modeling. A method was developed for the Data dRiven Engine for Archetype Models of Schools (DREAMS), an EnergyPlus-based inventory modeling structure that models different class typologies and takes into account not only indoor environmental quality criteria but also achievements, health, and healthcare costs. The results of the dynamic IEQ simulations showed that the impact of the construction age on learning performance rates existed and that it could be stronger in warmer regions.

In Milan, field studies were carried out in the Politecnico di Milano building in 16 air-conditioned and naturally ventilated rooms [51]. Thermal preference studies were also conducted among 985 students to assess the predictions given by the Ranger and adaptive comfort models. The authors confirmed the suitability of the Ranger model for predicting thermal sensations in air-conditioned classrooms with reasonable accuracy. In naturally ventilated classrooms, the adaptive model proved to be suitable for predicting students' comfort zones according to ASHRAE 55, while the adaptive comfort temperatures recommended by EN 15,251 were unacceptable for large numbers of students. There were no significant differences in the perception of thermal comfort between the sexes, except for two classes with natural ventilation, in which women's voices were closer to neutrality compared to men, who expressed warmer thermal sensations. Comparing the students' thermal sensation ratings with their thermal satisfaction showed that not all voices outside the comfort range always mean dissatisfaction, and vice versa.

A study of 583 primary school students in Galicia, Spain [52] confirmed the existence of a link between learning space and students' mathematical and artistic performance. It was also confirmed that the adaptation of iPEP scale (Indoor Physical Environment Perception scale) to measure students' perception of variables of the physical learning space in primary school is accurate and reliable. It was determined that the factorial structure of the learning-space construct consists of three factors: workplace comfort, natural environment, and building environment. The prognostic analysis of the results in mathematics confirmed a direct relationship with ventilation, room size, views, and attachment to the place.

A comprehensive review of research over the last 50 years on indoor air quality in classrooms in schools from over 40 countries is presented in [53]. The general conclusion that has been drawn from the analysis of the available literature is that students feel comfortable in rooms with a cooler climate than environments in which adults feel thermally neutral. Poor classroom air quality will result in the cognitive deterioration of students, with negative consequences for progressive learning while increasing short-term sick leave. Exposure to various air pollutants in school buildings threatens serious damage to the health of students, as they inhale a larger volume of air corresponding to their body weight than adults. In a number of studies the authors reviewed, [53] showed higher concentrations of pollutants in schools than in residential and commercial buildings. Volatile organic compound (VOC) pollutants are among the main indoor air pollutants, causing serious health problems for children and adults. In many schools, they have identified particulate matter pollution as the main source of indoor air pollution. In addition, Penicillium, Cladosporium, Aspergillus, and Alternaria were the most common fungi in school premises, and their prevalence varies according to climate and location, whether in the countryside or in the city. The authors note that there is a lack of research on the correlation between classroom CO_2 concentration and learning ability. Their analysis suggests that keeping classroom CO_2 levels below 900 ppm (absolute level) reduces the negative impact on learning.

Table 2 presents a list of publications on thermal comfort, ventilation and learning in schools.

Table 2. List of publications on thermal comfort, ventilation, and learning in school.

Authors	Year	Title	Journal DOI	Keywords
[36] Manca S.; Cerina V.; Tobia V.; Sacchi S.; Fornara F.	2020	The effect of school design on users' responses: A systematic review (2008–2017)	Sustainability 10.3390/SU12083453	learning space; psychological responses; school architectural features; students' performance; users' well-being

Table 2. *Cont.*

Authors	Year	Title	Journal DOI	Keywords
[40] Dudzińska A.; Kisilewicz T.	2021	Alternative ways of cooling a passive school building in order to maintain thermal comfort in summer	Energies 10.3390/en14010070	design builder; discomfort; energy efficiency; overheating; thermal comfort; weighted measure of discomfort
[41] Shrestha M.; Rijal H.B.	2023	Investigation on Summer Thermal Comfort and Passive Thermal Improvements in Naturally Ventilated Nepalese School Buildings	Energies 10.3390/en16031251	operative temperature; passive design; school building; simulation; thermal comfort; thermal environment
[42] Haddad S.; Synnefa A.; Ángel Padilla Marcos M.; Paolini R.; Delrue S.; Prasad D.; Santamouris M.	2021	On the potential of demand-controlled ventilation system to enhance indoor air quality and thermal condition in Australian school classrooms	Energy and Buildings 10.1016/j.enbuild.2021.110838	air quality; indoor environmental quality; school buildings; thermal comfort; ventilation
[44] Kapoor N.R.; Kumar A.; Alam T.; Kumar A.; Kulkarni K.S.; Blecich P.	2021	A review on indoor environment quality of Indian school classrooms	Sustainability (Switzerland) 10.3390/su132111855	acoustic comfort; artificial intelligence; classroom; COVID-19; indoor air pollution; indoor air quality; sick building syndrome; thermal comfort; ventilation; visual comfort
[45] Lala B.; Kala S.M.; Rastogi A.; Dahiya K.; Yamaguchi H.; Hagishima A.	2022	Building Matters: Spatial Variability in Machine Learning Based Thermal Comfort Prediction in Winters	Proceedings—2022 IEEE International Conference on Smart Computing, SMARTCOMP 2022 10.1109/SMART-COMP55677.2022.00078	classrooms; energy efficiency; IoT; machine learning; multi-class classification; natural ventilation; sensors; spatial variability; students; thermal comfort
[37] Peters T.; D'Penna K.	2020	Biophilic design for restorative university learning environments: A critical review of literature and design recommendations	Sustainability (Switzerland) 10.3390/su12177064	critical analysis; design method; education; learning; literature review; questionnaire survey; student; university sector
[38] Suradhuhita P.P.; Setyowati E.; Prianto E.	2022	Influence of a facade design on thermal and visual comfort in an elementary school classroom	IOP Conference Series: Earth and Environmental Science 10.1088/1755-1315/1007/1/012013	not provided
[43] Mazlan A.N.; Saad S.; Yahya K.; Haron Z.; Abang Hasbollah D.Z.; Kasiman E.H.; Rahim N.A.; Salehudddin A.M.	2020	Thermal comfort study for classroom in urban and rural schools in Selangor	IOP Conference Series: Materials Science and Engineering 10.1088/1757-899X/849/1/012016	not provided

Table 2. *Cont.*

Authors	Year	Title	Journal DOI	Keywords
[46] Lan K.; Chen Y.	2022	Air Quality and Thermal Environment of Primary School Classrooms with Sustainable Structures in Northern Shaanxi, China: A Numerical Study	Sustainability (Switzerland) 10.3390/su141912039	air quality; classroom; numerical simulation; sustainable building; thermal environment
[39] Talarosha B.; Satwiko P.; Aulia D.N.	2020	Air temperature and CO_2 concentration in naturally ventilated classrooms in hot and humid tropical climate	IOP Conference Series: Earth and Environmental Science 10.1088/1755-1315/402/1/012008	ASHRAE; CO_2 concentration; comfort temperature; natural ventilation; school classroom
[47] Chang L.-Y.; Chang T.-B.	2023	Air Conditioning Operation Strategies for Comfort and Indoor Air Quality in Taiwan's Elementary Schools	Energies 10.3390/en16052493	air-conditioning; elementary school; IAQ; thermal comfort; ventilation
[49] Tagliabue L.C.; Accardo D.; Kontoleon K.J.; Ciribini A.L.C.	2020	Indoor comfort conditions assessment in educational buildings with respect to adaptive comfort standards in European climate zones	IOP Conference Series: Earth and Environmental Science 10.1088/1755-1315/410/1/012094	adaptive comfort standards; climate zone; educational buildings; indoor comfort conditions
[51] Fabozzi M.; Dama A.	2020	Field study on thermal comfort in naturally ventilated and air-conditioned university classrooms	Indoor and Built Environment 10.1177/1420326X19887481	adaptive model; fanger model; field study; gender; natural ventilation; thermal comfort
[52] López-Chao V.; Lorenzo A.A.; Saorín J.L.; De La Torre-Cantero J.; Melián-Díaz D.	2020	Classroom indoor environment assessment through architectural analysis for the design of efficient schools	Sustainability (Switzerland) 10.3390/su12052020	acoustics; environmental quality; learning space; occupant comfort; sustainable architecture; sustainable building; thermal comfort; ventilation comfort; visual comfort
[50] Grassie D.; Karakas F.; Schwartz Y.; Dong J.; Milner J.; Chalabi Z.; Mavrogianni A.; Mumovic D.	2022	Modelling UK school performance by coupling building simulation and multi-criteria decision analysis	17th International Conference on Indoor Air Quality and Climate, INDOOR AIR 2022	building simulation; cognitive performance; decision analysis; indoor air quality; school buildings; thermal comfort
[53] Sadrizadeh S.; Yao R.; Yuan F.; Awbi H.; Bahnfleth W.; Bi Y.; Cao G.; Croitoru C.; de Dear R.; Haghighat F.; Kumar P.; Malayeri M.; Nasiri F.; Ruud M.; Sadeghian P.; Wargocki P.; Xiong J.; Yu W.; Li B.	2022	Indoor air quality and health in schools: A critical review for developing the roadmap for the future school environment	Journal of Building Engineering 10.1016/j.jobe.2022.104908	classroom air quality; energy use in schools; exposure risk; particle matter; ventilation; volatile organic compounds
[48] Catalina T.; Damian A.; Vartires A.; Nița M.; Racovițeanu V.	2023	Long-term analysis of indoor air quality and thermal comfort in a public school	IOP Conference Series: Earth and Environmental Science 10.1088/1755-1315/1185/1/012008	not provided

A review of the current literature on thermal comfort and students' ability to learn showed that there is no universal range of comfort temperature for all schools, as this temperature depends on the location of the building (rural/urban) and climatic zone. Pupils and teachers need about a week to adapt to the conditions in the school.

There is a direct relationship between ventilation, the size of the room, the view from the window, and attachment to the place occupied in the classroom and the student's performance. In addition to the air parameters, the environment also has an impact on the ability to learn: colors, the presence of plants, natural lighting, or the ability to open a window to take advantage of natural ventilation.

Children have difficulty in determining their thermal preferences and the level of thermal comfort but can assess their thermal sensations.

5. Thermal Comfort in Terms of Ventilation of Hospital Buildings

Hospital facilities are a category of buildings in which ventilation affects the feeling of thermal comfort and is particularly important from the point of view of hygienic safety: airborne disease transmission. The users of such facilities are people with special thermal requirements—sick people, people with impaired functions of the immune system, those undergoing convalescence, generally weakened people, and those representing a rather limited level of physical activity. On the other hand, hospitals employ staff who perform intense physical and/or intellectual work, are on the move (showing a high level of physical activity), and stay in the same rooms as patients. Expectations of these two groups, patients and personnel, regarding thermal parameters in rooms, are therefore divergent. The above-described aspects related to the ventilation of hospital rooms mean that they are still the subject of research and analysis, the latest of which is discussed in this chapter.

In the years 2020–2023, six review papers were published on the subject of hospital ventilation and thermal comfort, which were searched in the Scopus database by entering the phrase: "thermal comfort ventilation hospitals review". Two papers from the search results were excluded due to being less related to the subject of this review: "Engineering solutions for preventing airborne transmission in hospitals with resource limitation and demand surge" and "Biophilic design for restorative university learning environments: A critical review of literature and design recommendations".

In the review paper [54], hospitals are presented in the introduction as particularly energy-consuming in the comparison to other types of buildings. The main reason for this situation is the nature of the use of hospitals: 24/7 use as well as the constant need to use artificial lighting and technological processes. An interesting comparison was made in the context of average annual electric and thermal energy consumption per gross floor area, presented in Table 3.

Table 3. Electric and thermal energy consumption per gross floor area in various countries, based on [54].

Country	Electric Energy [kWh/m^2]	Thermal Energy [kWh/m^2]
Switzerland (CH)	75	195
Belgium (BE)	80	275
Netherlands (NL)	80	320
Sweden (SE)	100	175
United Kingdom (UK)	110	500
Greece (GR)	115	300
Austria (AU)	180	210
Unated States (US)	240	700
Canada (CA)	340	720

The high demand for both heat and electricity causes great interest in the possibility of reducing energy consumption in hospitals. The authors of the review [54] refer to studies that show that the very structure of the ventilation system can significantly affect both energy consumption and comfort, and that there are solutions that are beneficial in both respects: "when exhaust opening was set at the ceiling level (rear-middle-level)". In the conclusions, the authors point to the possibility of saving energy through the use of energy-saving ventilation control systems, which is also indicated by the authors of the review on ventilation systems in general [1] or the work comparing the operation of earth-to-air heat exchangers operating according to different schedules of use [10] (adapting to user needs).

Operating theaters are special rooms in hospital buildings, to which separate studies are devoted. The specificity of technical equipment systems in operating rooms reflects the specific technological needs of various categories of procedures and the risk of infection (from the operated patient or from the patient). An overview of ventilation systems in operating theaters only was published in [55] in 2021. In the highlights of the paper, the authors point out that the cleanliness and hygienic safety of the surgical zone is mainly due to adequate ventilation. At the same time, they note that the choice of the ventilation system is ambiguous and, despite many studies, it remains controversial. The problem related to the ventilation of operating rooms in the context of thermal comfort, already mentioned in the introduction, is the discrepancy between the needs of the staff and the patient—the first group expects a lower temperature for the comfort of intensive work, and the patient should not become too cold during the procedure when he lies still without clothes in a state of complete anesthesia. For this reason, surgical staff and HVAC engineers should cooperate and mutual understanding between them seems to be crucial, which indicates that a lot in promoting the knowledge is waiting to do it. In the conclusions, the following issues were raised as key in the context of ventilation systems:

(a) Design of the surgical lamp that will not block the air supply to the operation zone or cause air recirculation under the lamp;

(b) Broad laminar air flow diffusers play a great role in reducing the sensitivity of the laminar ventilation system to door opening;

(c) The use of an appropriate personnel dressing system may, on the one hand, reduce contamination from clothing, and at the same time serve to achieve a compromise in terms of thermal comfort of the surgical staff vs. the patient.

A review of works on thermal comfort in naturally ventilated patient rooms in the tropics is presented in [56]. The authors argue that natural ventilation is a solution that can provide adequate airflows in a tropical climate and provide thermal comfort at a low financial cost. Maintaining the proper air flow can be supported by additional devices and controlled by BMS (building management system) automation systems. The article points out that the thermal comfort of patients depends to a large extent on their health and many other factors, citing research conducted in a tropical climate. At the same time, research results have been found which show that in this climate, hospital staff experience discomfort more often than patients. The authors put emphasis on the review of works on the use of heat pumps and photovoltaic panels to power systems for maintaining thermal comfort in a building, including ventilation. In their conclusions, they concluded that this combination may be the answer to the need to ensure comfort with a high potential for energy savings, resulting from the high coefficient of performance of heat pumps (COP = 6.14) and the use of solar radiation as a renewable energy source, thanks to which photovoltaic (PV) panels produce electricity for heat pump power supply.

Studies of thermal comfort in naturally ventilated patient rooms in the tropics are described in [57] by an extended group of authors from [56]. Comfort was assessed using simulation, in situ measurements and field survey in Kepala Batas Hospital (Malaysia). One can read in the paper that "simulation results presented that more than half of the total occupants in the ward feel discomfort, with a predicted mean vote (PMV) between 1.0 and 1.6 and a predicted percentage of dissatisfied between 40% and 56%". The results of the simulations carried out in the work were consistent with the results obtained on the basis of measurements but differed from the questionnaire survey. Significantly more people voted that the conditions in the sick room are uncomfortable. The authors indicate that this may be due to a variety of diseases. For example, chronically ill patients with "Alzheimer's disease, Parkinson's disease, cardiovascular disease, diabetes, respiratory problems and kidney disease, are the most susceptible to heat". The results presented in the paper also showed that "people in a tropical climate zone are well accustomed to hot-humid weather and would tolerate high temperatures".

The last of the discussed review articles [58] comes from 2022 and is the most comprehensive knowledge base on past research in the field of thermal comfort, ventilation, and energy efficiency in hospital rooms. The article reviews the field-surveys of thermal comfort, indicating the methods used in individual studies and characterizing the group of respondents (e.g., patients, staff, staff and patients, pregnant women, etc.). The authors point out that a large number of works in the literature concern hypothermia as one of the most important consequences of the lack of thermal comfort and poorly designed/maintained HVAC systems. They cite the results of research in which an attempt was made to separate comfort zones—separate for the patient and staff—using a nozzle (a type of air curtain). The review concludes with the authors' thoughts on key themes that present major challenges today:

- Continuation of the study of factors inside and outside the building affecting the thermal comfort of users;
- Adjusting the temperature in the room to the condition of patients and the type of diseases, as well as research on adaptive thermal comfort in hospitals;
- The use of "self-warming blankets, prototype thermal compression devices, and in-line intravenous fluid warming" to improve the patient's thermal comfort during surgery while maintaining the thermal comfort of surgeons;
- Refining the guidelines regarding the time needed to achieve full functionality of operating rooms for various ventilation systems—so as to be able to save energy without sacrificing air quality and thermal comfort.

Table 4 presents list of discussed review publications in the field of thermal comfort and ventilation in hospitals.

Table 4. List of discussed review publications in the field of thermal comfort and ventilation in hospitals.

Authors	Year	Title	Journal DOI	Keywords
[54] Anwer A. Gatea et al.	2020	Energy Efficiency and Thermal Comfort in Hospital Buildings: A Review	International Journal of Integrated Engineering 10.30880/ijie.2020.12.03.005	Energy efficient, hospital, thermal comfort
[55] Sasan Sadrizadeh et al.	2021	A systematic review of operating room ventilation	Journal of Building Engineering 10.1016/j.jobe.2021.102693	Hospital operating room, indoor air quality, thermal comfort, infection control, surgical clothing system, source strength
[56] Abd Rahman et al.	2021	A Literature Review of Naturally Ventilated Public Hospital Wards in Tropical Climate Countries for Thermal Comfort and Energy Saving Improvements	Energies 10.3390/en14020435	thermal comfort; building energy; naturally ventilated ward; hybrid system; tropical climate
[58] Feng Yuan et al.	2022	Thermal comfort in hospital buildings—A literature review	Journal of Building Engineering 10.1016/j.jobe.2021.103463	Thermal comfort, hospital buildings, improvement measures, energy efficiency

Paper [59] presents the investigations of temperature, relative humidity, and carbon dioxide in four hospitals in Pakistan. Four types of rooms were assessed in every hospital: emergency room, operation room, intensive care unit, and medical ward. "The results show that occupancy rate, ambient thermal conditions, type of HVAC system, and building orientation are vital drivers of IEQ".

In the paper [60], the temperature and humidity in the operation rooms were investigated for (i) "conventional heating, ventilation, and air conditioning system" (ii) "liquid desiccant air conditioning (LDAC) system", and (iii) rotary desiccant air conditioning (RDAC) systems. The energy usage was monitored in parallel with airborne bacterial concentrations. The humidity for conventional, LDAC, and RDAC systems was respectively 66.7%, 60.8%, and 60.5%, while energy consumption was 14.1 kWh/m^2, 11.8 kWh/m^2, and 10.1 kWh/m^2, indicating RDAC as an energy-efficient and preferable system.

In [61], the following parameters were investigated in the newly built hospital in Bristol (United Kingdom): "temperature, RH, CO_2, particulate matter (PM2.5, PM10), and NO_2" and concentration levels of benzene, formaldehyde, and trichloroethylene. User surveys were used to assess lighting and acoustics comfort. The presented results show that mechanical ventilation ensures satisfactory air quality in the tested building; however, the authors of the article point to the possibility of improving the energy efficiency of the system, the possibility of an energy audit thanks to the BMS (building management system) technology and present suggestions for improving air quality and thermal comfort in rooms existing in hospitals and also pay attention to aspects that may be useful in the case of designing new buildings—e.g., to ensure adequate filtration.

The article [62] presents multi-criteria methods for assessing the suitability of a given ventilation system in an operating room. This method is based on the "analytic hierarchy process and fuzzy comprehensive evaluation", which takes into account seven parameters in the context of the three evaluation criteria "ventilation effectiveness, energy consumption and users' satisfaction". The proposed method was verified during a simulated operation in two rooms in St. Olavs Hospital in Norway, which were equipped with different ventilation systems: laminar or mixing. During the pilot tests, the laminar system obtained an acceptable result, and the mixing system an inadequate one. The authors suggest that this assessment method can be used during commissioning of operating rooms in order to verify the achievement of both comfort and energy efficiency parameters.

The adaptive comfort model was the subject of research described in the article [63] for nursing homes. In order to verify its applicability, surveys were conducted in five nursing homes in Spain (Mediterranean climate). The results showed that in the case of naturally ventilated rooms, the inhabitants presented a greater ability to adapt, understood as tolerance to comfort parameters deviating from the desired ones. This observation led to the conclusion that the use of natural ventilation without air conditioning can reduce energy consumption without significantly affecting the thermal comfort of residents. At the same time, the research confirmed that the ASHRAE 55:2020 thermal comfort model can be considered as close to the expectations of nursing home residents. Such dependence was not obtained for non-residents—service employees. Studying a larger group of these people is the goal of further studies planned in the future.

Parameters in the operating room: predicted mean vote (PMV) and predicted percentage of dissatisfied (PPD) were investigated by means of CFD simulation in [64]. The influence of air inlet temperature on the flow pattern and thus on the thermal comfort sensation was investigated. The results show that a temperature of about 18 °C is needed. Nevertheless, in the analyzed cases, the recirculation zones which can be repaired with usage of some air curtains occurred, but it was not simulated in the paper.

Table 5 presents a list of publications on thermal comfort and ventilation in hospitals.

Table 5. List of other publications on ventilation and thermal comfort in hospitals.

Authors	Year	Title	Journal DOI	Keywords
[57] Abd Rahman, Noor Muhammad et al.	2021	Thermal comfort assessment of naturally ventilated public hospital wards in the tropics	Building and Environment 10.1016/j.buildenv.2021.108480	naturally ventilated ward, public hospital, thermal comfort tropical climate
[59] Khan M. et al.	2020	Thermal comfort and ventilation conditions in healthcare facilities—Part 1: An assessment of indoor environmental quality (IEQ)	Environmental Engineering and Management Journal 10.30638/eemj.2020.087	CO_2 concentration; healthcare facilities; HVAC system; indoor environmental quality; thermal comfort
[60] Tsung-Yi Chien et al.	2020	Comparative Analysis of Energy Consumption, Indoor Thermal–Hygrometric Conditions, and Air Quality for HVAC, LDAC, and RDAC Systems Used in Operating Rooms	Applied Sciences 10.3390/app10113721	dehumidification air conditioning system; humidity; energy consumption; air quality; operating room; ventilation
[61] Nishesh Jain e tal.	2021	Building Performance Evaluation of a New Hospital Building in the UK: Balancing Indoor Environmental Quality and Energy Performance	Atmosphere 10.3390/atmos12010115	building performance evaluation; indoor environmental quality (IEQ); indoor air quality (IAQ); energy performance; hospitals
[62] Fan, Minchao et al.	2021	Suitability evaluation on laminar airflow and mixing airflow distribution strategies in operating rooms: A case study at St. Olavs Hospital	Building and Environment 10.1016/j.buildenv.2021.107677	air distribution strategy, operating room suitability evaluation
[63] N. Forcada et al.	2021	Field study on adaptive thermal comfort models for nursing homes in the Mediterranean climate	Energy and Buildings 10.1016/j.enbuild.2021.111475	adaptive thermal comfort model, nursing homes, elderly, naturally ventilated, air-conditioned, thermal comfort standards
[64] Gutierrez, Albio D. et al.	2022	Required thermal comfort conditions inside hospital operating rooms (ORS): a numerical assessment	Frontiers in Heat and Mass Transfer 10.5098/hmt.18.4	computational fluid dynamics (CFD); heat, ventilation, and air conditioning (HVAC); indoor environmental quality (IEQ); operating rooms (ORs); thermal comfort

A review of recent articles on ventilation in the context of thermal comfort did not show clear trends. Various areas of research are constantly assessed: from the issue of air distribution through methods of heat and moisture recovery and methods of controlling the operation of the system to the quality of filtration (PM2.5 PM10) or the issues of adaptive comfort of patients and hospital staff as two groups with opposite expectations. Access to real operating rooms for research purposes is limited, and the results are "sensitive". For these reasons, it can be expected that the works found in the literature do not fully reflect all activities that are taking place in this field in the world.

6. Thermal Comfort, Ventilation, and Quality of Sleep

People spend about one-third of their lives sleeping. Therefore, it is extremely important to provide a suitable indoor environment in bedrooms. The appropriate indoor environment consists of air quality, thermal comfort, and acoustic comfort. Research on the indoor environment in bedrooms focuses on these three aspects. The researchers consider the evaluation of indoor air quality and its relationship to room ventilation, thermal comfort, or acoustic comfort, with the inclusion of sleep quality measurements in addition to measuring these elements of the indoor environment.

Between 2020 and 2023, according to the Scopus database, with the search criteria selected for this review (title, abstract, or keywords including "thermal" and "comfort" and "ventilation" and "sleep quality"), 12 original studies were found and reviewed in this section.

Studies conducted in Africa focused on evaluating the feasibility of retrofitting buildings to reduce the spread of malaria [65]. At the same time, thermal comfort during the night was noted as an important factor for a good night's sleep. Special attention should be paid to this aspect. The authors recommended the introduction of increased airflow through buildings, i.e., an increase in the exchange of air. The analysis covered buildings in the Gambia—a traditional building with leaks at the junction of the roof and external walls, as well as more modern buildings with metal roofs with tight construction. Due to the construction of houses characterized by low airtightness, the authors were unable to assess ventilation using a typical blower door test. The effectiveness of ventilation was assessed by measuring carbon dioxide concentrations in a traditional (less airtight), and a modern building was analyzed. It was estimated that in the case of houses located in Africa, it is essential to provide airflow through which pollutants can be diluted, which reduces the risk of contracting malaria but also improves thermal comfort. Thermal comfort is influenced by the speed of air movement. By creating additional openings in buildings or introducing additional windows, the authors obtained a greater multiplicity of air exchanges. The study also concluded that for old-type buildings or traditional houses, the African method of assessing ventilation by measuring carbon dioxide concentrations could give good results. This is a certain indication that can be used to assess the indoor environment of existing buildings, especially leaky ones.

For the authors of [65], it was also important to draw attention to the fact that the analysis of thermal comfort to ensure appropriate conditions, especially for falling asleep, is often discussed while the importance of ventilation and air movement is omitted. Similar conclusions were reached by other authors of [66], who studied ventilation scenarios for tiny sleeping spaces. Tiny sleeping spaces are increasingly popular spaces that can be found in the form of capsule hotels, sleeping boxes, or sleeper compartments. The influence of air velocity in the capsule on the feeling of comfort was analyzed. Scenario 1 included the air velocity in the capsule in the range of 0.1 to 0.24 m/s and the provision of conditions in which the PMV would be between −0.5 and +0.5. In Scenario 2, the minimum amount of air was introduced and the speed was significantly higher (0.5 m/s). In Scenario 3, the DR value was limited to 5.3% to reduce the draught. The authors evaluated the effects of temperature and air movement in the capsule. They showed that it is more important to ensure adequate air exchange than to focus on the temperature of the supply air. The most comfortable conditions were obtained for Scenario 1. It is recommended to use air exchange in the sleeping environment in such a way that there is air movement but in the range of up to 0.24 m/s.

Analyses in the summer using a climate chamber were carried out in China [67]. The authors examined the sleep of 24 people and the influence of changing climatic conditions in terms of air temperature (28/30/32 °C), relative humidity (50/80%), and air velocity (0.39/0.69/1.17 m/s). Sleep was assessed using the Fitbit app, which records the duration of different phases of sleep. Analysis of the results showed that in the case of climatic comfort during sleep, the air movement is important, due to which even at higher air temperature and higher humidity a favorable PMV range can be obtained. At air temperature above

32 °C, the movement of air does not cause a feeling of comfort. It was estimated that feelings of comfort may be related to sleep phases, as it was noticed that the surface temperature of the skin varies before falling asleep, during sleep, and after waking up. The authors also carried out an energy assessment, indicating that obtaining high air velocities can cause increased electricity consumption. Therefore, they recommend considering the introduction of local elements in bedrooms that force air movement without the need to cool it. This can contribute to reducing the energy demand for air conditioning systems.

In Denmark [68], surveys were carried out in which information was collected on the configuration of bedrooms and the ventilation systems in which the buildings are equipped. Information was collected from 517 people. Questions were asked about room ventilation, how the air is supplied, the location of the building and bedroom, noise and the thermal environment. Sleep was assessed using the Pittsburg Sleep Quality Index (PSQI). Surveys have shown that only 33% of bedrooms are equipped with mechanical ventilation, 14% with exhaust air systems, and the rest are naturally ventilated. Respondents also rated disturbance and the largest resulted from the lack of mechanical ventilation, the presence of carpets, and feelings heat. People who felt the thermal environment was too hot were more likely to ventilate rooms during the day and before bedtime. The results showed that there is a link between ventilation and the feeling of thermal comfort and that people who experience discomfort try to remedy this by ventilating rooms. The authors also pointed out that field studies are necessary to link people's feelings with objective measurement results.

Subjective surveys and measurements were also conducted in China among students living in dormitories [69]. The survey asked about their own feelings about thermal comfort, humidity environment, acoustic environment, indoor air quality, and satisfaction level. Each answer was assigned points on a scale of -3 to $+3$. From cold/very poor/very noisy/very dry/very satisfied with -3 to extremely different answers with $+3$ (hot/very humid/very quiet/very good/very satisfied). Temperature, relative humidity, the concentration of carbon dioxide and particulate matter, formaldehyde concentration, and noise were also measured. Measurements were taken for one month in two small bedrooms with four beds. The authors found a correlation between survey responses and measured air parameter values. The air temperature was 26–28 °C, which significantly exceeds the comfort values by 4–5 °C. Relative humidity was about 40% and was at an acceptable level. In turn, the CO_2 concentration on average was about 4000 ppm, which means that air quality was very poor. The authors assessed the internal environment is uncomfortable and does not meet commonly accepted values that indicate appropriate conditions. The students in their survey responses also described the prevailing conditions as unsatisfactory, which only confirmed the measured values. Although the study was conducted in a short period of time and only in two bedrooms, the authors indicate further research directions, which should include a larger number of subjects. They decided that the results could help implement the guidelines for dorm users: ventilate frequently, especially before bedtime, lower the temperature with local devices, and use earplugs.

In Romania [70], tests of CO_2 concentration in the bedrooms of buildings that had been refurbished 2 years earlier were undertaken. Measurements of the CO_2 concentration were carried out in three rooms with natural ventilation. The concentration of CO_2 after 8 h of sleep reached values of up to 5680 ppm (when two people slept in the bedroom). This is a very high value that should not occur in any room where people are. Spending time in rooms with such high concentration can negatively affect not only the quality of sleep but even health. The authors noted that the bedrooms were aired before bedtime, but due to the significant increase in CO_2 concentration, this was ineffective. Additionally, the authors estimated that in winter, airing the room for 10 min causes a decrease in internal temperature by 3.3 °C (at an outside air temperature of 1.5 °C), so it is disadvantageous due to energy consumption for heating. The authors recommend the use of sensor-controlled mechanical ventilation in bedrooms.

The problem of heating costs, in terms of providing comfortable conditions in bedrooms, was raised in research in New Zealand [71]. The authors conducted a long-term study evaluating thermal comfort in children's bedrooms. They estimated that more than 50% of children sleep in uncomfortable conditions, which in these studies were determined by low indoor air temperatures. They identified that the reason for this condition is poor insulation of external partitions, insufficient ventilation, and lack of financial resources for heating buildings. They proposed to use a heat pump for heating. At the same time, they assumed that there should be good air quality in bedrooms, determined by the concentration of PM2.5. They ran CFD simulations that included the introduction of ventilation into bedrooms, assumed indoor temperatures, and determined PMV and PPD. In their conclusions, they stated that it is possible to achieve thermal comfort (PMV index of 0.72) when the set temperature is 22 °C. Research was carried out in the first weeks of winter. An interesting observation, which required further research, was the assessment that at the temperature of 19 °C in the second month of the study, when the energy consumption for heating was higher than at a temperature of 22 °C. The authors assessed that this was due to the higher humidity of the outside air.

Research on thermal comfort during sleep, in connection with ventilation, was conducted by authors from China [72]. They assessed that the bedrooms are heated in different ways: radiation, ceiling heating, wall heating, and others. The distribution of heating affects the formation of zones with different parameters, which can affect the feeling of comfort during sleep. The authors proposed a combination of heating and ventilation systems, stratum ventilation heating, thanks to which comfortable conditions can be obtained. This system is characterized by blowing air from the side of the head of a sleeping person. The temperature distribution in this system ensures appropriate thermal conditions. Thanks to this system, it is also possible to maintain adequate air quality. By preparing the right amount of air, this system is energy efficient. Further research by the same authors [73] concerned a more accurate assessment of air distribution in bedrooms using the aforementioned stratum ventilation system and mechanical mixing ventilation. The stratum system is recommended because by supplying air to the zone occupied by people, with the same value of the PMV comfort index, 10% energy savings can be achieved.

The definition of a ventilation strategy for bedrooms in single-family buildings was carried out in research on the characteristics of naturally ventilated and mechanically ventilated apartments [74]. CO_2 concentrations and air exchange rates were assessed in nine variants, including bedroom door/open, natural/mechanical ventilation, and various combinations of these cases. Due to a series of many measurements, the authors found that when the bedroom has natural ventilation, its door should be open during sleep. An even better effect can be achieved when a mechanical air extraction is used simultaneously with the door open (e.g., in the kitchen). In the case of mechanical ventilation, it was determined that air exchange with a multiplicity of $0.6 \, h^{-1}$ allowed maintaining of the CO_2 concentration below 1000 ppm. The authors also determined the CO_2 emission of a sleeping person at the level of 7.9–10.9 L/h, which is less than the standards (12.8–13.9 L/h). The authors assessed that in terms of thermal comfort, both in the case of natural and mechanical ventilation, conditions within the recommended values were obtained.

Table 6 presents a list of publications in the field of thermal comfort, ventilation in bedrooms with regard to sleep quality.

Table 6. List of discussed review publications in the field of thermal comfort and ventilation in bedrooms.

Authors	Year	Title	Journal DOI	Keywords
[67] Chenqiu Du et. al.	2022	A model developed to predict thermal comfort during sleep in response to appropriate air velocity in warm environments	Building and Environment 10.1016/j.buildenv.2022.109478	sleep quality; thermal comfort; warm environment; appropriate air velocity; prediction model
[69] Dan Mio et al.	2022	Associations of indoor environmental quality parameters with students' perceptions in undergraduate dormitories: a field study in Beijing during a transition season	International Journal of Environmental Research and Public Health 10.3390/ijerph192416997	dormitory; indoor environmental quality; indoor air quality; noise; carbon dioxide; well-being
[66] Haiguo Yin et al.	2020	Performance evaluation of three attached ventilation scenarios for tiny sleeping spaces	Building and Environment 10.1016/j.buildenv.2020.107363	tiny sleeping spaces; sleepin environment; double side-attached ventilation; air quality
[68] Chenxi Liao et al.	2021	A survey of bedroom ventilation types and the subjective sleep quality associated with them in Danish housing	Science of the Total Environment 10.1016/j.scitotenv.2021.149209	not provided
[65] Jakob B. Knudsen et al.	2020	Measuring ventilation in different typologies of rural Gambian houses: a pilot experimental study	Malaria Journal 10.1186/S12936-020-03327-0	airflow; housing, malaria; the Gambia
[72] Jian Liu and Zhang Lin	2020	Performance of stratum ventilated heating for sleeping environment	Building and Environment 10.1016/j.buildenv.2020.107072	heating; sleeping environment; horizontal temperature distribution; stratum ventilation; operative temperature
[73] Jian Liu and Zhang Lin	2020	Energy and exergy analyses of different air distribution in residential buildings	Energy and Buildings 10.1016/j.enbuild.2020.110694	sleeping environment; air distribution; stratum ventilation; energy and exergy analysis; computational fluid dynamics
[71] Mohammad Al-Rawi	2021	The thermal comfort sweet-spot: A case study in residential house in Waikato, New Zealand	Case Studies in Thermal Engineering 10.1016/j.csite.2021.101530	indoor environmental quality; computational fluid dynamics; Waikato; New Zealand
[70] D.A. Adincu et al.	2020	Experimental measurements of CO_2 concentrations in sleeping rooms	IOP Conference Series: Materials Science and Engineering 10.1088/1757-899X/997/1/012137	not provided
[74] Chandra Sekhar et al.	2020	Detailed characterization of bedroom ventilation during heating season in a naturally ventilated semi-detached house and a mechanically ventilated apartment	Science and Technology for the Build Environment 10.1080/23744731.2020.1845019	not provided

Paper [75], although meeting the search criteria, was not included in the review because it did not address the subject of sleep quality or bedroom issues. The study looked at thermal comfort in dormitories, and the recommendations were about thermal comfort during the day rather than at night. Also not included is [76], which is an article that was a review of research in Indonesia.

Sleep quality can be defined in different ways. The lengths of individual sleep phases are assessed or other indicators are used, e.g., the Pittsburg Sleep Quality Index (PSQI). Although there are no strict norms or guidelines for sleep quality, by analyzing these selected methods, it is possible to conclude whether sleep was valuable. Other methods of assessing sleep quality are conducting survey. Although these are rather subjective studies, taking into account the objective measurements of various parameters, scientists conclude about the impact of selected factors on people's feelings. In the analyzed works, the research was conducted in different groups of people (children, students) and different types of sleeping rooms (dormitories, tiny sleeping spaces and residential buildings). The choice of parameters to assess thermal comfort, ventilation, or air quality parameters, as well as the assessment of the impact of other factors that may affect the quality of sleep in bedrooms, depends on the location of the building and local conditions, as well as the season or the way the bedroom is used. The research was conducted in China, Denmark, Romania, New Zealand, and the Gambia. Because sleep is one of the elements of everyday life, its quality may be influenced not only by ventilation and thermal sensations, which should be borne in mind when analyzing the results.

7. Ventilation Impact on Thermal Comfort of Atrium Buildings

A modern atrium is a covered space surrounded by a building, usually equipped with a glazed roof, skylights, or large windows. The atrium can significantly improve building occupants' comfort due to providing a large common space and additional daylight, giving an impression of a connection with an exterior environment. However, the thermal comfort of the atrium space depends on many factors, including interaction with solar radiation and thermal buoyancy forces related to the height of the space. Proper ventilation can significantly affect thermal comfort by providing comfortable air temperature and velocity in the occupied zone, dissipating solar heat gains in the summer, and influencing the air temperature gradient. In addition, as the atrium can impact the energy balance of the building significantly, the optimal ventilation strategy can improve energy performance and thermal comfort in the atrium. Aspects related to atrium ventilation and thermal comfort are still the subject of recent research.

From January 2020 to May 2023, according to the Scopus database, with the search criteria selected for this review (title, abstract, or keywords including "thermal" and "comfort" and "ventilation" and ("atrium" or "atria")), no review papers were published, but 14 original studies were found. These studies are reviewed in this paper.

The paper [77] presents the investigation of energy performance and thermal comfort of the atrium. It is found that geometric parameters, especially height, presence and location of the platform, shape, and opaque-to-transparent surfaces are the most important factors. The research was conducted for the winter and summer in the atrium of a hotel located in China. Orthogonal experiments and CFD simulations were conducted. Although the energy aspect is the main focus of this research, some conclusions are drawn regarding thermal comfort and ventilation, i.e., the presence of the platform significantly impacts the thermal environment of the space under it. The platform shields the floor from solar radiation in the summer, stabilizes and unifies the temperature distribution in winter, and forces the ventilation air under the platform to flow with a cross path, which allows for maintaining better thermal comfort under the platform.

Winter thermal characteristic of the central atrium with skylights is studied in [78] for a teaching building in the cold climate of China. Field measurements of indoor thermal environment parameters (i.e., air temperature, humidity, atmospheric pressure, PMV, PPD) and occupants' thermal comfort rate questionnaire survey were implemented. An analysis of vertical and horizontal temperature distribution was conducted with the use of measurement instruments such as a thermal infrared imager, and illuminance UV recorder. It was proved, that the atrium thermal environment was affected by the fenestration airtightness, chimney effect, and skylight radiation. Based on the results of 70 surveys on thermal comfort, the formula was developed, describing the measured thermal sensation and PMV index as a function of operative temperature. Furthermore, it was concluded that the measured neutral temperature is lower than predicted with the Fanger model (about 17 °C in comparison to about 18.4 °C, respectively), and the measured comfort temperature range is wide, ranging from about 13 °C to about 21 °C. The formula for temperature in the atrium as a function of altitude was proposed. As the wide indoor air temperature range (reaching over 8 °C) impacted occupants' rating for thermal comfort, some solutions for improving thermal comfort were proposed, i.e., floor heating and cross ventilation, to provide more heterogeneous indoor thermal conditions.

Paper [79] is an analysis of various design parameters of the atrium in a commercial building on thermal comfort, daylighting, and natural ventilation. A central atrium was assessed with a field study, and a quantitative analysis was conducted to study the importance of various natural ventilation and daylighting strategies related to thermal comfort. The window-to-wall ratio and the glazing type were selected as parameters impacting the thermal comfort and daylighting the most. It was concluded that the natural ventilation, if designed optimally, can significantly decrease the energy consumption of the building. The window opening area and schedule are the most important design parameters.

An effect of the atrium ceiling shape on catching sea breeze and its impact on indoor thermal comfort and natural ventilation is the purpose of an original study [80], conducted for a building located in Bushehr (Iran) with a BWh (hot and humid coastal) climate. A CFD simulation method with a model validated with empirical measurements in wind tunnels was implemented. Twelve shapes of the roof were simulated for reference summer conditions, and Fanger's PMV/PPD indexes were used to determine thermal comfort. The authors concluded that in a hot climate, an atrium should be designed with particular emphasis on natural ventilation to prevent overheating and to improve indoor thermal comfort. In the extensive literature review, the authors indicate examples of research on the positive impact of high air velocity (e.g., 3 m/s) on thermal comfort, as well as the impact of atrium geometry on the efficiency of natural ventilation and indoor temperature stratification. Some studies confirming the possibility of shape and opening optimization to increase the effectiveness of wind-assisted natural ventilation were listed, as well as studies describing the impact of various ventilation strategies (e.g., night cooling, the use of chimney draft, and solar chimneys) on thermal comfort in the atrium. The authors confirmed with their research that the appropriate shape of the atrium roof in windy areas (the indicated average wind speed in Bushehr is 5 m/s) can cause additional negative pressure at the opening, increasing the effectiveness of natural ventilation. A significant impact of the roof shape on the ventilation effectiveness and resulting thermal comfort was observed, with the PMV index ranging from +2.4 to +3.0.

The performance evaluation of a vertical linear supply slot diffuser in a ring-shaped atrium located in a building in Kuala Lumpur (Malaysia), shaded from solar radiation by surrounding, taller buildings is the purpose of the research paper [81]. The CFD simulation method was implemented with the Ansys Fluent software. The conducted research proved that the vertical linear supply diffusers with high capacity (about 450 m^3/h/m) and high supply air velocity (6.56 m/s) and located around the perimeter of the atrium at the low height (bottom level about 0.5 m above the floor) with exhaust grilles located under them can provide good airflow distribution, with a stratified thermal plume in the occupied, lower zone of the atrium. Acceptable thermal comfort was estimated for the supply diffusers located along the sidewalls of the atrium, according to calculated PMV and PPD indicators. It should be noted, however, that the presented results are adequate for atriums without the influence of solar radiation, which is typically an important factor determining airflow and thermal comfort in this type of space.

In [82], a study of energy and thermal characteristics of an atrium in a moderate climate of Poland is conducted. A large atrium with a glazed roof, giving a high glazed-to-opaque surface ratio, was studied in two variants, as a zone independent of the adjacent spaces and as a passive heater for fresh air for ventilation of adjacent spaces. Incorporating the atrium can significantly lower the use of artificial lighting in adjacent spaces and result in fewer hours of artificial lighting usage, which can reduce energy demand directly. In addition, reducing the time of use of built-in lighting impacts the HVAC systems' operational characteristics by lowering internal heat gains. The authors stated that the atrium makes it difficult to maintain the temperature in adjacent spaces in the comfortable range of 20–25 °C throughout the whole calendar year; however, in the case of the educational building unoccupied during the summer holiday, the comfortable temperature can be maintained for about 90% of the usage time. It was proved that the use of the atrium as a preliminary heat exchanger for fresh air for adjacent space ventilation can provide a comfortable temperature in the atrium's occupied zone (near the floor) but is related to the limitation of overheating the atrium in the summer.

Occupants' thermal comfort perception in two different, naturally ventilated central atrium spaces of public markets in Malaysia is the topic of research in [83]. This study included questionnaire surveys and field measurements of air temperature, humidity, and velocity. For the permanent occupants of the buildings (sellers), excessive temperature and humidity as well as excessively still air velocity were repeated feedbacks, and the use of individual mechanical fans or occupying zones with the best thermal conditions were the most common responses for the perceived thermal dissatisfaction. It was emphasized that there were ceiling fans located under the roofs of both atriums, which were unable to generate a perceptible air velocity in the occupied zone. The small sample size for the surveys should be noted (40 and 20 responders in both cases, respectively).

An open-plan, naturally ventilated multi-story office building with an atrium is studied in [84]. The CFD and WBM (water bath modeling) methods were used to examine ventilation efficiency reflected by the ACH (air changes per hour) factor and internal room air temperature above ambient. A total number of 36 cases varying in the design of the atrium, openings, and core configuration were examined, for which the abovementioned parameters were calculated for different floors of the building. According to the results, the ventilation efficiency is most sensitive to the opening design varying between 2.32 and 2.48 ACH for the mean floors, while the average horizontal thermal stratification is most sensitive to the atrium design, varying between 0.24 and 0.33 °C. The mean room air temperature above ambient at various office depths (locations between the external and atrium façade) also varies depending on the building configuration, which justifies the use of complex computational methods to predict airflow and thermal conditions in atrial buildings.

In [85], a multistory commercial building with a central atrium is studied. Thermal comfort in the atrium was to be provided by connection with neighboring mechanically air-conditioned office rooms, which allows hybrid ventilation of the atrium with the cooled air from the office rooms flowing through the public area to the atrium. The displacement and stack ventilation of the building were studied. The CFD simulation method was used to evaluate how external temperature, supply air inlet locations, flow paths between the atrium and surrounding public areas, ceiling height, and location of heat sources in offices affect the ventilation of the atrium. The PMV and PPD indices were used as a measure of thermal comfort. The authors found that displacement ventilation can be more effective than stack ventilation when the ambient temperature is high due to the impact of indoor–outdoor temperature difference on the pressure distribution. In high buildings with air intakes on each floor, reversed pressure distribution may occur in the upper floors that will cause the airflow from the atrium to the public area, which is a path opposite to the desired one. According to the results, this effect may happen when the ambient temperature is more than 3 °C higher than the indoor temperature, and the supply air is provided with inlets around the perimeter of the building. Ambient conditions have a limited effect on the indoor environment when fresh air is not provided by the external air intakes (i.e., displacement ventilation is designed instead of stack ventilation). The authors found that the air inlets between offices and public areas located at a low height provide better thermal comfort expressed in PMV than when the air inlets are located high (i.e., averaged PMV for supply air inlet at 0.4 m or 2.0 m above the floor was calculated as +1.13 and +1.34, respectively). Furthermore, the higher the handrails were (partitions between the atrium and surrounding public areas), the better averaged thermal comfort was noticed (i.e., resulting PMV was from +0.70 to +0.98 for handrail height of 1.9 m, and from +0.78 to +1.09 for 1.1 m). The combination of ceiling height, heat sources, and air inlets impact displacement ventilation system efficiency, and neutral PMV values can be achieved with a proper combination of the abovementioned parameters.

The paper [86] presents the CFD analysis of a subway station with the atrium space, ventilated using a piston ventilation pattern. As it is a specific case not related to typical buildings, this paper is not reviewed.

The paper [87] presents an analysis of cooling load reduction due to nighttime natural ventilation. A central atrium of a non-residential building, with an opaque roof and two exterior window facades on opposite sides of the building, equipped with a cooling floor, was studied to verify that opening opposite windows provide effective natural ventilation of the building and accumulation of cold in its thermal mass through the night, to be used during the day. The efficiency of nighttime precooling was simulated with the EnergyPlus model, while the CFD simulations with ANSYS Fluent were implemented to predict the wind-inducted pressure on the façade with window openings located near the floor on the windward side and near the roof on the leeward side to increase stack effect. The thermal and airflow dynamic models were validated by full-scale measurements. The authors determined that natural ventilation driven by stack and wind can be efficiently covered by modern simulation tools to predict night cooling potential. The potential of passive night cooling to reduce the daytime cooling load by about 27% in the hot climate of Portugal was proven, even with relatively small openings used (the total area of the operable windows in comparison to the floor area of the atrium was only about 1%). Thermal comfort was assessed with a set of surveys and indoor environmental parameters measured on two days with similar weather conditions. In the first case, only passive nighttime cooling was used, while additional active day cooling was implemented in the second. Air and mean radiant temperature, air speed, and relative humidity were measured, while about 30 occupants were asked to vote for their thermal sensation in both cases. According to the results, it was proved that nighttime natural cooling does not over-cool the building, has no negative effect on occupant's thermal sensation and allows the maintaining of good thermal comfort conditions without activation of the supplementary cooling floor system, as more than 80% of the occupants felt comfortable in that case, which is a decrement in comparison to the over 90% satisfied in the case of ventilation night cooling + day floor cooling case, but is still an acceptable threshold of people satisfied with the thermal environment.

The central atrium with roof fenestration, located in a library in Cyprus characterized by a coastal, windy climate with high air temperatures and relative humidity is an object of research in [88]. The atrium with skylights was supposed to provide natural lighting and improve the thermal comfort of the building, equipped with operable windows located around its perimeter. The authors emphasized the problem of summer overheating caused by the atrium skylights, which are impossible to eliminate by shading elements without loss of natural illumination, so the study investigated the implementation of a double-skin skylight to improve the thermal performance of the atrium without significant loss of natural lighting. The simulation method with EDSL TAS software was implemented, and the adaptive thermal comfort model was adopted to assess predicted occupants' thermal sensation. It was found, that implementation of a double-skin façade increases the air temperature in the upper part of the atrium, and enhances the effect of thermal buoyancy, stimulating natural ventilation during the summer season. Additionally, additional thermal insulation is provided for the winter season, as the original skylights were only 6 mm clear glass. Several operation strategies with different ratios of the opened windows and the double-skin façade materials were simulated for the winter and summer. According to the results, opening the atrium roof leads to an almost doubling of thermally comfortable working hours in the summer. Implementation of double-skin skylights improves thermal comfort during working hours even more significantly depending on the total area of fenestration opened, lowering the temperature in the building by enhancing natural ventilation. Importantly, no improvement in winter thermal performance was deducted. In addition to the original research, a wide literature review on the topic of atrium thermal performance and double-skin envelopes was presented in the paper.

The review paper [89] presents a systematic review of the literature on the daylighting performance of atria and fulfills the search criteria of this review but does not apply to issues related to ventilation and thermal comfort, so this paper is not reviewed.

A mechanically ventilated atrium of a Malaysian theme park is the subject of CFD (Ansys Fluent) simulation studies of heat and airflow in [90]. The authors conducted a simulation of natural ventilation driven by the buoyancy force and compared the resulting thermal comfort with those provided by fully-mechanical ventilation systems. According to their results, natural ventilation is ineffective and leads to worsening indoor thermal conditions in comparison to fully-mechanical ventilation, but hybrid ventilation with a fresh air inlet fan supplying exterior air to the lower part of an atrium to supplement the mechanical ventilation and cooling system can improve air distribution and maintain thermal comfort inside the atrium.

The articles are summarized in Table 7.

Table 7. List of other publications on ventilation and thermal comfort in atrium.

Authors	Year	Title	Journal DOI	Keywords
[77] J. Pang et al.	2023	Effects of complex spatial atrium geometric parameters on the energy performance of hotels in a cold climate zone in China	Journal of Building Engineering 10.1016/j.jobe.2023.106698	building energy performance; complex spatial atrium; design parameters; atrium platform position; slope of trapezoidal atrium profile
[78] C. Xu et al.	2023	Study on winter thermal environmental characteristics of the atrium space of teaching building in China's cold region	Journal of Building Engineering 10.1016/j.jobe.2023.105978	atrium space; cold region; measurement; teaching building; thermal environment
[79] H. Sahu, J. Vijayalaxmi	2023	Optimization of the Integrated Daylighting and Natural Ventilation in a Commercial Building	Lecture Notes in Civil Engineering 10.1007/978-981-19-9139-4_9	atrium optimization; daylighting; natural ventilation; thermal comfort; visual comfort
[80] J. Shaeri et al.	2023	Effects of sea-breeze natural ventilation on thermal comfort in low-rise buildings with diverse atrium roof shapes in BWh regions	Case Studies in Thermal Engineering 10.1016/j.csite.2022.102638	atrium roof shapes; sea-breeze; natural ventilation; thermal comfort; bushehr; BWh
[81] Y.H. Yau, U.A. Rajput	2023	Performance evaluation of an architecturally-designed vertical high capacity linear slot diffuser in a tropical atrium	Architectural Science Review 10.1080/00038628.2022.2140988	CFD simulation; tropical atrium; slot diffuser; turbulence model; deflector angle

Table 7. *Cont.*

Authors	Year	Title	Journal DOI	Keywords
[82] K. Ratajczak et al.	2022	Incorporating an atrium as a HVAC element for energy consumption reduction and thermal comfort improvement in a Polish climate	Energy and Buildings 10.1016/j.enbuild.2022.112592	atrium; daylighting; ventilation; building simulations; energy savings; university buildings; IES VE
[83] A. Ghazali et al.	2022	Indoor Thermal Comfort Perception at Atrium Zone: Case Study of Naturally Ventilated Public Market	Journal of Advanced Research in Applied Sciences and Engineering Technology 10.37934/araset.29.1.1329	thermal comfort; atrium zone; natural ventilation
[84] T. Corbett et al.	2020	Sensitivity analysis of proposed natural ventilation IEQ designs for archetypal open-plan office layouts in a temperate climate	Advances in Building Energy Research 10.1080/17512549.2020.1813197	buoyancy-driven ventilation; building design; computational fluid dynamics; internal environmental quality; natural ventilation; thermal comfort; water-bath modeling
[85] H.-H. Hsu et al.	2021	Hybrid Ventilation in an Air-Conditioned Office Building with a Multistory Atrium for Thermal Comfort: A Practical Case Study	Buildings 10.3390/buildings11120625	computational fluid dynamics (CFD); stack ventilation; displacement ventilation; atrium design; energy saving
[86] Y. Wen et al.	2020	Integrated design for underground space environment control of subway stations with atriums using piston ventilation	Indoor and Built Environment 10.1177/1420326X20941349	architectural design; atrium; piston effect; subway station; underground space; ventilation
[87] D.P. Albuquerque et al.	2020	Full-scale measurement and validated simulation of cooling-load reduction due to nighttime natural ventilation of a large atrium	Energy and Buildings 10.1016/j.enbuild.2020.110233	night cooling; natural ventilation; cooling load measurement; thermal comfort; EnergyPlus; CFD
[88] R. Sokkar, H.Z. Alibaba	2020	Thermal Comfort Improvement for Atrium Building with Double-Skin Skylight in the Mediterranean Climate	Sustainability 10.3390/su12062253	atrium; natural ventilation; thermal comfort; passive design strategy; Mediterranean climate; double skin skylight
[89] H. Omrany et al.	2020	Is atrium an ideal form for daylight in buildings?	Architectural Science Review 10.1080/00038628.2019.1683508	atrium; daylight; energy efficiency; visual comfort; sustainable design; atrium design
[90] N.A.M. Fohimi et al.	2020	CFD Simulation on Ventilation of an Indoor Atrium Space	CFD Letters 10.37934/cfdl.12.5.5259	CFD simulation; hybrid ventilation; atrium space

A review of recent research papers on ventilation and thermal comfort in atriums confirms the relevance of this topic and proves the need for further research. According to the presented studies, comprehensive analysis and optimization of architectural and engineering solutions allow the development of an atrium that improves not only the building's energy characteristic but also thermal comfort. As atriums are typically high spaces with many glazings, their interior thermal conditions are affected by many factors with great importance for solar heat gains and stack effects. Due to the high complexity of atrium thermal environments, a general conclusion can be drawn regarding the research methods used in the revised papers:

- The CFD analysis and in situ measurements are recognized as the most popular research methods to predict air movement (ventilation) patterns;
- Questionnaire surveys or calculations based on steady-state Fanger or adaptive thermal comfort models are used widely to assess thermal comfort.

8. The Future Directions of Research

This article reviews the latest publications on thermal comfort and ventilation in relation to various buildings present in public space. The results presented by the authors, the methods used, and the conclusions were analyzed.

The future directions of research, in analyzed articles relevant to buildings, include developing more advanced HVAC control systems which take into account online learning or predictive control. More buildings in different climate zones could be explored in the context of the transfer learning performance.

There are many more buildings and activities that also require special conditions, like sports buildings [91–95], including swimming pools [96–98] and fitness centers [99], as well as kindergartens [100–104], nurseries [105–107], museums [108–110], and churches [111–113].

The other group of papers on thermal comfort deals with methods of assessing thermal comfort [114–117] or on the subject of the personalization of comfort through ventilation [118,119]. Due to the assumed criteria for the selection of publications for the review and the extensiveness of the subject, the above issues have not been included. But the authors believe these are topics worth considering for future research.

9. Summary

A literature review of 53 articles on thermal comfort and ventilation for different types of buildings and spaces was conducted. In the articles, the conclusions, research methods, and limitations that could set directions for engineering activities that could bring benefits to society were sought. The most important findings are summarized below.

9.1. Office Buildings

Summarized findings of reviewed works in the field of thermal comfort and ventilation in office buildings are presented in Table 8.

In studies on offices, large, open spaces are most often studied [28,30–34]; sometimes the same offices in different locations are studied [31]. In the case of the research in [28], 52 spaces located in 26 cities from "ASHRAE RP-884 Database", 57 cities located in 30 countries from "The Scales Project Dataset", and 1 building (24 participants: 16 females, 8 males) from Medium US Office were numerically analyzed, and in [30] three different offices were analyzed. In terms of ventilation, there was both natural [31–33] and controlled [28–30,34] ventilation, with virtually no predominance of either system, which indicates the commonness of both types of ventilation in office buildings. In the analyses, measurements of air parameters in spaces [31,33] and measurements along with surveys on the feelings of their users [30,34] prevailed, but research works based on simulations [32], transfer learning [29], and a combination of simulations in EnergyPlus and machine learning algorithms [28] appeared. The discussed studies have many limitations given by their authors, among which one can specify the study of individual thermal comfort optimizing [28] or the study of only one office [29,30,32,33]. The authors also indicated that the results of the research should be supplemented with the results of questionnaires [31,32]. The main conclusions from the research are as follows.

Table 8. Summarized findings of reviewed works in the field of thermal comfort and ventilation in office buildings.

Paper	Space Type	Ventilation Type	Research Method	Number of Subjects	Limitations	Country, Climate	Main Findings
[28] Peng B., Hsieh S.-J.	Open-space office room	Controlled	Co-simulation (EnergyPlus and MATLAB))	1 open-space office room 4 citis (16 main batch	There is a high air exchange between cubes with individual volume flow rate of vent (formed in the middle of the room), so optimizing for the average of 6 cubes equals local discomfort (e.g., cool local climate for one objective).	Chicago, Phoenix, College Station, Tampa, United States of America	Novel hybrid data-driven thermal comfort model (PMV-SVM-ANN model) and HVAC control system which improve the thermal comfort of occupants in an open-space office room by 43.4–69.9% with slightly lower energy consumption (1–3.6%) were designed.
[29] Gao N et al.	Medium office	Controlled	Transfer learning with Algorithm TL-MLP-C * and TL-MLP	25,623 observations from ASHRAE RP-884 Database, 8225 participants from The Scales Project Dataset and 24 participants from Medium US Office Dataset (Philadelphia city, PA, USA).	Only the one office located in the 'temperate' climate zone was used as the target building. The best performance of prediction model is when at least six factors are provided and if there are lower (than six) then performance drops. Only MLP models were investigated.	Algorithm TL-MLP-C *:—Philadelphia, USA Algorithm TL-MLP: different climate zones all over the world	Developing of transfer-learning-based multilayer perceptron model in order to make thermal comfort prediction more precise (accuracy 54.5%)
[30] Yong, N.H. et al.	Open office	mechanical ventilation system	Measurments Surveys	3 office areas	Thermal environment and air quality were studied only at three office spaces.	Malaysia, tropics	Thermal comfort indices were indicated: the neutral temperature was 23.9 °C (that is 1.3 K lower than predicted 25.2 °C). About 81% of the occupants found their thermal environment comfortable with high rate of adaptation Phisical parameters, air quality, and ventilation rate per person were acceptable.
[31] M. Badeche, Bouchahm Y.	Office building with large glazing	Natural	Measurments According a Post Occupancy Evaluation (POE) method	4 office rooms located in diverse orientations (NE), (SE), (SW)	Thermal and visual comfort measurements only were conducted. For example, there is no information about occupant survey questionnaire, which is part of POE.	Algeria, Semi-arid climate	Impact of large glazing areas on thermal comfort and visual comfort. Natural ventilation was unable to release of accumulated heat gains. 86–100% of working-time was overheating, so adaptive solar façade, smart glazing, and moving shading devices was suggested.

Table 8. *Cont.*

Paper	Space Type	Ventilation Type	Research Method	Number of Subjects	Limitations	Country, Climate	Main Findings
[32] Ortiz, M.C., Morales, S., Cabrera, V.	Admission office at campus	Natural	CFD simulation (ANSYS Fluent 2019)	1 admission office	In order to improve the research and increase the number of satisfied users, it would be necessary to change the time step to hourly, conduct field surveys and introduce an adaptive approach.	Quito in Ecuador	Analysis of opening area of skylight, distributing louvers along the skylight, door position, impact of furniture in context of air movement in room, PMV, and PPD. Opening the door increases ventilation rates by 30%. Indoor air temperature equaled 18 °C and air speed in range: 0.25–0.75 m/s represented cool sensation (clothing level = 0.65) and slightly cool sensation (clothing level = 1.0). Indoor air temperature equaled 25 °C met thermal comfort.
[33] Lahji K., Puspitasari P.	Architecture studio room	Natural	Measurements:	1 architecture studio room	Only one room was analysed.	Jakarta in Indonesia	PMV value in range 2.23–3.27 (warm–hot thermal sensation) was observed and was correlated with wind velocity. In order to meet natural sensation, mechanical aids or hybrid air conditioning system is required.
[34] Borsos, Á. et al.	Open-plan office	Controlled	Measurments of IEQ Comfort survey	1 open-plan office, 216 respondents	1 There are no data on the IEQ parameters for air conditioning or no heating and no validation of the comfort map. Used A-weighted sound pressure level (LA) method underestimates the degree of noise produced by activities or sources characterized by a low-frequency contribution.	Budapest Hungary	65%, 50%, 47% of respondents were dissatisfied with the adjustability of noises and sounds, ventilation and thermal conditions, respectively, in the work environment. An office 'comfort map' based on occupants' individual IEQ preferences was developed.

In order to maintain a thermal comfort while minimizing office buildings' energy usage it is necessary to create an accurate thermal comfort model. Thermal comfort models based on thermal comfort data from room occupants achieve better performance than the PMV (predicted mean vote) model and adaptive model. On the other hand, it requires the collecting and processing of large data sets. Control of indoor air exchange in natural and mechanical ventilation systems is crucial to increase work productivity.

Thermal comfort in office buildings depends on the glazed area. In semi-arid climates, large glazing areas affect the high energy consumption of cooling systems to maintain comfortable temperature. Night ventilation absence is a reason for heat accumulation in the occupied zone.

In tropical climates, air extraction in buildings with natural ventilation (using skylight) is difficult because indoor and outdoor temperatures are similar. However, open doors increase ventilation rates (up to 30%).

Understanding office workers' individual preferences for IEQ affects increase of productivity. "Comfort maps" provide the opportunity to choose the optimal workplace.

Although it seems that office buildings are well researched, studies conducted in recent years indicate that there is further potential in their research, especially in terms of the thermal comfort of users. It should be borne in mind that research on thermal comfort in offices should use surveys and questionnaires, because feedback from people staying in offices may be important for recommending new models or guidelines for air parameter ranges. Simulations can be a response to performing research on a small research sample. Thanks to the measurement of the object and the execution of a digital twin, the data can be used for further simulations, thanks to which the number of tested cases will increase. The most common conclusion when inadequate thermal comfort of office users is detected is the recommendation to use air conditioning and/or controlled ventilation systems. According to the authors of this review paper, future research should also focus on the issues of improving controlling accuracy of the HVAC systems in offices in order to match the current needs of users while minimizing energy consumption for air treatment and its circulation.

9.2. Schools

Summarized findings of reviewed works in the field of thermal comfort and ventilation in schools are presented in Table 9.

The review included 18 articles. Most studies were conducted in primary schools [38–40,43,45–47]. In more than 80% of the analyzed cases, the subject of the analysis was natural ventilation, and in two (which are not review articles), air conditioning was included. The research methods used by the authors included in situ measurements [38,39,41–43,45,46,51], simulations [40,41,45–47], surveys: [42,43,51,52], and various combinations of those. Articles [36,44,53] are review articles.

The review of the literature on the impact of thermal comfort and ventilation on the ability to learn led to the following conclusions. Temperature is a very important factor affecting children's' performance. High temperatures cause a loss of concentration; students prefer cooler classroom temperatures than temperatures that adults feel as thermally neutral. In the tropical climate, cross-ventilation allows improving thermal comfort and maintaining the level of CO_2 concentration below 1000 ppm.

CO_2 concentration below 900 ppm avoids negative impact on learning ability. However, the exact correlation between CO_2 levels and learning ability should be further explored.

72

Table 9. Summarized findings of reviewed works in the field of thermal comfort and ventilation in school buildings.

Paper	Space Type	Ventilation Type	Research Method	Number of Subjects	Limitations	Country, Climate	Main Findings
[36] Manca S.; Cerina V.; Tobia V.; Sacchi S.; Fornara F. (2020)	School	Not specified	Systematic review;	68 empirical papers	Not specified	Various	A friendly environment (colors, ergonomic furniture…) as well as acoustic, thermal, and natural light comfort are important for the well-being and productivity of the users. The amount of ventilation affects student performance. Future research should emphasize individual variables of age, gender, specific needs, and interactions or coexistence of different variables to capture the high level of complexity of phenomena.
[40] Dudzińska A.; Kisilewicz T. (2021)	Primary school	Four variants: mechanical ventilation, forced night ventilation, natural night cooling, natural during the day and night	Simulation (Design Builder)	1 school 1 digital model	Results are valid for summertime	Budzów, Poland Summer	The role of mechanical ventilation and the possibility of night ventilation in reducing discomfort were investigated. Suggestion of the "aggregated measure of overheating" method for estimating hours of discomfort.
[41] Shrestha M.; Rijal H.B. (2023)	Secondary school	Natural	Field survey, simulation (DesignBuilder)	3 schools (one in Kathmandu, two in Dhading) Survey of 246 students, 737 responses (three-time votes)	Results are valid for summertime	Katmandu, Nepal Summer	Use of simulation in the design process to assess how the building behaves in terms of comfort and thermal environment. 1. Summer average comfort temperature: 26.9 °C. 2. Further studies should be conducted in different seasons and climates.
[42] Haddad S.; Synnefa A.; Ángel Padilla Marcos M.; Paolini R.; Delrue S.; Prasad D.; Santamouris M. (2021)	Secondary school	Mechanical ventilation (DCV)	Measurements, field survey	1 school, 2 adjacent classrooms The survey took place in late August to early September 2018 (305 responses) and in early April (377 responses)	Not specified	Sydney, Australia	Survey studies and Indoor Air Quality (IAQ) measurements confirm the influence of indoor temperature and CO_2 concentration on students' fatigue levels. It takes one week for students to adapt to a sudden change in the average outdoor temperature.

Table 9. *Cont.*

Paper	Space Type	Ventilation Type	Research Method	Number of Subjects	Limitations	Country, Climate	Main Findings
[44] Kapoor N.R.; Kumar A.; Alam T.; Kumar A.; Kulkarni K.S.; Blecich P. (2021)	School	Natural Controlled	Systematic Review Meta-Analyses (PRISMA);	37 articles	Not specified	India	A systematic review of Indoor Environmental Quality (IEQ) parameters related to research conducted in Indian school classrooms. The need for a standardized IEQ testing method in Indian schools.
[45] Lala B.; Kala S.M.; Rastogi A.; Dahiya K.; Yamaguchi H.; Hagishima A. (2022)	Primary school	Natural	Measurments, machine learning	5 schools, 14 classrooms, 512 students	Only one month-long field experiment	Dehradun, India	A comparative analysis of spatial variability in the performance of the machine learning model for children (in situ measurements) and adults (ASHRAE-II database).
[37] Peters T.; D'Penna K. (2020)	University	Natural	Semi-systematic review	32 articles	Not specified	Not specified	Openable windows should be used in study rooms. Enhancing ventilation improves cognitive performance. Each classroom and office should have windows on the exterior of the building, and if this is not possible, opening into the atrium to let daylight into the room. Natural ventilation reduces carbon dioxide levels, so air quality closer to serviced windows may be best to study.
[38] Suradhuhita P.P.; Setyowati E.; Prianto E. (2021)	Elementary school	Natural	Measurements	2 classrooms	Field measurements for the room temperature were carried out in the time range from 07:00–16:00 WIB on Wednesday, 2 June 2021	Indonesia (Jawa Tengah)	Openings in facades and the support of mechanical ventilation are necessary to ensure thermal comfort in accordance with the Indonesia National Standard.

Table 9. *Cont.*

Paper	Space Type	Ventilation Type	Research Method	Number of Subjects	Limitations	Country, Climate	Main Findings
[43] Mazlan A.N.; Saad S.; Yahya K.; Haron Z.; Abang Hasbollah D.Z.; Kasiman E.H.; Rahim N.A.; Salehudddin A.M. (2020)	School	Not specified	Measurement survey (Likert's scale rating)	2 schools in urban area, 2 schools in rural area	Measurments took place from 10 to 22 January 2019	Selangor, Malysia	On-field measurement and questionnaire survey revealed that urban areas experienced a rapid increase in temperature, while rural areas remained stable. The survey results confirmed that both students and teachers experienced reduced concentration when the indoor temperature was too high.
[46] Lan K.; Chen Y. (2022)	Primary school	Underground ventilation pipes and double-sided corridors	On-site monitoring in the classroom + numerical simulations based on finite element software	1 typical classroom and numerical model	Measurements five times a day from December to January	Shaanxi Province, China	Dual-sided corridors, underground ventilation ducts, and windows with a height/width ratio of 1 can provide energy-efficient and habitable building structures for primary school classrooms in the northern Shaanxi region of China.
[39] Talarosha B.; Satwiko P.; Aulia D.N. (2020)	Primary school	Natural cross -ventilation	Measurments	1 classroom	Measurment lasts only 4 days in April	Medan Helvetia, Indonesia	Despite the overheating in the classroom (indoor air temperature ranged from 30.5 °C to 34.5 °C), the median CO_2 concentration remained below 1000 ppm (maximum value: 637 ppm).
[47] Chang L.-Y.; Chang T.-B. (2023)	Elementary school	Natural ventilation, mechanical ventilation	Simulations (Python)	1 class of the school building; 4 variant A/C all day; delaying the turn-on time of the A/C system, turning the A/C system off during the lunch time period; turning the A/C system off during one class each afternoon.	Not specified	Taiwan	In connection with the plan to install air conditioning systems in all schools in Yuan, Taiwan, the authors conducted simulation studies on various air conditioning control strategies to achieve the average 8 h CO_2 concentration target in classrooms. The optimal strategy identified was operating the air conditioning throughout the school day in conjunction with a mechanical ventilation system providing a fresh air exchange rate of 5 l/s per person in the room.

Table 9. *Cont.*

Paper	Space Type	Ventilation Type	Research Method	Number of Subjects	Limitations	Country, Climate	Main Findings
[49] Tagliabue L.C.; Accardo D.; Kontoleon K.J.; Ciribini A.L.C. (2020)	Classroom	Natural	Simulation (EnergyPlus, Comsol Multiphysics)	Building energy model of the standard classroom; "Modulo Didactico" building model	Not specified	European climate zones (Italy, Greece, United Kingdom)	A dynamic simulation was conducted to assess the thermal performance of school buildings and explore possibilities for their improvement in several representative locations across Europe. It was demonstrated that thermal comfort in educational buildings during the summer can be achieved in certain temperate climate regions without the use of mechanical systems. In addition to this, informing users about building monitoring and management is crucial to ensuring thermal comfort conditions within the spaces.
[51] Fabozzi M.; Dama A. (2019)	University	Natural air-conditioning	Measurments Survey (ASHRAE 55 Standard)	16 classroom survey of 985 students	Evaluation took place in summer time	Milan, Italy Summer	The authors confirmed the utility of the Ranger model for predicting thermal sensations in air-conditioned classrooms with reasonable accuracy. In naturally ventilated classrooms, the adaptive model proved suitable for predicting students' comfort zones according to the ASHRAE 55 standard, while the adaptive comfort temperatures recommended by the EN 15,251 standard were unacceptable for a large number of students. There were no significant differences in the perception of thermal comfort between genders.

Table 9. *Cont.*

Paper	Space Type	Ventilation Type	Research Method	Number of Subjects	Limitations	Country, Climate	Main Findings
[52] López-Chao V.; Lorenzo A.A.; Saorín J.L.; De La Torre-Cantero J.; Melián-Díaz D.	Primary school	Natural	Survey based on the Indoor Physical Environment Perception scale (iPEP scale)	9 schools, 27 classrooms, 583 primary school students	There are no data on the IEQ parameters for air	Spain (Galicia)	The prognostic analysis of students' mathematics results confirmed their direct correlation with ventilation, room size, views, and attachment to place. It was also validated that the adaptation of the iPEP scale in the primary school is accurate and reliable.
[50] Grassie D.; Karakas F.; Schwartz Y.; Dong J.; Milner J.; Chalabi Z.; Mavrogianni A.; Mumovic D.	School	Natural	Simulations (EnergyPlus)	111 different combinations of naturally ventilated schools, from 5 different eras of construction, in 13 geographical regions	Modeling approach developed for naturally ventilated schools only	United Kingdom	A method called Data-driven Engine for Archetype Models of Schools (DREAMS) has been developed, based on the EnergyPlus stock modeling framework, which models different typologies of classrooms and takes into account not only indoor environmental quality criteria but also academic achievements, health, and healthcare costs.
[53] Sadrizadeh S.; Yao R.; Yuan F.; Awbi H.; Bahnfleth W.; Bi Y.; Cao G.; Croitoru C.; de Dear R.; Haghighat F.; Kumar P.; Malayeri M.; Nasiri F.; Ruud M.; Sadeghian P.; Wargocki P.; Xiong J.; Yu W.; Li B. (2022)	School	Different types	Critical review	304 publications	Not specified	Various	A comprehensive review of research from the last 50 years concerning indoor air quality in classrooms across more than 40 countries.
[48] Catalina T.; Damian A.; Vartires A.; Nița M.; Racovițeanu V. (2023)	Secondary school	Natural	Measurments	2 clasrooms	Not specified	Romania	Ventilation in classrooms is mandatory to achieve a satisfactory level of thermal comfort and indoor air quality, and even natural ventilation through window opening remains a simple and effective tool to reduce pollutant concentrations in indoor spaces.

It is not possible to give a one-size-fits-all comfort temperature range for all educational buildings because it depends on many factors, including the climate zone and students' previous exposure to air conditioning. But night ventilation allows reducing thermal discomfort during the day.

The publications often lack information about the limitations of research, but they can mainly be defined as conducting research only for one period of the year [40,41,51] or covering only a short measurement period [38,39,45,46].

Educational buildings are designed for different age groups: children, teenagers, adults, as well as for various types of activities—science, artistic expression, sports, but also office work. In studies conducted in schools, it is good to use a holistic approach, because a lot of elements affect the effectiveness of the learning process [36,37,52]. The planned research may take into account the different purpose of the rooms and the different ages of the participants, which may affect the responses to the survey. It is worth noting that in the cited studies, schools are mainly equipped with natural ventilation, which means that thermal comfort is often not achieved, not to mention the quality of indoor air, which is often poor in educational buildings. This indicates a new direction of research on ventilation systems that could be used in existing buildings for such a purpose. When designing new educational buildings or during the renovation process, computer simulations can be an important tool. However, for the simulations to reflect reality, it is very important to introduce parameters that precisely define the characteristics of the building along with its technical equipment, surroundings, and users, which is why expanding/defining simulation algorithms and machine learning so that they take into account the specificity of individual age groups (children, youth, adults) seems to be important.

9.3. Hospitals

Summarized findings of reviewed works in the field of thermal comfort and ventilation in hospitals are presented in Table 10.

Research on hospitals is conducted for various spaces in hospitals: hospital ward [57,61], emergency room [59], operation room [59,60,62,64], intensive care unit [59], or nursing home [63]. It would be expected that healthcare facilities should have controlled ventilation through which a better indoor environment can be achieved, and this is indeed the case. Natural ventilation/controlled ventilation/both types of ventilation were present in correspondingly 29%/57%/14% of the reviewed papers. Most studies use controlled ventilation [59–62,64]. There are various research methods: simulations [57,60,62,64], in situ measurements [57,59,61,62], surveys [57,61–63], and their various combinations. The main limitations are the inability to generalize the results due to the test conditions in only one climate [59,61,63] or the limited number of analyzed systems [60,64]. The main the main insights after the literature review are summarized below.

A review of research papers and review articles on ventilation and thermal comfort in hospitals shows that this is a subject that still requires continuous research. The multitude of parameters affecting the final effect does not allow for simple and universal conclusions and recommendations. Climate, tightness of the building envelope, ventilation system, control of its operation, staff attire, type of room (operating room, hospital ward, others), and type of patient illness are just some of them. To this day, it is not settled which ventilation system should be used and when to ensure adequate air quality and thermal comfort. New technologies are being developed to ensure thermal comfort for both patients and hospital staff (divergence of interests), aiming at ad hoc (temporary) warming of the patient during surgery in order to avoid hypothermia, also with the use of intravenous warming substances.

Table 10. Summarized findings of reviewed works in the field of thermal comfort and ventilation in hospitals.

Paper	Space Type	Ventilation Type	Research Method	Number of Subjects	Limitations	Country, Climate	Main Findings (General)
[57] Abd Rahman, Noor Muhammad et al.	Hospital wards	Natural	Simulation, in situ measurement, surveys	1 hospital ward located on the 3rd floor (last floor)	Results are not presented for the similar rooms from the 1st and 2nd floors, which are air-conditioned. A comparison in terms of thermal comfort and energy consumption would be interesting.	Malaysia	More than half of the total occupants in the ward feel discomfort, with a predicted mean vote (PMV) between 1.0 and 1.6 and a predicted percentage of dissatisfied between 40% and 56%. "People in a tropical climate zone are well accustomed to hot-humid weather and would tolerate high temperatures."
[59] Khan M. et al.	Emergency room, operation room, intensive care unit medical ward	Natural and controlled	in situ measurements	4 hospitals and 4 space types in each	The results and conclusions are accurate for the hot climate.	Pakistan	Temperature, relative humidity, and carbon dioxide were measured. "The results show that occupancy rate, ambient thermal conditions, type of HVAC system, and building orientation are vital drivers of IEQ".
[60] Tsung-Yi Chien et al.	Operating room	Controlled	Simulation	3 variants of HVAC system: (i) "conventional"; (ii) "liquid desiccant air conditioning system (LDAC)"; (iii) rotary desiccant air conditioning system (RDAC)	Three systems were analyzed and the best one was selected. There are more possibilities and they are related to the type of operation and its specific requirements. The conclusions may therefore not be appropriate for every situation.	Taiwan	RDAC (rotary desiccant air-conditioning system) is preferable as the most energy-efficient for the operating room.

Table 10. *Cont.*

Paper	Space Type	Ventilation Type	Research Method	Number of Subjects	Limitations	Country, Climate	Main Findings (General)
[61] Nishesh Jain et al.	Hospital wards	Controlled	In situ measurement site visits semi-structured interviews with the facility managers	3 hospital wards located at 3rd, 4th, and 7th floor; 47 responses on the survey	Measurements and surveys were carried out in the new building to draw conclusions for buildings designed in the future. Three rooms in a single building in the UK climate were analyzed. The conclusions are not universal for other room types, other HVAC systems, or other climate zones.	United Kingdom	Temperature, RH, CO_2, PM2.5, PM10, NO_2, VOCs, electricity and gas use were measured. "The study highlights the need for an integrated and holistic approach to building performance to ensure that healthy environments are provided while energy efficiency targets are met."
[62] Fan, Minchao et al.	Operating room	Controlled	Simulation, in situ measurement, surveys	2 types of ventilation system: laminar or mixing for two real-size operating rooms	The presented method seems to be universal for a comprehensive assessment of the quality of the HVAC system and the energy consumption of operating rooms. It has been validated in only two operating theatres: laminar and mixing ventilation systems.	Norway	The method to assess ventilation systems in hospitals based on the "analytic hierarchy process and fuzzy comprehensive evaluation" is presented. Taking into account 7 parameters in the context of 3 evaluation criteria: "Ventilation effectiveness, energy consumption, and users' satisfaction". Multi-criteria decision making. Problem was solved. Laminar system was assessed as satisfying and mixed as dissatisfying.

Table 10. *Cont.*

Paper	Space Type	Ventilation Type	Research Method	Number of Subjects	Limitations	Country, Climate	Main Findings (General)
[63] N. Forcada et al.	Nursing home	Natural	Surveys	5 nursing homes	The results and conclusions are accurate for Mediterranean climate.	Spain	The use of natural ventilation without air conditioning can reduce energy consumption without significantly affecting the thermal comfort of residents. ASHRAE 55:2020 thermal comfort model can be considered as close to the expectations of nursing home residents.
[64] Gutierrez, Albio D. et al.	Operating room	Controlled	Simulation	Single operating room	Simulations were carried out for one variant of the operating room with one of the many possible HVAC systems.	Colombia, USA	CFD simulations to calculate predicted mean vote (PMV) and predicted percentage of dissatisfied (PPD) in the operating room. The temperature and airflow patterns were investigated to show that proper design is needed to obtain comfortable conditions.

Ventilation in hospital buildings plays an important role in ensuring sanitary safety, removing pollutants and maintaining thermal comfort. However, the choice of the ventilation system remains controversial—although laminar systems seem to be the most advantageous, there is no one universal solution that will work in every case nor guidelines that would show when to use which type of system. In addition, most works emphasize that ventilation systems in hospitals, in particular in operating rooms, are energy-intensive and should be energy optimized.

Research in hospitals is difficult to conduct because the hospital consists of many different spaces with different purposes and requirements. There are patients with different health conditions, so it can be difficult to survey them. The conditions of care are not conducive to carrying out measurements. It seems that simulations can be a good way to find new, profitable solutions. Validation measurements should be used as well as those that allow the creation of a digital model for simulation purposes. Simulations allow us to take into account more various cases and make the possibility of comparing results for different locations. A limitation in generalizing conclusions may also be the different local regulations in force in each country.

9.4. Bedrooms

Summarized findings of reviewed works in the field of thermal comfort and ventilation while sleeping are presented in Table 11.

The presented articles concerned several aspects related to ventilation and thermal comfort in bedrooms. Although the authors presented valuable results, it can be noted that the described studies contain small samples of the studied objects. The largest research sample involved surveying 517 people in Denmark [68]. Thanks to this, the authors obtained a picture of the ventilation systems used in this country. A greater number of analyzed variants is also found when simulations are used for research, mainly CFD [66,72,73]. Thanks to this, the authors analyzed from 3 to 24 different situations. In 100% of such works, the CFD models were validated against experimental data carried out in a climatic chamber or using a laboratory stand. Ansys Fluent or Solidworks software were used for simulations, whereas in cases where experiments were conducted, the authors performed field tests in one building up to a maximum of three rooms within one study [65,69–71,74]. Cases in which the measurements took place in a climatic chamber prevail [67].

Research on thermal comfort in bedrooms and while sleeping is carried out all over the world, and their results usually refer to specific climatic conditions. The ones discussed concerned the summer period [65,67] and heating seasons [68–73] or did not specify the period of the year due to the nature of the analyses conducted (influence of air movement). Both types of ventilation were present in correspondingly reviewed papers, but part of the research took place in a laboratory.

Good indoor air quality and proper thermal conditions are a key factors that may provide good sleep quality. To achieve it, there are some recommendations that may be followed. In the case of buildings located in countries with warm climates, the lack of air movement in the room may cause a problem with sleep. The guideline addressed to designers and implementers of building modernization resulting from the conducted research is the recommendation for air movement. When designing systems to ensure proper sleeping conditions, it is more important to pay attention to ventilation and air movement than to the temperature of the supplied air. This may be achieved by introducing local devices in bedrooms. Thanks to this, the air conditioning will only be able to work in the event of a very high air temperature.

Table 11. Summarized findings of reviewed works in the field of thermal comfort and ventilation while sleeping.

Paper	Space Type	Ventilation Type	Research Method	Number of Subjects	Limitations	Country, Climate	Main Findings
[65] Jakob B. Knudsen et al.	Bedroom in traditional Gambian building	Natural	Measurements	2 experiments (1) Measurements of five methods for assessing ventilation in two houses; (2) Measurements of CO_2 concentration in 5 similar houses.	External parameters between the night vary (weather, wind). Future CFD modeling study designed to improve airflow in buildings to keep the occupants comfortable while keeping out mosquitoes.	Gambia (Africa) Climate: hot and humid Summer season	Recommendations of providing air movements in bedrooms and air exchange. Recommendation for airtightness tests for old-type buildings. Recommendation of the different construction of houses—adding screened doors and windows to the house. The conclusion that proper ventilation may protect against malaria.
[66] Haiguo Yin et al.	Climatic chamber Tiny sleeping spaces in full scale	Controlled	CFD simulations Validation measurements	Three scenarios of data: (1) Airflow change; (2) Supply air temperature change; (3) Air distribution change.	Not specified	China	Recommendation of providing ventilation with air movements (0.24 m/s) to keep thermal comfort during falling asleep.
[67] Chenqiu Du et. Al.	Climatic chamber Representation of bedroom	Controlled	Measurements (experiments)	24 people 3 sets of air parameters (temperature, relative humidity, and air velocity) Total Sleep Time (TST), Sleep Onset Latency (SOL), and Slow-Wave Sleep (SWS)	The values used for the bedding system are not able to reflect accurately the actual thermal resistance when sleeping. Conclusions are based only on one limited experiment and need more experimental studies and tests to describe typical scenarios. The use of wired temperature sensors in this study may cause discomfort for subjects during sleep. Other factors like air quality, CO_2 concentration, lighting, noise, etc., were not included but may have an impact.	China Summer season	A PMV model was proposed considering three aspects of the body heat exchanges: the parts in contact with the mattress (e.g., back, buttocks), the parts covered by the quilts or clothes (e.g., abdomen, chest), and the parts directly exposed to the air (e.g., head, arms, legs) Recommendation of providing air movement at air temperature 28–32 °C. Guidelines for adopting air movements to improve thermal environments whilst saving energy by the installation of local fans will be beneficial for thermal comfort and energy.

Table 11. *Cont.*

Paper	Space Type	Ventilation Type	Research Method	Number of Subjects	Limitations	Country, Climate	Main Findings
[68] Chenxi Liao et al.	Existing bedrooms in Denmark	35.3%% mechanical ventilation 24.6% exhaust ventilation 49.9% natural ventilation	Survey (1) Type of ventilation, bedroom, surroundings, and location (2) Subjective sleep quality—Pittsburgh Sleep Quality Index (PSQI)	517 people	Results are for the heating season, and to generalize them, more data from non-heating seasons are needed. Results are based on a survey, so field measurements are necessary to validate them.	Denmark Winter—heating season	People respond with their behavior to thermal conditions in the bedroom, e.g., by opening the window when the air feels stuffy. In most residential buildings in Denmark, there is natural ventilation. In bedrooms with mechanical ventilation, people feel more comfortable conditions. Recommendation of a combination of objective and subjective research will allow future solutions to be proposed—e.g., opening windows before sleeping.
[69] Dan Mio et al.	Student dormitories	Natural	Survey Maesurments: IEQ parameters	Two 4-bed bedrooms	Results are limited to similar room types and climates. The sample of respondents was small. The research should be repeated in a larger sample of similar objects and models should be created. Recommendations of providing more field studies with measurements and surveys should be done in the future to help implement guidelines.	Beijing, China Heating season	Analyzed dormitories were of poor air quality. Correlation between subjective and objective results exists. Recommendation of providing local air movement devices, using earplugs.
[70] D.A. Adincu et al.	Bedrooms in a refurbished building	Natural	Measurements (temperature, relative humidity, CO_2)	3 bedrooms with different numbers of people.	Results are limited to the pollution concentration inside specific rooms. The authors plan for extended further experiments.	Bucharest, Romania Winter season	Analyzed bedrooms were of poor air quality because of a lack of mechanical ventilation. Airing is not effective for providing proper indoor air quality and during winter it may lower room temperature. Use of sensor-controlled mechanical ventilation. Use of sensor-controlled mechanical ventilation.

Table 11. *Cont.*

Paper	Space Type	Ventilation Type	Research Method	Number of Subjects	Limitations	Country, Climate	Main Findings
[71] Mohammad Al-Rawi	Residential house Children's bedroom	Natural	Measurements CFD simulations (SOLIDWORKS)	1 bedroom A short period of time	The authors stated that further research should be performed on the impact of humidity of outside air the on performance of a heat pump.	Waikato, New Zeland Winter season	Air temperature during sleep is low in New Zealand houses because of poor insulation, insufficient ventilation, and lack of financial resources for heating bills. The use of heat pumps for heating by which a PMV index of 0.72 may be reached is proposed.
[72] Jian Liu and Zhang Lin	Climatic chamber Laboratory stand	Controlled	Simulations CFD (Fluent)	17 cases	Not specified	China Winter season	A new type of heating system—stratum ventilation heating—allows for blowing air from the side of the head, which provides an appropriate distribution system for proper thermal comfort in the bedroom.
[73] Jian Liu and Zhang Lin	Climatic chamber Laboratory stand	Controlled	Simulations CFD (Fluent)	3 air-distribution systems 7 supply air temperatures 3 levels of PMV	Not specified	China Winter season	By supplying air to the occupied zone, with the same value of PMV, 10% energy savings may be achieved in comparison of other air distribution and heating systems.
[74] Chandra Sekhar et al.	Semi-detached house (SDH) Apartment (A)	Natural (SDH) Controlled (A)	Measurements (t, RH, CO_2) Simulations	2 bedrooms 9 variants of different use of a bedroom	Not specified	Denmark Heating season	To ensure air quality at the right level (CO_2 concentration of 1000 ppm), the use of the bedroom should be adapted to the type of ventilation. In bedrooms with natural ventilation, door should be opened during sleep. For mechanical ventilation, air exchange of $0.6\ h^{-1}$ allows for maintaining CO_2 concentration below 1000 ppm.

Still, there is a need to conduct field studies on the quality of sleep and the quality of the indoor environment in order to be able to better recommend various beneficial design solutions for people to improve their well-being at home. Each measurement should be supplemented with surveys, which may be helpful in future works. Surveys may give reliable results that can be obtained without the need to mount sensors. But additional, even local measurements, can show the effectiveness of certain actions.

The authors most often give limitations that apply to the conducted research. Among them, the most common limitation, which can also be attributed to cases where the authors themselves did not take it into account, is usually a small research sample [65,69–71]. An additional limitation is conducting the research in a selected season of the year [68–70,74], which is important, especially when there is natural ventilation in the analyzed building [65,69–71,74]. A certain limitation is also the use of only simulations and conducting research on experimental stands, which may not reflect the real conditions of the bedroom.

In order to overcome the defined limitations of the cited studies, when planning research work on the subject of sleep quality and air parameters in bedrooms, one should follow the authors of already published studies. Research should be planned covering the period of the heating season as well as the time outside the heating season. The research sample should be expanded, although this may be difficult due to the nature of the research. Survey research (subjective) should be supplemented with objective research (measurement).

Regardless of the planned research, even in the case of a small research sample, field studies on sleep parameters and air parameters in bedrooms are important and should be continued due to the fact that good sleep quality is extremely important for well-being and health, and people spend one-third of their lives sleeping.

9.5. Atria

Summarized findings of reviewed works in the field of thermal comfort and ventilation in buildings with atriums are presented in Table 12.

The atrium is a building element that occurs all over the world. The analyzed studies concerned various types of buildings: offices [84,85], educational [78,82,88], commercial [77,79,83,87], and others [80,81,90]. In most of the analyses, the authors use computational simulations [77,79–82,84,85,87,88,90] to simulate the thermal state of the zone and calculate thermal comfort identifiers. Mostly, the CFD tools were used. Only in the case of two studies, the authors performed in situ measurements and used questionnaires [78,83] to draw some scientific conclusions directly, although in some papers the in situ measurements were used as additional research tools to validate the accuracy of the developed computational models [79,87]. The central atrium is the most studied type, and mainly only one selected atrium case was analyzed [77–82,85,87,88,90], except for [83], in which two atrial spaces were researched. However, the use of simulation software allowed one to analyze multiple cases of the same atrium differing in particular physical and operational parameters.

In terms of the type of ventilation occurring in the analyzed atrial buildings, natural ventilation is the most common [79,80,83–85,87,88,90] as the atrium may increase its efficiency (natural ventilation was presented in about 67% of reviewed papers). The controlled ventilation is less popular in the reviewed papers, but the interaction of the atrium with this type of HVAC system was also studied in [78,81,82,90] (which is about 33% of the reviewed papers).

Table 12. Summarized findings of reviewed works in the field of thermal comfort and ventilation in buildings with atrium.

Paper	Space Type	Ventilation Type	Research Method	Number of Subjects	Limitations	Country, Climate	Main Findings
[77] J. Pang et al.	Hotel	Not specified	Simulations (Ansys Fluent, Design Builder),	Single central atrium, 27 simulation cases with different atrium shape and size	The findings may be useful only for atriums with platform at the lower levels, as it greatly affects thermal conditions.	China	Height, platform location, and the opaque-to-transparent surface ratio of the atrium are factors of high importance for atrium energy performance.
[78] C. Xu et al.	College	Controlled	In situ measurement, surveys	Single central atrium, 70 questionnaires	The findings are based on a single, specific case study (winter, central atrium, specific ventilation system, radiant heating).	China, Winter	Fanger model overestimates thermal comfort for an atrium with central skylights. An original formula for PMV for that kind of space is provided.
[79] H. Sahu, J. Vijayalaxmi	Commercial	Natural	Simulations	Single central atrium, 11 design variables simulated	The findings are based on the computational model designed and validated for the selected single-type of central atrium.	India	The window-to-wall ratio and the glazing type are selected as important parameters affecting the thermal comfort in the central atrium with natural ventilation. The ventilation strategy, including window opening area and schedule, allow a decrease in the energy consumption for cooling.
[80] J. Shaeri et al.	Not mentioned	Natural	CFD simulations (Ansys Fluent)	Central atrium in low-rise building, 12 roof shapes	Simulations take into account the sea breeze impact on the atrium. Natural ventilation efficiency, which is valid for coastal cases.	Iran	The roof shape impacts the natural ventilation of the atrium and thermal comfort.
[81] Y.H. Yau, U.A. Rajput	Anonymous skyscraper	Controlled	CFD simulations	Single hexagonal atrium	The study investigates specific air distribution system with linear diffusers and low-level cool air supply.	Malaysia	A specific solution of ventilation air supply through diffusers with high air supply velocities (>6.5 m/s) and exhaust air from under the supply diffusers ensures good thermal comfort in the ring-shaped atrium.

Table 12. *Cont.*

Paper	Space Type	Ventilation Type	Research Method	Number of Subjects	Limitations	Country, Climate	Main Findings
[82] K. Ratajczak et al.	University	Controlled	Simulations (IES-VE)	Single central atrium	The study focuses on the energy efficiency and does not quantitatively analyze thermal comfort in detail.	Poland	Incorporating a central atrium with a glazed roof provides natural light to the adjacent spaces, lowering the internal heat gain and impacting the HVAC system's performance. An atrium can be used as a passive heater for fresh air in a moderate climate.
[83] A. Ghazali et al.	Public market	Natural	In situ measurements, surveys	2 central atriums, measurements conducted for 7 days in May, traders surveyed	As the surveys are addressed to traders, conducted thermal comfort analysis is valid for the long exposure and relatively permanent location of the subjects. The results for visitors would likely be different.	Malaysia	Thermal comfort is not provided with central ceiling fans, and the local fans are used to neutralize the effect of high temperature and humidity.
[84] T. Corbett et al.	Open-plan office	Natural	CFD simulations (ANSYS-CFX), water bath modeling	9 cases, with 3 types of the central atrium and 3 types of the building core	Simulations performed with simplified, constant climatic conditions not corresponding to winter and summer.	United Kingdom	The natural ventilation efficiency in the building with a central atrium is very sensitive to the design of the perimeter openings.
[85] H.-H. Hsu et al.	Office	Natural, hybrid	CFD simulations	Case study of single central atrium	The results are obtained for the specific tall building with two atriums central atriums opened to the perimeter air-conditioned spaces: first in the lower part, with the wall-opening, and the second one in the upper part, with the roof-opening	Taiwan	Stack ventilation is unsuitable when the external temperature is higher than the internal temperature, which is typical for air-conditioned buildings. The hybrid ventilation scheme is developed for the atrium surrounded by air-conditioned spaces, in which the air is transferred to the atrium from neighboring spaces.

Table 12. *Cont.*

Paper	Space Type	Ventilation Type	Research Method	Number of Subjects	Limitations	Country, Climate	Main Findings
[87] D.P. Albuquerque et al.	Non-residential	Natural	Simulations (EnergyPlus), CFD (Ansys Fluent), in situ measurements	Single central atrium, analyzed on several summer days	The study covers interaction between passive cooling and radiant cooling floor, which is a specific cooling system impacting significantly the air flow pattern in the zone.	Portugal	The passive cooling provided by nighttime ventilation allows reduction of the cooling load for active cooling systems in atrial spaces by about 27%. Relatively small openings can be used for passive cooling.
[88] R. Sokkar, H.Z. Alibaba	Library	Natural	Simulations (EDSL Tas)	Single central atrium, several dozen cases	Authors noted that their results are influenced by large transparent facades, but they found it an average model.	Cyprus	The central atrium with skylights provides good natural lighting; however, it also causes summer overheating that is impossible to be eliminated by shading without loss of natural lighting. Double-skin facade for the skylights connected with the operable skylights is the solution for enhancing natural ventilation and improving thermal comfort during the summer season.
[90] N.A.M. Fohimi et al.	Indoor theme park	Natural, controlled	CFD simulations	Single atrium	The study investigates single, specific case. Although some conclusions are drawn, they cannot be generalized.	Malaysia	Mechanically-driven ventilation provides better thermal comfort than the natural, buoyancy driven ventilation in the air-conditioned space.

The main conclusions from the study of buildings with atriums are as follows.

An atrium designed taking into account the naturally occurring stack forces can significantly improve the conditions of thermal comfort related to the air temperature, humidity, and velocity, but it requires an analysis of many geometric parameters, not only of the atrium itself but also of the whole building, as well as a ventilation system and local climatic conditions, as various synergistic effects can significantly affect ventilation efficiency and thermal comfort.

A properly selected ventilation strategy and configuration of the ventilation system together with the architectural parameters of the atrium allow for a significant improvement in thermal comfort indicators in the atrium and adjacent spaces and can lead to a reduction in the energy demand of the building's HVAC systems, which is proved by many reviewed case studies.

In general, atrial spaces allow taking advantage of the stack effect and implementation of natural, non-mechanically-driven ventilation, which is the most commonly seen potential for improving energy efficiency and thermal comfort, being verified in many research papers. According to the complexity of the atrium's thermal environment, the CFD simulation tools are widely used, with satisfactory accuracy confirmed by many in situ measurements.

The main limitations in research on atrium buildings are mainly related to the number of research facilities, as mentioned above. The limitations are related to the analysis of the selected special architecture of the facility [77,85,88] and specific boundary conditions, e.g., climate [78–81,83,84,87,90]. As the atrial spaces can be, according to reviewed papers, very sensitive to small changes in individual design parameters (both physical and operational parameters of the atrium itself) as well as the properties of the external environment (i.e., temperature, solar radiation, wind), there are plenty of possible boundary conditions. This makes it difficult to relate the results obtained for a particular atrium to another, not identical one.

To overcome the defined limitations of the cited studies, when planning future research work on the subject of atrial buildings, it seems appropriate to study other types of specific atrial buildings, taking into account their individual physical and operational parameters to further understand the thermal characteristics and thermal comfort of the atrium and expand the research case database. Research taking into account not only one particular set of boundary conditions (including exterior climate parameters, at least in the winter/summer season) may contribute greatly to the field and allow drawing more universal conclusions.

Author Contributions: Conceptualization, K.R., F.P., Ł.A., J.S. and K.P., methodology, K.R., F.P., Ł.A., J.S. and K.P.; writing—original draft preparation, K.R., F.P., Ł.A., J.S. and K.P.; writing—review and editing, K.R., F.P., Ł.A., J.S. and K.P.; supervision: K.R. and Ł.A.; funding acquisition, K.R. and Ł.A. All authors have read and agreed to the published version of the manuscript.

Funding: This research was funded by the Polish Ministry of Science and Higher Education, grant number 0713/SBAD/0981.

References

1. Amanowicz, Ł.; Ratajczak, K.; Dudkiewicz, E. Recent Advancements in Ventilation Systems Used to Decrease Energy Consumption in Buildings—Literature Review. *Energies* **2023**, *16*, 1853. [CrossRef]
2. Dudkiewicz, E.; Szałański, P. A review of heat recovery possibility in flue gases discharge system of gas radiant heaters. *E3S Web Conf.* **2019**, *116*, 00017. [CrossRef]
3. Orman, L. Enhancement of pool boiling heat transfer with pin-fin microstructures. *J. Enhanc. Heat Transf.* **2016**, *23*, 137–153. [CrossRef]
4. Hussain, L.; Khan, M.M.; Masud, M.; Ahmed, F.; Rehman, Z.; Amanowicz, Ł.; Rajski, K. Heat Transfer Augmentation through Different Jet Impingement Techniques: A State-of-the-Art Review. *Energies* **2021**, *14*, 6458. [CrossRef]
5. Niemierka, E.; Jadwiszczak, P. Experimental investigation of a ceramic heat regenerator for heat recovery in a decentralized reversible ventilation system. *Int. Commun. Heat Mass Transf.* **2023**, *146*, 106899. [CrossRef]

6. Ratajczak, K.; Amanowicz, Ł.; Szczechowiak, E. Assessment of the air streams mixing in wall-type heat recovery units for ventilation of existing and refurbishing buildings toward low energy buildings. *Energy Build.* **2020**, *227*, 110427. [CrossRef]

7. Zender, E. Improvement of indoor air quality by way of using decentralised ventilation. *J. Build. Eng.* **2020**, *32*, 101663. [CrossRef]

8. Zender-Świercz, E.; Telejko, M.; Galiszewska, B.; Starzomska, M. Assessment of Thermal Comfort in Rooms Equipped with a Decentralised Façade Ventilation Unit. *Energies* **2022**, *15*, 7032. [CrossRef]

9. Zender, E. Assessment of Indoor Air Parameters in Building Equipped with Decentralised Façade Ventilation Device. *Energies* **2021**, *14*, 1176. [CrossRef]

10. Amanowicz, Ł.; Wojtkowiak, J. Comparison of Single- and Multipipe Earth-to-Air Heat Exchangers in Terms of Energy Gains and Electricity Consumption: A Case Study for the Temperate Climate of Central Europe. *Energies* **2021**, *14*, 8217. [CrossRef]

11. Michalak, P. Hourly Simulation of an Earth-to-Air Heat Exchanger in a Low-Energy Residential Building. *Energies* **2022**, *15*, 1898. [CrossRef]

12. Michalak, P.; Szczotka, K.; Szymiczek, J. Audit-Based Energy Performance Analysis of Multifamily Buildings in South-East Poland. *Energies* **2023**, *16*, 4828. [CrossRef]

13. Kordana-Obuch, S.; Starzec, M.; Słyś, D. Assessment of the Feasibility of Implementing Shower Heat Exchangers in Residential Buildings Based on Users' Energy Saving Preferences. *Energies* **2021**, *14*, 5547. [CrossRef]

14. Piotrowska, B.; Słyś, D.; Kordana-Obuch, S.; Pochwat, K. Critical Analysis of the Current State of Knowledge in the Field of Waste Heat Recovery in Sewage Systems. *Resources* **2020**, *9*, 72. [CrossRef]

15. Kordana-Obuch, S.; Starzec, M.; Wojtoń, M.; Słyś, D. Greywater as a Future Sustainable Energy and Water Source: Bibliometric Mapping of Current Knowledge and Strategies. *Energies* **2023**, *16*, 934. [CrossRef]

16. Wojtkowiak, J.; Amanowicz, Ł.; Mróz, T. A new type of cooling ceiling panel with corrugated surface—Experimental investigation. *Int. J. Energy Res.* **2019**, *43*, 7275–7286. [CrossRef]

17. Baborska, M.; Kostka, M. Seasonal Air Quality in Bedrooms with Natural, Mechanical or Hybrid Ventilation Systems and Varied Window Opening Behavior-Field Measurement Results. *Energies* **2022**, *15*, 9328. [CrossRef]

18. Wojtkowiak, J.; Amanowicz, Ł. Effect of surface corrugation on cooling capacity of ceiling panel. *Therm. Sci. Eng. Prog.* **2020**, *19*, 100572. [CrossRef]

19. Sinacka, J.; Szczechowiak, E. An Experimental Study of a Thermally Activated Ceiling Containing Phase Change Material for Different Cooling Load Profiles. *Energies* **2021**, *14*, 7363. [CrossRef]

20. Dudkiewicz, E.; Fidorów, N.; Jeżowiecki, J. Wpływ sprawności promienników podczerwieni na koszt zużycia energii. *Rocz. Ochr. Śr.* **2013**, *15*, 1804–1817.

21. Amanowicz, Ł. Controlling the Thermal Power of a Wall Heating Panel with Heat Pipes by Changing the Mass Flowrate and Temperature of Supplying Water—Experimental Investigations. *Energies* **2020**, *13*, 6547. [CrossRef]

22. Gojak, M.; Bajc, T. Thermodynamic sustainability assessment for residential building heating comparing different energy sources. *Sci. Technol. Built Environ.* **2022**, *28*, 73–83. [CrossRef]

23. Kaczmarczyk, J.; Ferdyn-Grygierek, J. Thermal Comfort and Energy Use with Local Heaters. *Energies* **2020**, *13*, 2912. [CrossRef]

24. Cichowicz, R.; Dobrzański, M. Spatial Analysis (Measurements at Heights of 10 m and 20 m above Ground Level) of the Concentrations of Particulate Matter (PM10, PM2.5, and PM1.0) and Gaseous Pollutants (H2S) on the University Campus: A Case Study. *Atmosphere* **2021**, *12*, 62. [CrossRef]

25. Cichowicz, R.; Wielgosiński, G. Analysis of Variations in Air Pollution Fields in Selected Cities in Poland and Germany. *Ecol. Chem. Eng. S* **2018**, *25*, 217–227. [CrossRef]

26. Snyder, H. Literature review as a research methodology: An overview and guidelines. *J. Bus. Res.* **2019**, *104*, 333–339. [CrossRef]

27. Roelofsen, P. The impact of office environments on employee performance: The design of the workplace as a strategy forproductivity enhancement. *J. Facil. Manag.* **2002**, *1*, 247–264. [CrossRef]

28. Peng, B.; Hsieh, S.J. Cyber-Enabled Optimization of HVAC System Control in Open Space of Office Building. *Sensors* **2023**, *23*, 4857. [CrossRef]

29. Gao, N.; Shao, W.; Rahaman, M.S.; Zhai, J.; David, K.; Salim, F.D. Transfer learning for thermal comfort prediction in multiple cities. *Build. Environ.* **2021**, *195*, 107725. [CrossRef]

30. Yong, N.H.; Kwong, Q.J.; Ong, K.S.; Mumovic, D. Post occupancy evaluation of thermal comfort and indoor air quality of office spaces in a tropical green campus building. *J. Facil. Manag.* **2022**, *20*, 570–585. [CrossRef]

31. Badeche, M.; Bouchahm, Y. A study of Indoor Environment of Large Glazed Office Building in Semi Arid Climate. *J. Sustain. Archit. Civ. Eng.* **2021**, *29*, 175–188. [CrossRef]

32. Ortiz, M.C.; Morales, S.; Cabrera, V. Bioclimaic optimization: Skylight ground floor new building, Udla Park Torre II. *Proc. Int. Struct. Eng. Constr.* **2021**, *8*, AAE-18-2. [CrossRef]

33. Lahji, K.; Puspitasari, P. Thermal comfort analysis by adjusting the tipping window opening distance using PMV. In Proceedings of the Symposium on Advance of Sustainable Engineering 2021 (Simase 2021): Post Covid-19 Pandemic: Challenges and Opportunities in Environment, Science, and Engineering Research, Bandung, Indonesia, 18–19 August 2021; p. 050121. [CrossRef]

34. Borsos, Á.; Zoltán, E.; Pozsgai, É.; Cakó, B.; Medvegy, G.; Girán, J. The Comfort Map—A Possible Tool for Increasing Personal Comfort in Office Workplaces. *Buildings* **2021**, *11*, 233. [CrossRef]

35. McGee, B.L.; Couillou, R.J.; Maalt, K. Work from Home: Lessons Learned and Implications for Post-pandemic Workspaces. *Interiority* **2023**, *6*, 91–114. [CrossRef]

36. Manca, S.; Cerina, V.; Tobia, V.; Sacchi, S.; Fornara, F. The Effect of School Design on Users' Responses: A Systematic Review (2008–2017). *Sustainability* **2020**, *12*, 3453. [CrossRef]
37. Peters, T.; D'Penna, K. Biophilic Design for Restorative University Learning Environments: A Critical Review of Literature and Design Recommendations. *Sustainability* **2020**, *12*, 7064. [CrossRef]
38. Suradhuhita, P.P.; Setyowati, E.; Prianto, E. Influence of a facade design on thermal and visual comfort in an elementary school classroom. *IOP Conf. Ser. Earth Environ. Sci.* **2022**, *1007*, 012013. [CrossRef]
39. Talarosha, B.; Satwiko, P.; Aulia, D.N. Air temperature and CO_2 concentration in naturally ventilated classrooms in hot and humid tropical climate. *IOP Conf. Ser. Earth Environ. Sci.* **2020**, *402*, 012008. [CrossRef]
40. Dudzińska, A.; Kisilewicz, T. Alternative Ways of Cooling a Passive School Building in Order to Maintain Thermal Comfort in Summer. *Energies* **2020**, *14*, 70. [CrossRef]
41. Shrestha, M.; Rijal, H.B. Investigation on Summer Thermal Comfort and Passive Thermal Improvements in Naturally Ventilated Nepalese School Buildings. *Energies* **2023**, *16*, 1251. [CrossRef]
42. Haddad, S.; Synnefa, A.; Marcos, M.Á.P.; Paolini, R.; Delrue, S.; Prasad, D.; Santamouris, M. On the potential of demand-controlled ventilation system to enhance indoor air quality and thermal condition in Australian school classrooms. *Energy Build.* **2021**, *238*, 110838. [CrossRef]
43. Mazlan, A.N.; Saad, S.; Yahya, K.; Haron, Z.; Hasbollah, D.Z.A.; Kasiman, E.H.; Rahim, N.A.; Salehudddin, A.M. Thermal comfort study for classroom in urban and rural schools in Selangor. *IOP Conf. Ser. Mater. Sci. Eng.* **2020**, *849*, 012016. [CrossRef]
44. Kapoor, N.R.; Kumar, A.; Alam, T.; Kumar, A.; Kulkarni, K.S.; Blecich, P. A Review on Indoor Environment Quality of Indian School Classrooms. *Sustainability* **2021**, *13*, 11855. [CrossRef]
45. Lala, B.; Kala, S.M.; Rastogi, A.; Dahiya, K.; Yamaguchi, H.; Hagishima, A. Building Matters: Spatial Variability in Machine Learning Based Thermal Comfort Prediction in Winters. In Proceedings of the 2022 IEEE International Conference on Smart Computing (SMARTCOMP), Helsinki, Finland, 20–24 June 2022; IEEE: Piscataway, NJ, USA, 2022; pp. 342–348. [CrossRef]
46. Lan, K.; Chen, Y. Air Quality and Thermal Environment of Primary School Classrooms with Sustainable Structures in Northern Shaanxi, China: A Numerical Study. *Sustainability* **2022**, *14*, 12039. [CrossRef]
47. Chang, L.-Y.; Chang, T.-B. Air Conditioning Operation Strategies for Comfort and Indoor Air Quality in Taiwan's Elementary Schools. *Energies* **2023**, *16*, 2493. [CrossRef]
48. Catalina, T.; Damian, A.; Vartires, A.; Ni, M.; Racovi, V. Long-term analysis of indoor air quality and thermal comfort in a public school. *IOP Conf. Ser. Earth Environ. Sci.* **2023**, *1185*, 012008. [CrossRef]
49. Tagliabue, L.C.; Accardo, D.; Kontoleon, K.J.; Ciribini, A.L.C. Indoor comfort conditions assessment in educational buildings with respect to adaptive comfort standards in European climate zones. *IOP Conf. Ser. Earth Environ. Sci.* **2020**, *410*, 012094. [CrossRef]
50. Grassie, D.; Karakas, F.; Schwartz, Y.; Dong, J.; Milner, J.; Chalabi, Z.; Mavrogianni, A. Modelling UK school performance by coupling building simulation and multi-criteria decision analysis. In Proceedings of the Indoor Air 2022, Kuopio, Finland, 13–16 June 2022. Available online: https://discovery.ucl.ac.uk/id/eprint/10146883/1/20220403-IA2022_Full_Paper_v3_FINAL%20%2 81%29.pdf (accessed on 1 July 2023).
51. Fabozzi, M.; Dama, A. Field study on thermal comfort in naturally ventilated and air-conditioned university classrooms. *Indoor Built Environ.* **2020**, *29*, 851–859. [CrossRef]
52. López-Chao, V.; Lorenzo, A.A.; Saorín, J.L.; De La Torre-Cantero, J.; Melián-Díaz, D. Classroom Indoor Environment Assessment through Architectural Analysis for the Design of Efficient Schools. *Sustainability* **2020**, *12*, 2020. [CrossRef]
53. Sadrizadeh, S.; Yao, R.; Yuan, F.; Awbi, H.; Bahnfleth, W.; Bi, Y.; Cao, G.; Croitoru, C.; de Dear, R.; Haghighat, F.; et al. Indoor air quality and health in schools: A critical review for developing the roadmap for the future school environment. *J. Build. Eng.* **2022**, *57*, 104908. [CrossRef]
54. Gatea, A.; Batcha, M.F.M.; Taweekun, J. Energy Efficiency and Thermal Comfort in Hospital Buildings: A Review. *Int. J. Integr. Eng.* **2020**, *12*, 33–41.
55. Sadrizadeh, S.; Aganovic, A.; Bogdan, A.; Wang, C.; Afshari, A.; Hartmann, A.; Croitoru, C.; Khan, A.; Kriegel, M.; Lind, M.; et al. A systematic review of operating room ventilation. *J. Build. Eng.* **2021**, *40*, 102693. [CrossRef]
56. Rahman, N.M.A.; Haw, L.C.; Fazlizan, A. A Literature Review of Naturally Ventilated Public Hospital Wards in Tropical Climate Countries for Thermal Comfort and Energy Saving Improvements. *Energies* **2021**, *14*, 435. [CrossRef]
57. Rahman, N.M.A.; Haw, L.C.; Fazlizan, A.; Hussin, A.; Imran, M.S. Thermal comfort assessment of naturally ventilated public hospital wards in the tropics. *Build. Environ.* **2022**, *207*, 108480. [CrossRef]
58. Yuan, F.; Yao, R.; Sadrizadeh, S.; Li, B.; Cao, G.; Zhang, S.; Zhou, S.; Liu, H.; Bogdan, A.; Croitoru, C.; et al. Thermal comfort in hospital buildings—A literature review. *J. Build. Eng.* **2022**, *45*, 103463. [CrossRef]
59. Khan, M.; Thaheem, M.J.; Khan, M.; Maqsoom, A.; Zeeshan, M. Thermal comfort and ventilation conditions in healthcare facilities—Part 1: An assessment of indoor environmental quality (IEQ). *Environ. Eng. Manag. J.* **2020**, *19*, 917–933.
60. Chien, T.Y.; Liang, C.C.; Wu, F.J.; Chen, C.T.; Pan, T.H.; Wan, G.H. Comparative Analysis of Energy Consumption, Indoor Thermal–Hygrometric Conditions, and Air Quality for HVAC, LDAC, and RDAC Systems Used in Operating Rooms. *Appl. Sci.* **2020**, *10*, 3721. [CrossRef]
61. Jain, N.; Burman, E.; Stamp, S.; Shrubsole, C.; Bunn, R.; Oberman, T.; Barrett, E.; Aletta, F.; Kang, J.; Raynham, P.; et al. Building Performance Evaluation of a New Hospital Building in the UK: Balancing Indoor Environmental Quality and Energy Performance. *Atmosphere* **2021**, *12*, 115. [CrossRef]

62. Fan, M.; Cao, G.; Pedersen, C.; Lu, S.; Stenstad, L.-I.; Skogås, J.G. Suitability evaluation on laminar airflow and mixing airflow distribution strategies in operating rooms: A case study at St. Olavs Hospital. *Build. Environ.* **2021**, *194*, 107677. [CrossRef]
63. Forcada, N.; Gangolells, M.; Casals, M.; Tejedor, B.; Macarulla, M.; Gaspar, K. Field study on adaptive thermal comfort models for nursing homes in the Mediterranean climate. *Energy Build.* **2021**, *252*, 111475. [CrossRef]
64. Gutierrez, A.D.; Sezer, H.; Ramirez, J.L. Required thermal comfort conditions inside hospital operating rooms (ORS): A numerical assessment. *Front. Heat Mass Transf.* **2022**, *18*, 1–12. [CrossRef]
65. Knudsen, J.B.; Pinder, M.; Jatta, E.; Jawara, M.; Yousuf, M.A.; Søndergaard, A.T.; Lindsay, S.W. Measuring ventilation in different typologies of rural Gambian houses: A pilot experimental study. *Malar. J.* **2020**, *19*, 273. [CrossRef]
66. Yin, H.; Li, Y.; Deng, X.; Han, Y.; Wang, L.; Yang, B.; Li, A. Performance evaluation of three attached ventilation scenarios for tiny sleeping spaces. *Build. Environ.* **2020**, *186*, 107363. [CrossRef]
67. Du, C.; Lin, X.; Yan, K.; Liu, H.; Yu, W.; Zhang, Y.; Li, B. A model developed for predicting thermal comfort during sleep in response to appropriate air velocity in warm environments. *Build. Environ.* **2022**, *223*, 109478. [CrossRef]
68. Liao, C.; Akimoto, M.; Bivolarova, M.P.; Sekhar, C.; Laverge, J.; Fan, X.; Lan, L.; Wargocki, P. A survey of bedroom ventilation types and the subjective sleep quality associated with them in Danish housing. *Sci. Total Environ.* **2021**, *798*, 149209. [CrossRef] [PubMed]
69. Miao, D.; Cao, X.; Zuo, W. Associations of Indoor Environmental Quality Parameters with Students' Perceptions in Undergraduate Dormitories: A Field Study in Beijing during a Transition Season. *Int. J. Environ. Res. Public Health* **2022**, *19*, 16997. [CrossRef]
70. Adîncu, D.A.; Popescu, A.; Atanasiu, M. Experimental measurements of CO_2 concentrations in sleeping rooms. *IOP Conf. Ser. Mater. Sci. Eng.* **2020**, *997*, 012137. [CrossRef]
71. Al-Rawi, M. The thermal comfort sweet-spot: A case study in a residential house in Waikato, New Zealand. *Case Stud. Therm. Eng.* **2021**, *28*, 101530. [CrossRef]
72. Liu, J.; Lin, Z. Performance of stratum ventilated heating for sleeping environment. *Build. Environ.* **2020**, *180*, 107072. [CrossRef]
73. Liu, J.; Lin, Z. Energy and exergy analyze of different air distributions in a residential building. *Energy Build.* **2021**, *233*, 110694. [CrossRef]
74. Sekhar, C.; Bivolarova, M.; Akimoto, M.; Wargocki, P. Detailed characterization of bedroom ventilation during heating season in a naturally ventilated semi-detached house and a mechanically ventilated apartment. *Sci. Technol. Built Environ.* **2021**, *27*, 158–180. [CrossRef]
75. Abdollahzadeh, N.; Farahani, A.V.; Soleimani, K.; Zomorodian, Z.S. Indoor environmental quality improvement of student dormitories in Tehran, Iran. *Int. J. Build. Pathol. Adapt.* **2023**, *41*, 258–278. [CrossRef]
76. Budiawan, W.; Tsuzuki, K.; Prastawa, H. Bibliometric Analysis of Thermal Comfort and Sleep Quality Research Trends in Indonesia. *IOP Conf. Ser. Earth Environ. Sci.* **2022**, *1098*, 012025. [CrossRef]
77. Pang, J.; Fan, Z.; Yang, M.; Liu, J.; Zhang, R.; Wang, W.; Sun, L. Effects of complex spatial atrium geometric parameters on the energy performance of hotels in a cold climate zone in China. *J. Build. Eng.* **2023**, *72*, 106698. [CrossRef]
78. Xu, C.; Wang, Y.; Hui, J.; Wang, L.; Yao, W.; Sun, L. Study on winter thermal environmental characteristics of the atrium space of teaching building in China's cold region. *J. Build. Eng.* **2023**, *67*, 105978. [CrossRef]
79. Sahu, H.; Vijayalaxmi, J. Optimization of the Integrated Daylighting and Natural Ventilation in a Commercial Building. In *Building Thermal Performance and Sustainability*; Springer Nature: Singapore, 2023; Volume 316, pp. 129–149. [CrossRef]
80. Shaeri, J.; Mahdavinejad, M.; Vakilinejad, R.; Bazazzadeh, H.; Monfared, M. Effects of sea-breeze natural ventilation on thermal comfort in low-rise buildings with diverse atrium roof shapes in BWh regions. *Case Stud. Therm. Eng.* **2023**, *41*, 102638. [CrossRef]
81. Yau, Y.H.; Rajput, U.A. Performance evaluation of an architecturally-designed vertical high capacity linear slot diffuser in a tropical atrium. *Archit. Sci. Rev.* **2023**, *66*, 53–69. [CrossRef]
82. Ratajczak, K.; Bandurski, K.; Płóciennik, A. Incorporating an atrium as a HAVC element for energy consumption reduction and thermal comfort improvement in a Polish climate. *Energy Build.* **2022**, *277*, 112592. [CrossRef]
83. Ghazali, A.; Zin, W.M.Z.W.; Ismail, M. Indoor Thermal Comfort Perception at Atrium Zone: Case Study of Naturally Ventilated Public Market. *J. Adv. Res. Appl. Sci. Eng. Technol.* **2022**, *29*, 13–29. [CrossRef]
84. Corbett, T.; Spentzou, E.; Eftekhari, M. Sensitivity analysis of proposed natural ventilation IEQ designs for archetypal open-plan office layouts in a temperate climate. *Adv. Build. Energy Res.* **2022**, *16*, 171–201. [CrossRef]
85. Hsu, H.H.; Chiang, W.H.; Huang, J.S. Hybrid Ventilation in an Air-Conditioned Office Building with a Multistory Atrium for Thermal Comfort: A Practical Case Study. *Buildings* **2021**, *11*, 625. [CrossRef]
86. Wen, Y.; Leng, J.; Yu, F.; Yu, C.W. Integrated design for underground space environment control of subway stations with atriums using piston ventilation. *Indoor Built Environ.* **2020**, *29*, 1300–1315. [CrossRef]
87. Albuquerque, D.P.; Mateus, N.; Avantaggiato, M.; Da Graça, G.C. Full-scale measurement and validated simulation of cooling load reduction due to nighttime natural ventilation of a large atrium. *Energy Build.* **2020**, *224*, 110233. [CrossRef]
88. Sokkar, R.; Alibaba, H.Z. Thermal Comfort Improvement for Atrium Building with Double-Skin Skylight in the Mediterranean Climate. *Sustainability* **2020**, *12*, 2253. [CrossRef]
89. Omrany, H.; Ghaffarianhoseini, A.; Berardi, U.; Ghaffarianhoseini, A.; Li, D.H.W. Is atrium an ideal form for daylight in buildings? *Archit. Sci. Rev.* **2020**, *63*, 47–62. [CrossRef]
90. Fohimi, N.A.M.; Asror, M.H.; Rabilah, R.; Mohammud, M.M.; Ismail, M.F.; Ani, F.N. CFD Simulation on Ventilation of an Indoor Atrium Space. *CFD Lett.* **2020**, *12*, 52–59. [CrossRef]

91. Zacharko, M.; Cichowicz, R.; Andrzejewski, M.; Chmura, P.; Kowalczuk, E.; Chmura, J.; Konefał, M. Air Pollutants Reduce the Physical Activity of Professional Soccer Players. *Int. J. Environ. Res. Public. Health* **2021**, *18*, 12928. [CrossRef]

92. Salonen, H.; Salthammer, T.; Morawska, L. Human exposure to air contaminants in sports environments. *Indoor Air* **2020**, *30*, 1109–1129. [CrossRef] [PubMed]

93. Ramos, C.A.; Wolterbeek, H.T.; Almeida, S.M. Exposure to indoor air pollutants during physical activity in fitness centers. *Build. Environ.* **2014**, *82*, 349–360. [CrossRef]

94. Ramos, C.A.; Reis, J.F.; Almeida, T.; Alves, F.; Wolterbeek, H.T.; Almeida, S.M. Estimating the inhaled dose of pollutants during indoor physical activity. *Sci. Total Environ.* **2015**, *527–528*, 111–118. [CrossRef]

95. Hurnik, M.; Ferdyn-Grygierek, J.; Kaczmarczyk, J.; Koper, P. Thermal Diagnosis of Ventilation and Cooling Systems in a Sports Hall—A Case Study. *Buildings* **2023**, *13*, 1185. [CrossRef]

96. Sobhi, M.; Fayad, M.A.; Al Jubori, A.M.; Badawy, T. Impact of spectators attendance on thermal ambience and water evaporation rate in an expansive competitive indoor swimming pool. *Case Stud. Therm. Eng.* **2022**, *38*, 102359. [CrossRef]

97. Ciuman, P.; Lipska, B. Experimental validation of the numerical model of air, heat and moisture flow in an indoor swimming pool. *Build. Environ.* **2018**, *145*, 1–13. [CrossRef]

98. Lee, L.T.; Blatchley, E.R. Long-Term Monitoring of Water and Air Quality at an Indoor Pool Facility during Modifications of Water Treatment. *Water* **2022**, *14*, 335. [CrossRef]

99. Slezakova, K.; Peixoto, C.; Pereira, M.D.C.; Morais, S. Indoor Air Quality Under Restricted Ventilation and Occupancy Scenarios with Focus on Particulate Matter: A Case Study of Fitness Centre. In *Occupational and Environmental Safety and Health III*; Arezes, P.M., Baptista, J.S., Carneiro, P., Branco, J.C., Costa, N., Duarte, J., Guedes, J.C., Melo, R.B., Miguel, A.S., Perestrelo, G., Eds.; Springer International Publishing: Cham, Switzerland, 2022; Volume 406, pp. 345–354. [CrossRef]

100. Majewski, G.; Telejko, M.; Orman, Ł.J. Preliminary results of thermal comfort analysis in selected buildings. *E3S Web Conf.* **2017**, *17*, 00056. [CrossRef]

101. Krawczyk, D.A.; Rodero, A.; Teleszewski, T.J. Examples of the HVAC Systems' Modernization in the Existing Schools and Kindergartens. *IOP Conf. Ser. Mater. Sci. Eng.* **2020**, *809*, 012009. [CrossRef]

102. Telejko, M.; Zender-Świercz, E. Attempt to Improve Indoor Air Quality in Kindergartens. *Procedia Eng.* **2016**, *161*, 1704–1709. [CrossRef]

103. Swiercz, E.Z.; Telejko, M. Indoor Air Quality in Kindergartens in Poland. *IOP Conf. Ser. Mater. Sci. Eng.* **2019**, *471*, 092066. [CrossRef]

104. Gładyszewska-Fiedoruk, K. Analysis of stack ventilation system effectiveness in an average kindergarten in north-eastern Poland. *Energy Build.* **2011**, *43*, 2488–2493. [CrossRef]

105. Taneichi, S.; Tanaka, I. Study on ventilation issues in urban nursery facilities: Long-term field survey in Yokohama, Japan. *E3S Web Conf.* **2023**, *396*, 01084. [CrossRef]

106. Jung, C.; Salameh, M.; Sherzad, M.; Arar, M. The Changes in Indoor Air Pollutant Concentration in Nursery in Dubai, United Arab Emirates. *Int. J. Adv. Res. Eng. Innov.* **2023**, *5*, 47–56. [CrossRef]

107. Ratajczak, K. Ventilation Strategy for Proper IAQ in Existing Nurseries Buildings—Lesson Learned from the Research during COVID-19 Pandemic. *Aerosol Air Qual. Res.* **2022**, *22*, 210337. [CrossRef]

108. Ferdyn-Grygierek, J.; Kaczmarczyk, J.; Blaszczok, M.; Lubina, P.; Koper, P.; Bulińska, A. Hygrothermal Risk in Museum Buildings Located in Moderate Climate. *Energies* **2020**, *13*, 344. [CrossRef]

109. Indrie, L.; Oana, D.; Ilies, M.; Ilies, D.C.; Lincu, A.; Ilies, A.; Baias, S.; Herman, G.; Onet, A.; Costea, M.; et al. Indoor air quality of museums and conservation of textiles art works. Case study: Salacea Museum House, Romania. *Ind. Textila* **2019**, *70*, 88–93. [CrossRef]

110. Ilie, D.C.; Marcu, F.; Caciora, T.; Indrie, L.; Ilie, A.; Albu, A.; Costea, M.; Burtă, L.; Baias, S.; Ilie, M.; et al. Investigations of Museum Indoor Microclimate and Air Quality. Case Study from Romania. *Atmosphere* **2021**, *12*, 286. [CrossRef]

111. Caciora, T.; Herman, G.V.; Ilieș, A.; Baias, Ș.; Ilieș, D.C.; Josan, I.; Hodor, N. The Use of Virtual Reality to Promote Sustainable Tourism: A Case Study of Wooden Churches Historical Monuments from Romania. *Remote Sens.* **2021**, *13*, 1758. [CrossRef]

112. Albu, A.V.; Caciora, T.; Berdenov, Z.; Ilies, D.C.; Sturzu, B.; Sopota, D.; Herman, G.V.; Ilies, A.; Kecse, G.; Ghergheles, C.G. Digitalization of garment in the context of circular economy. *Ind. Textila* **2021**, *72*, 102–107. [CrossRef]

113. Marcu, F.; Ilieș, D.C.; Wendt, J.A.; Indrie, L.; Ilieș, A.; Burta, L.; Caciora, T.; Herman, G.V.; Todoran, A.; Baias, S.; et al. Investigations regarding the biodegradation of the cultural heritage. Case study of traditional embroidered peasant shirt (Maramures, Romania). *Rom. Biotechnol. Lett.* **2020**, *25*, 1362–1368. [CrossRef]

114. Bajc, T.; Kerčov, A.; Gojak, M.; Todorović, M.; Pivac, N.; Nižetić, S. A novel method for calculation of the CO_2 concentration impact on correlation between thermal comfort and human body exergy consumption. *Energy Build.* **2023**, *294*, 113234. [CrossRef]

115. Kercov, A.; Bajc, T.; Gojak, M.; Todorovic, M.; Pivac, N.; Nizetic, S. Comparison between different thermal comfort models based on the exergy analysis. In Proceedings of the 2022 7th International Conference on Smart and Sustainable Technologies (SpliTech), Split/Bol, Croatia, 5–8 July 2022; IEEE: Piscataway, NJ, USA, 2022; pp. 1–4. [CrossRef]

116. Koelblen, B.; Psikuta, A.; Bogdan, A.; Annaheim, S.; Rossi, R.M. Thermal sensation models: Validation and sensitivity towards thermo-physiological parameters. *Build. Environ.* **2018**, *130*, 200–211. [CrossRef]

117. Psikuta, A.; Kuklane, K.; Bogdan, A.; Havenith, G.; Annaheim, S.; Rossi, R.M. Opportunities and constraints of presently used thermal manikins for thermo-physiological simulation of the human body. *Int. J. Biometeorol.* **2016**, *60*, 435–446. [CrossRef] [PubMed]
118. Chludzińska, M.; Bogdan, A. The effect of temperature and direction of airflow from the personalised ventilation on occupants' thermal sensations in office areas. *Build. Environ.* **2015**, *85*, 277–286. [CrossRef]
119. Harrouz, J.P.; Karam, J.; Ghali, K.; Ghaddar, N. Personalized ventilation with embedded air treatment system for simultaneous cooling and sorption-based carbon and humidity capture. *Energy Convers. Manag.* **2023**, *291*, 117290. [CrossRef]

energies

MDPI

Article

A New Approach to the Economic Evaluation of Thermomodernization: Annual Assessment Based on the Example of Production Space

Orest Voznyak [1], Edyta Dudkiewicz [2,*], Marta Laska [2], Ievgen Antypov [3], Nadiia Spodyniuk [3], Iryna Sukholova [1] and Olena Savchenko [1]

[1] Department of Heat, Gas Supply and Ventilation, Lviv Polytechnic National University, St. Bandery, 12, 79013 Lviv, Ukraine; orest.t.vozniak@lpnu.ua (O.V.); iryna.y.sukholova@lpnu.ua (I.S.); olena.o.savchenko@lpnu.ua (O.S.)

[2] Faculty of Environmental Engineering, Wroclaw University of Science and Technology, Wybrzeże Wyspiańskiego 27, 50-370 Wroclaw, Poland; marta.laska@pwr.edu.pl

[3] Heat and Power Engineering Department, Education and Research Institute of Energetics, Automation and Energy Efficiency, National University of Life and Environmental Sciences of Ukraine, Heroiv Oborony 12, 03041 Kyiv, Ukraine; ievgeniy.antypov@nubip.edu.ua (I.A.); n_spoduniuk@nubip.edu.ua (N.S.)

* Correspondence: edyta.dudkiewicz@pwr.edu.pl

Abstract: Energy and economic assessments are of great relevance in the context of decision processes for the most optimal solutions for building renovations. Following the method recommended by UNIDO, economic analyses of thermal modernization options are carried out based on the Simple Payback Time (SPBT), Net Present Value Ratio (NPVR) and Internal Rate of Return (IRR) indices. Incorporating these indicators and a new approach that involves aggregating thermomodernization activities not only in the cold and warm seasons separately, but throughout the whole year, an economic evaluation of the thermomodernization of a production space was carried out. In this case study, the renovation options included wall insulation, window replacement, the installation of infrared heater, a two-flow air diffuser (TFAD) and variable air volume. The economic effect indicated by the highest NPVR over a normative period of 15 years was obtained for the installation of an infrared heater and a TFAD with a variable mode ventilation system. The SPBT for this case was also the lowest.

Keywords: UNIDO; electric radiant heater; two-flow air diffuser (TFAD); variable air volume (VAV)

check for updates

Citation: Voznyak, O.; Dudkiewicz, E.; Laska, M.; Antypov, I.; Spodyniuk, N.; Sukholova, I.; Savchenko, O. A New Approach to the Economic Evaluation of Thermomodernization: Annual Assessment Based on the Example of Production Space. *Energies* **2024**, *17*, 2105. https://doi.org/10.3390/en17092105

Academic Editors: David Borge-Diez and Jay Zarnikau

Received: 13 February 2024
Revised: 3 April 2024
Accepted: 19 April 2024
Published: 28 April 2024

1. Introduction

A significant policy priority of European countries is saving energy [1–8]. In order to effectively save energy, technical innovations must be applied in a rational economic way. However, in order to provide sufficient indoor conditions in industrial premises, a large amount of energy is consumed for heating and cooling in the cold and warm seasons, respectively [9–13].

Today, reducing energy consumption and moving away from fossil fuels is a priority. This can be addressed through the widespread use of renewable and waste energy [14]. However, in certain European countries, like Ukraine and Poland, gas is still one of the main sources of energy. This raises questions about the diversification of the gas supply and pricing policies. Therefore, in these countries, policies to reduce the need for thermal energy consumption and its economic justification have been widely implemented. Energy saving problems in all European countries include topics such as heat supply, internal HVAC systems in buildings, heating and ventilation in the cold season [13,15], ventilation and air conditioning in the warm season [16,17], and hot water supply throughout the year [18,19]. Other considerations include building envelope, indoor environment and building operation. In order to effectively reduce energy consumption, the thermal modernization of buildings primary involves lowering the coefficients of heat transfer of walls and windows [20]. A reduction in heating and cooling demand is achieved by the modernization of existing systems, thus maximizing the effect of both thermal and economic savings. The heat transfer coefficient of envelopes and windows in new buildings must not exceed the maximum values given in the standards and regulations defined in particular countries [21,22].

An analysis of the literature clearly indicates that thermal modernization activities bring economic benefits, social benefits, environmental benefits and benefits for the energy system. There is even evidence for modernizing a building to a passive and nearly zero energy building standard [23–25].

The thermal diagnostics of a building can reveal factors influencing its heat consumption. By integrating various partial diagnoses, one can determine the overall heat consumption of the entire structure and gather data for devising improved solutions. When evaluating energy consumption, it is important to consider the indoor environment's quality, ensuring that energy-saving measures do not compromise it. The diagnostic procedure involves two phases: inspections and diagnostic measurements. Inspections aim to assess the technical conditions of systems and compare their performances with the designer's specifications. Diagnostic measurements are conducted in order to evaluate system operation under real conditions. A comprehensive method for assessing the physical envelope, HVAC systems and indoor environment was proposed in the literature [26,27].

When making decisions, including those related to improving the energy efficiency of a building, the decision-maker, consciously or unconsciously, most often uses the single-criterion method, looking for a solution that best meets one selected criterion. The single-criterion method gives a clear result, which decision should be made, knowing that many important parameters of the analyzed solutions will not be taken into account at all [28].

The basic single-criterion analytical methods include Cost–Benefit Analysis (CBA) and Cost-Effectiveness Analysis (CEA) [29]. The CBA analysis is carried out in three stages. In the first stage, after defining which variants of solutions should be compared, all costs and benefits of each variant are given a monetary value. During the second stage, costs and benefits are compared for all analyzed variants. The comparison is made using different measures of economic efficiency [28].

The economic criterion can be described using many indicators known from the literature [30]. The most frequently used are investment and operating costs related to energy consumption. Importantly, the operating costs index may combine many elements: energy consumption for heating, hot water preparation and cooling, electricity for lighting, operation of equipment, machines, technologies, water consumption and sewage disposal. They can be analyzed together or individually. The most popular indicators of economic efficiency include Net Present Value (NPV), Simple Payback Time (SPBT) and Internal Rate of Return (IRR) of an investment. Additionally, the economic uncertainty and service life of the installation are assessed. However, a tool based on the assumptions of life cycle costing (LCC) can be used to compare the economic effectiveness of alternative investment solutions and the profitability of a product over the entire life cycle. Depending on the detail and purpose of the analysis, there are three types of life cycle costing: conventional LCC (also called traditional or business LCC), environmental LCC and social LCC. This tool requires a large amount of data and complements the multi-criteria analysis, which eliminates unacceptable solutions that are not subject to detailed analyses, e.g., LCC.

However, the literature review shows that the most popular methods include net present value or internal rate of return of an investment. The third stage of the decision-making process is the selection of the best solution, i.e., the one that best meets the adopted objective function. This evaluation mechanism has also been recommended by the United Nations Industrial Development Organization (UNIDO).

The motivation for scientific research and studies in this area is the search for optimal solutions that not only yield results in reducing energy consumption, but are also economically justified. As indicated by numerous articles [20,29,31–33], the renovation of buildings, despite unquestionable benefits in reducing energy consumption, is an undertaking requiring significant investment. The authors took the initiative to identify the optimal solution for another case and thus contributed to research towards cost optimization using a new approach in the selection of thermal modernization options and measures, because in many studies on lowering energy demand in buildings, thermal modernization measures are considered seasonally (Table 1). In the cold season, thermal modernization of the buildings focuses mainly on the building envelope [20,34–36], the heating systems [21,22] and the heat sources, for example, gas supply [37]. In the aforementioned publications, the technical merits and the economic efficiency were investigated.

Table 1. Thermal modernization measures undertaken in case of thermal modernization activities.

Authors, Year	Title; Journal	Thermal Modernization Measure	Season
Krawczyk D.A., 2004 [38]	The optimum variants of warming up walls, roof, windows change and a heating system modernization in the typical school according to its localization (in Polish); Instal.	Reducing heat losses through building partitions, improving the efficiency of the heating system.	1
Przesmycka N. et al., 2023 [39]	Modernisation of hospital buildings built in the 20th century in the context of architectural, functional and operational problems; Architectus.	Modernization of installation, thermomodernization, replacement of window and door joinery.	1
Krawczyk D.A., 2014 [40]	Theoretical and real effect of the school's thermal modernization—A case study; Energy and Buildings.	Reducing heat losses through building partitions, improving the efficiency of the heating system.	1
Gładyszewska-Fiedoruk K. et al., 2014, [41]	The possibilities of energy consumption reduction and a maintenance of indoor air quality in doctor's offices located in north-eastern Poland; Energy and Buildings.	Replace radiators and pipes; pipe insulation; wall insulation; window replacements.	1
Ołtarzewska A. et al., 2022 [20]	Analysis of the influence of selected factors on heating costs and pollutant emissions in a cold climate on the example of a service building located in Bialystok; Enegies.	Thermal insulation of the building partitions, heat source replacement.	1
Lis A., 2020 [8]	Renewable Energy Sources and Rationalisation of Energy Consumption in Buildings as a Way to Reduce Environmental Pollution; Renewable Energy	Thermal insulation of building envelope; renewable energy sources; replacement of the windows and doors; modernize the central heating system; hot water heating; installation of heating elements with low thermal inertia and thermostatic valves; installation lagging on the central heating pipes; ineffective electric heaters with a centralized heating from their own boiler room replacement.	1
Lis A. et al., 2019 [42]	The quality of the microclimate in educational buildings subjected to thermal modernization	Insulation of building envelope and modernization of the heating system and hot water preparation.	1
Bøhm B., 2013 [43]	Production and distribution of domestic hot water in selected Danish apartment buildings and institutions. Analysis of consumption, energy efficiency and the significance for energy design requirements of buildings; Energy Conversion and Management Journal	Improving the DHW system.	1/2
Hałacz J. et al., 2020 [44]	Assessment of Reducing Pollutant Emissions in Selected Heating and Ventilation Systems in Single-Family Houses; Energies	Heat sources, mechanical ventilation with ground-coupled heat exchanger.	1/2

Table 1. *Cont.*

Authors, Year	Title; Journal	Thermal Modernization Measure	Season
Ferdyn-Grygierek J. et al., 2019 [45]	HVAC control methods for drastically improved hygrothermal museum microclimates in warm season; Building and Environment	HVAC control.	2
Ratajczak, K et al., 2020 [46]	Assessment of the air streams mixing in wall-type heat recovery units for ventilation of existing and refurbishing buildings toward low energy buildings; Energy and Buildings	Mechanical ventilation systems with heat recovery.	2
Zender-Świercz E. et al., 2013 [47]	Thermomodernization a building and its impact on the indoor microclimate; Structure and Environment	Sealing the roof.	1/2

Seasons: 1—cold season, 2—warm season.

The main aim of this paper is to propose a new approach to the economic evaluation of thermal modernization in a building. This new approach takes into account thermo-modernization measures from different seasons, creating a year-round assessment. This is especially important in the Central European area at the moment, where climate change means that cooling demand will play an increasingly important role in the summer season [26]. A case study of production space in Lviv in Ukraine is considered in which the options included walls insulation, window replacement and the utilization of devices like radiant heaters and TFAD working in variable mode. These measures led to the improvement of inner conditions throughout the year. The of economic calculation concepts are recommended by UNIDO.

2. Materials and Methods

2.1. Economic Indicators

Every investment process aimed at generating economic profits should undergo a thorough analysis. This is particularly crucial for investment processes where the returns are spread over an extended period, such as in the construction industry. Economic analysis provides information to determine whether a project is financially viable and if the proposed solutions are economically sound. Numerous examples from past practice demonstrate the use of economic analysis as a decision-support tool in investment decision-making [48].

Considering the operation of heating and ventilation systems throughout the year, the following abbreviations are used:

- The Thermal Modernization Measure (TMM) is an undertaking aimed at reducing the demand for heat or cooling energy in a building;
- The Seasonal Thermal Modernization Measure (STMM) separately considers the cold and warm seasons, and refers to the insulation of the building envelopes, as well as the heating and ventilation systems;
- The Year-Round Thermal Modernization Measure (YTMM) refers to the cumulative effects of STMMs in both seasons, i.e., throughout the year;
- The Thermal Modernization Option (TMO) describes the cumulative effect of several STMMs, considered separately for the cold and warm season. In other words, the TMO is the combination of several STMMs in the corresponding period of the year.

The economic indicators of STMMs and TMOs are determined in publications [16,49].

Savings due to the application of electric infrared heaters (K_d) in EUR/year were calculated following formula:

$$K_d = Q_H \cdot (C_H - C_E) \tag{1}$$

where

K_d—annual energy savings due to the difference in heat and electrical energy costs, EUR/year;

C_H and C_E—the costs of heat and electricity, respectively, EUR/GJ;

Q_H—the year-round energy demand of the heating system, GJ/year, which is calculated based on heat losses in W, determined by the formula:

$$q_h = U \cdot A \cdot (t_{in} - t_5) \tag{2}$$

where

U—heat transfer coefficient, $W/(m^2 K)$;

A—the area of the building element (window, wall, etc.), m^2;

t_{in}—indoor temperature during the cold season, °C;

t_5—the temperature of the coldest five days, °C;

According to duration of the heating season and the mean external temperature of the heating season t_{hs}, the temperature ratio is $(t_{in} - t_{hs})/(t_{in} - t_5)$.

The warm season calculation procedure contains the cooling demand calculations determined by the formula for heat gains in W:

$$q_c = c \cdot \rho \cdot L \cdot (t_{ex} - t_{in}) \tag{3}$$

where

c—air specific heat, $J/(kg \cdot K)$;

ρ—air density, kg/m^3;

L—volume flow rate of air, m^3/s;

t_{ex}—temperature of exhaust, °C;

t_{in}—temperature of supply air, °C.

It can be noted that the proposed approach has some limitations. There is a simple method of calculating the heat demand and cooling load of a building. In the future, building cooling loads can be estimated based on rough sets and deep extreme learning machines [50]. Furthermore, deep learning models based on EWKM, random forest algorithms, SSA and BiLSTM for building energy consumption prediction can be also implemented [51].

The value of savings K_{iCS} as a result of each thermomodernization process in the cold season is determined according to formula:

$$K_{iCS} = \Delta Q_{iCS} \cdot C_{HE} \tag{4}$$

where

K_{iCS}—savings for the cold season, EUR/year;

ΔQ_{iCS}—energy savings for one STMM, GJ/year;

C_{HE}—the cost of energy in the cold season; EUR/year.

The value of savings K_{iWS} as a result of each thermomodernization process in the warm season is determined according to formula:

$$K_{iWS} = \Delta Q_{iWS} \cdot C_{EC} \tag{5}$$

where

K_{iWS}—savings for the warm season, EUR/year;

ΔQ_{iWS}—energy savings for one STMM, GJ/year;

C_{EC}—the cost of energy in the warm season, EUR/year.

The equation to calculate the cost savings throughout the year K (EUR/year) as an effect of the YTMM takes the form:

$$K = \Delta Q_{CS} \cdot C_{HE} + \Delta Q_{WS} \cdot C_{EC} \tag{6}$$

where

ΔQ_{CS} and ΔQ_{WS}—saved heat energy in the cold season and cooling in the warm season, respectively, GJ/year;

C_{HE} and C_{EC}—the cost of energy in the cold and warm seasons, respectively, EUR/year.

Investment costs I_i are the sum of investment costs of thermal modernization measures I_{iCS} for the cold season and I_{iWS} for the warm season, and are determined by:

$$I_i = I_{iCS} + I_{iWS} \tag{7}$$

where

I_{iCS} and I_{iWS}—investment costs of STMMs in the cold and warm seasons, respectively, EUR.

An evaluation of the profitability of investments was carried out based on methods recommended by the United Nations Industrial Development Organization (UNIDO) [52]. This is a specialized agency of the United Nations (UN) that promotes industrial development for poverty reduction, inclusive globalization and environmental sustainability. It was established in 1966 and obtained the status of specialized UN agency in 1985. Currently, UNIDO has 170 member states. The organization's main areas of interest are divided into four strategic priorities—creating shared prosperity, increasing the competitiveness of the economy, environmental protection, and strengthening knowledge and institutions.

The decision to undertake a thermal modernization investment is made when the "sum" of positive values, benefits resulting from the project, is greater than the expenses related to the investment. Various types of economic efficiency indicators are used. In accordance with UNIDO recommendations, the values of such indicators should be calculated according to Simple Payback Time (SPBT) and common dynamic indices, namely Net Present Value Ratio (NPVR) and Internal Rate of Return (IRR). The methodology recommended by UNIDO presents the method for conducting financial analysis and evaluating investment projects. According to this methodology, the financial profitability of the project from the investor's point of view is the primary criterion in the investment assessment, more important than other advantages of the project.

Simple Payback Time (SPBT) is defined as the time needed to recover the capital expenses incurred for the investment. It is calculated from the moment the investment is launched until the sum of gross benefits obtained as a result of the investment balances the incurred expenditure. Since the SPBT for STMMs is determined from [2,5], the SPBT for YTMMs is calculated as follows:

$$SPBT = \frac{I_{iCS} + I_{iWS}}{K_{iCS} + K_{iWS}} \tag{8}$$

where

K_{iCS} and K_{iWS}—the cost of energy saved in the cold and warm seasons, respectively. These costs correspond to the STMM, EUR/year.

The net present value ratio is described by the following formula:

$$NPVR = \frac{K_i \cdot t}{(1 + r)^t} - I_i \tag{9}$$

where

K_i—the cost of energy saved throughout the year for the corresponding i-th YTMM, EUR/year, and is determined from (6);

t—current time in years (the maximum time is defined to be 15 years);

r—discount rate, r = 0.08;

I_i—determined by Formula (7).

The method is exposed to the typical limitations found in economic analysis, namely, the adoption of the value of the discount rate r-factor. The r-value is a bank value that is constantly adjusted; that is, it changes slightly depending on the economic situation in the country. The current economic situation in Ukraine is unstable for objective reasons, so the value of r is very unstable and difficult to predict. During the initial work on this article, the r-value was changed several times according to the NBU (National Bank of Ukraine). As a result, the calculations become much more complicated and require additional assumptions and simplifications. In this regard, a fixed value of r was adopted for the conditions of a stable economic situation. If we consider another arithmetic value of r, then this does not change the essence of the matter, but will only lead to obtaining another corresponding arithmetic value of NPVR.

The NPVR value takes into account the entire duration of the project. This is its basic advantage. It encourages risk by showing the full benefits of making a given investment. The disadvantage is the difficulty in selecting the observation period for which the benefits expressed in the NPVR value are calculated. One of the parameters for choosing the length of this period may be the period until the project will generate benefits, i.e., until its "economic death", or more clearly, the period of full depreciation of the purchased machinery and equipment. In some cases, the investor may impose different, shorter periods when assessing the benefits of the project in the light of their expectations as to when the benefits of the invested funds will be realized. In certain cases, the length of the analysis period may be determined by one of the interested parties, e.g., a bank or by law. In order for the thermal renovation option to be profitable, it is necessary that its Simple Payback Time is shorter than 15 years, and the Internal Rate of Return is greater than 8% [49,53]. Thus, in Poland and Ukraine, a period of 15 years is assumed for thermal modernization investments.

As per guidance from UNIDO, the net present value ratio stands out as the pivotal measure for evaluating the optimality of annual thermal renovation options. The NPVR not only serves as an indicator, but also plays a crucial role in discerning the financial viability of different thermal upgrading alternatives. A notable aspect is that a higher NPVR value is indicative of the specific thermal renovation choice that not only enhances energy efficiency but also promises the utmost financial gain over the anticipated operational lifespan of the building. This criterion thus offers a comprehensive perspective, combining both economic and thermal performance considerations, in order to guide decision-makers towards the most financially lucrative and sustainable thermal modernization strategy. It should be noted that other economic indexes are also important but are not a priority. In particular, the SPBT can be compared with the normative operating time of 15 years and the Internal Rate of Return, with the normative value r = 0.08 (8%), to draw appropriate conclusions.

The thermal modernization options that combine two (or more) energy-saving measures—one of which is profitable (NPVR > 0) and the other unprofitable (NPVR < 0) —are interesting. They can be both profitable but not optimal, and finally unprofitable. First of all, everything depends on the NPVR indicator, and the rest of the economic characteristics are its consequences. This is clear in the analyzed case, as window replacement generates a negative NPVR, but when used with other STMMs, it allows for a positive net present value.

The Internal Rate of Return (IRR) is a certain degree of discount when NPVR = 0. Since, by the definition t = 15 years, IRR with parameters described above is determined by the formula:

$$IRR = \left(\frac{15 \cdot K_i}{I_i} \right)^{\frac{1}{15}} - 1 \tag{10}$$

The year-round interest rates are proposed to be determined from Formulas (9) and (10).

Indicators of the economic criterion of investment are also the service life of the installation, defined as a guarantee of durability, resistance to corrosion and damage, and economic uncertainty, described as the probability that the actual economic parameters will differ significantly from the estimated ones, e.g., due to an unexpected increase in the cost of installing equipment or fuel.

2.2. New Approach

Thermal modernization activities include the adoption of assumptions for the selection and assessment of the thermal modernization package. The criteria for assessing the effects of their use are as follows [54]:

- Demand for usable energy, EU index;
- Capital expenditure;
- Saving the annual energy cost;
- Simple payback period;
- Net present value ratio.

According to the UNIDO recommendations, the optimality criterion of the annual thermal renovation option is the NPVR. The maximum value of the NPVR indicates the option of thermal upgrading that provides the maximum profit over the assumed period of operation of the building.

The new approach consider several thermomodernization measures from different seasons, creating a year-round assessment. In order to determine the YTMM, the authors propose to calculate it as the linked effect of both STMMs for the cold and warm seasons, and then to pinpoint the TMO with the highest NPVR value.

Preliminary activities require comprehensive in situ thermal diagnostics of buildings in order to determine the scope of thermal modernization of the facility and to estimate the energy consumption of a given facility.

Thermal modernization improvement options are proposed separately for the cold and warm seasons (STMM).

Next, two seasonal thermal modernization measures are combined as the YTMM. Their number is obtained according to the rule of combinatorics. For example, when the cold season includes three options, and the warm season two options, the number of YTMMs is six. Since there is a difference between energy costs in the cold and warm seasons, these values were considered separately. However, these values must be considered aggregate in order to estimate the annual effect.

In the next step, the annual energy savings are estimated, which allows us to calculate the SPBT, NPVR and IRR for each YTMM. In order to achieve the maximum saving effect level for the best options, namely thermal modernization options (TMOs) linked to several TMM from both seasons, the investment cost, the annual savings and the investment profitability indices are juxtaposed in a table (Table 2) using a special algorithm: annual energy-saving measures must be arranged vertically in order of increasing Simple Payback Time from minimum to maximum, while their cumulative effect are arrange horizontally, marked with a "+" sign. Generally, one should take into account the possibility of their simultaneous action, for example, the replacement of windows and sealing of windows. Such measures cannot be implemented simultaneously.

Table 2. Design of table for the optimization of combinations of STMMs.

YTMM	Thermal Modernization Options (TMO)				
	I	**II**	...	...	...
TMM for cold + TMM for warm season	+	+	+	+	+
TMM for cold + TMM for warm season		+	+	+	+
TMM for cold + TMM for warm season			+	+	+
TMM for cold + TMM for warm season				+	+
TMM for cold + TMM for warm season					+
Indicators					
Investment cost I Annual savings K SPBT, year NPVR, EUR IRR, %	$SPBT_I <$	$SPBT_{II} <$	$SPBT_i <$ $NPVR_{max}$	$SPBT_{ii} <$	$SPBT_{nmax}$

Two pluses in the column mean that TMO II includes the TMM for the winter season plus the TMM for the summer season (i.e., TMO I), plus an additional TMM, for the summer or winter season, selected from the next YTMM pair. The new TMM is, of course, additionally included in the investment costs, and the option is positioned in line with the increase in the SPBT. For TMO III, a new, previously unconsidered TMM is added from the new YTMM pair. If the TMM was already included in the previous option, the new TMO does not introduce new information and the cost indexes are equal to the previous TMO.

The scheme of activities within the designation of the TMO with the maximum value of NPVR is presented in Figure 1.

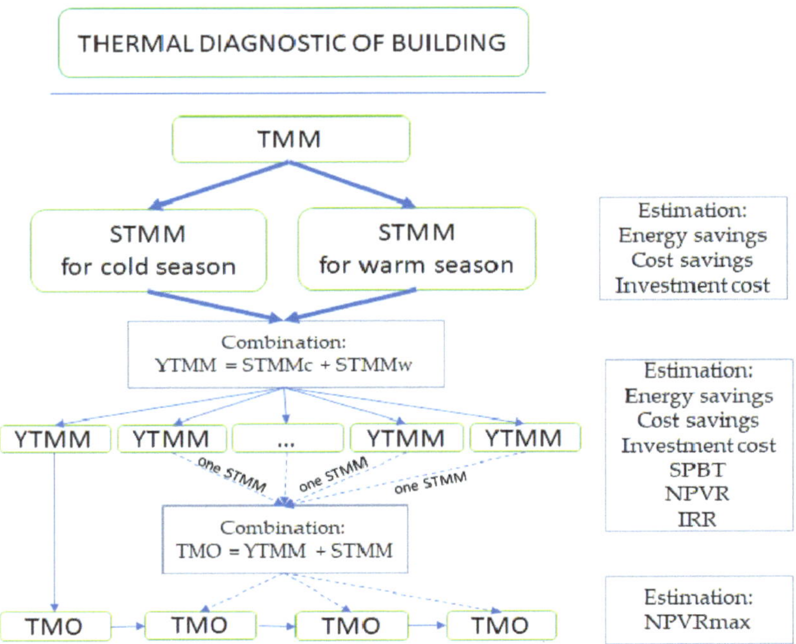

Figure 1. Scheme of activities within the designation of the TMO.

The use of a new approach to identify a thermal modernization option combining several thermal modernization measures for two seasons is shown as an example of a real production building as a case study.

3. Case Study

3.1. Building Characteristic and Its Location

The considered space is a production building, originally equipped with a traditional water heating system, located in Lviv city in Ukraine (Figure 2), an area with a Central European climate that is of interest due to the increasing need for cooling in summer and the large differences between summer and winter outdoor air temperatures. The naturally ventilated facility was built in 1995 and consists of one floor. The volume of the building is 245 m^3, while the usable area is 70 m^2 (5 m × 14 m). The window surface area was calculated as about 9.7 m^2. During the cold season, a temperature t_{in} of 18 °C is maintained indoors, while in warm season, a supply temperature of 23 °C and exhaust of 26 °C are designed. The layout of the investigated space, including the locations of two infrared heaters and two TFADs, proposed as a thermomodernization options, is presented in Figure 3. The internal walls are on axes 1 and 2, and the external walls are on axes A and B.

Figure 2. Lviv city in Ukraine [55].

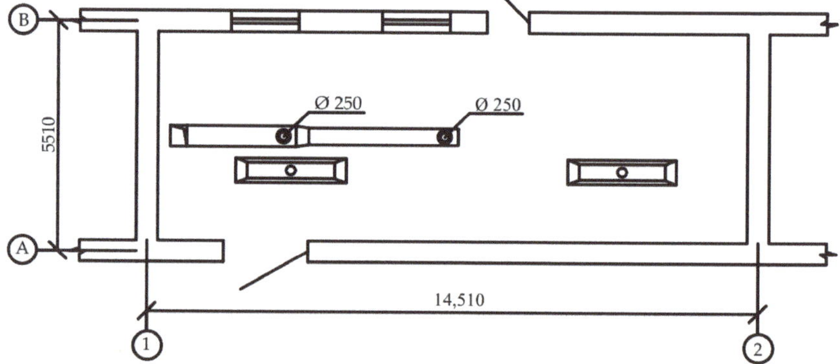

Figure 3. The layout of the production space.

The main climatic characteristics of Lviv city are as follows:

- First climate zone according to the DBN V.2.6-31 standard [56];
- Climate zone 5A according to the ASHRAE;
- The mean annual air temperature is 8.8 °C;
- The mean temperature of the coldest five days in the cold season is −19 °C;
- The duration of the heating season (cold season) is 179 days and the remaining days of the year are the warm season. There are no specific regulations governing the heating season, including when it starts and ends. Nevertheless, a certain period of time has been defined and confirmed by case law. Accordingly, the heating season starts in October, when the mean daily external temperature of three consecutive days is lower than 12 °C;
- The mean temperature of the heating season is +0.4 °C;
- The mean temperature of the five hottest days in the warm season is +23 °C.

3.2. Thermal Modernization Improvements Options

The wall and window heat transfer coefficients should adhere to the specified maximum values outlined in the standards and regulations set by individual countries. The thermal insulation requirements in Ukraine are considerably less stringent compared to other countries [21,22]. The minimum thermal insulation requirements for Ukraine are visible in Table 3.

Table 3. Heat transfer coefficients U for building envelopes.

Building Envelope	U [W/(m²·K)]
Exterior walls	0.30
Flat roof above heated spaces	0.20
Floor on the ground	0.20
Flat roof over unheated attic	0.20
Flat roof over unheated basement	0.27
Exterior windows	1.33
Exterior doors	1.67

For the analysis of thermomodernization improvements, window replacement and external wall insulation were proposed, as well as elements of the building installations dedicated to the cold (heating) and warm seasons (ventilation).

The following STMMs for cold season were implemented:

- Window replacement (from $R = 0.36$ m²K/W to $R = 0.9$ m²K/W). The U value can be simply calculated, i.e., old windows with U-value of 2.8 W/(m²K) were replaced by new windows with $U = 1.1$ W/(m²K);
- The insulation of the external walls was improved from $R = 1$ m²K/W to $R = 3.3$ m²K/W (achieved by polystyrene foam boards with a thickness of 100 mm), i.e., U before thermomodernization equaled 1 W/(m²K), while after $U = 0.3$ W/(m²K);
- Installation of the infrared heaters.

For the warm season, the STMMs are as follows:

- Installation of two-flow air diffuser TFAD;
- Use of variable air volume (VAV) in the ventilation system by application of an integrated actuator.

Based on the aforementioned STMMs, the following six YTMM combinations were proposed:

- Window replacement and installation of two-flow air diffuser TFAD,
- Window replacement and implementation of VAV,
- Insulating the walls and installation of two-flow air diffuser TFAD,
- Insulating the walls and implementation of VAV,
- Installation of infrared heaters and two-flow air diffuser TFAD,
- Installation of infrared heaters and implementation of VAV.

The economic performance indicators determined from Equations (1)–(10) can be applied to YTMO, which is the aggregate effect of different TMOs in different periods of the year.

The following energy costs were adopted in the analysis [49,53]:

- For electricity, $C_{EE} = 0.036$ EUR/kWh (10 EUR/GJ),
- For heating, $C_{HE} = 14$ EUR/GJ,
- For cooling, $C_{EC} = 40$ EUR/GJ.

During the warm season investigation, the conversion factor of the cooling device was taken into account in order to determine energy consumption. This value was calculated based on the operation data of the electric motor of the compressor of the refrigerating device and the ranges $\varepsilon = 3$–5. For further calculations, the average ε value of 4 was assumed to be optimal.

3.3. Modernization of Building Installation

Proposed thermomodernization measures comprising the modernization of the building installation consisted of application solutions dedicated the warm and cold seasons separately. For the cold season and industrial premises, the use of radiant heating systems with infrared heaters deserves attention [57]. Infrared heaters [58–60], by directing infrared radiation straight into the occupied zone, create comfort and uniform thermal conditions. These devices are powered by electricity or gas. They are safe and work in a wide range of temperatures with high efficiency, while the high inertia and high-quality regulation provide a reduction in the running costs of the heating system. The aforementioned features of infrared heaters and issues such as uniformity of temperature field, mounting height, thermal efficiency, thermal load, air temperature and the temperature of the surrounding surfaces, i.e., walls, ceilings, and floors, and the reduction of operating costs are described in [61,62]. These publications also sought ways to increase the technical and thermal efficiency of the devices.

The infrared heater type NL–12 R, adapted for analysis as a thermomodernization option for the cold season, provides local heating via electromagnetic waves and has the ability for the power settings to be quickly adjusted, thus creating dynamically changing indoor conditions in the production space. The height and angle position of the heater, which can be adjusted, affects the irradiation area, while the reflector, which is an integral part of the device, directs infrared rays straight into the heating zone, increasing the efficiency of the heater and the comfort of occupants.

In the warm season, the thermal and humidity conditions in the room must also meet the user's expectations and the legal and technological requirements [13,63]. Simultaneously, the mechanical ventilation and air conditioning systems must meet energy saving and efficiency requirements [64]. The positive cooling effect, through a greater convective heat exchange with the human body, without the input of additional energy for cooling, which is desirable in the warm season, can be achieved through increased airflow. The description of the solutions based on periodic change in the volume flow rate and the initial velocity of air diffusers are described in [65,66]. This current article proposes, as one of the thermomodernization measure, the utilization of the original, proprietary technical solution, namely, the two-flow air diffuser (TFAD) presented in Figure 4 [67]. A TFAD, together with an energy-efficient ventilation system, is able to form air flows with an intensive attenuation of air velocity and temperature. It has the ability to regulate air flow and create a prompt and dynamic change in indoor microclimate conditions [65]. The design of the TFAD (Figure 4) allows air flow geometry to be changed in two ways, namely, as a swirled air stream directed downward (vertical), and as a flat air stream directed to the surface of the ceiling (horizontal), depending on actual needs. The TFAD has an inlet nozzle installed against the inlet section of the device with a deflector that forms an adjustable annular gap. Additionally, it is equipped with blades that are attached to the TFAD by a rod with a control handle with a changeable angle. The twisting blades of the TFAD allow the angle of expansion of the stream to be changed and enable the formation of a swirled air stream with high-intensity mixing of the supply and surrounding air, thus changing the amount of air supply. The TFAD design allows air flow to be improved by reducing the air velocity attenuation coefficient. The adjustment screw of the gap brings the change in the amount of air that enters the adjustable annular gap. The integration of the TFAD with an LM24A electric actuator, manufactured by the company Belimo, allows the change of the inclination angle of the twisting blades, thus supplying air in the required direction according to the required parameters. Connecting the electric actuator, which is controlled from the automation unit, enables the smooth change in volumetric air flow and intensity. As a result, the device works with air volume, which can be increased with a stable air velocity maintained within regulatory limits without any disruptions, and thus, the TFAD is dedicated to variable air volume ventilation systems (VAV) [66].

Further improvement of the aerodynamic characteristics of the two-flow air diffuser and the properties of air flow (velocity and temperature fields, velocity and temperature attenuation coefficients, and the influence of the angles of inclination of the twist blades) in terms of aerodynamic resistance and acoustic characteristics of aerodynamic noise, can be achieved as proposed in publication [65]. In order to confirm the effectiveness of the devices implemented to reduce energy consumption, it would be necessary to conduct additional experimental studies applying mathematical models [68–71] and numerical modelling methods as described in [72–74].

Figure 4. A two-flow air diffuser (TFAD), equipped with electric actuator Belimo LM24A.

The application of a TFAD allows for energy savings in air conditioning systems for small-scale premises (such as the one presented in this article) and the maintenance of normative air velocity and temperature, creating dynamic and comfortable indoor conditions in a room. The energy efficiency of cooling in an air conditioning system equipped with a TFAD integrated with an electric actuator during the warm season is examined in [49].

3.4. Results and Discussion

Considering the climatic characteristics of Lviv, the heat loss of the building in the cold season before thermal modernization was calculated in order to determine the base value of the energy demand Q_0. After each thermal modernization measure was performed, a new heat loss Q_i was calculated. Energy saving ΔQ_{cs} and ΔQ_{WS} were determined as the differences between the energy demand before the thermal modernization process Q_0 and the energy demand after the thermal modernization process Q_i. More precisely, the state before and after the thermal modernization:

- Heat losses q_h were determined from the Formula (2), the temperature of the coldest five days $t_5 = -19\,°C$, $t_{in} = 18\,°C$, and the U coefficients of the thermal modernization measures;
- The determination of the building energy demand Q_0 in GJ during the cold season, in accordance with the average temperature of the heating season, that is, +0.4 °C, and taking into account the duration of cold season (179 days);
- Similarly, the determination of the building energy demand Q_i in GJ for each STMM separately;
- The determination of energy saving: $\Delta Q_{CS} = Q_0 - Q_i$ for each STMM separately;
- The determination of investments in EUR for each STMM separately, in accordance with Equations (1), (4) and (5);
- The determination of annual energy and savings by the combination of two seasonal thermal modernization measures (one each for the cold and warm seasons) as a sum of both savings.

While the thermal modernization measures in the cold season reduce the value of U W/(m²K), the need for energy for heating decreases. The use of the VAV mode allows for a reduction in the value of L; additionally, TFAD devices are efficient and cheap while maintaining comfortable indoor conditions in the warm season; thus, the need for energy for cooling is reduced.

In the cold season, three STMMs were considered. In Table 4, the obtained data are presented, including energy demand Q_0 and Q_i, energy saving ΔQ_{cs} and savings K_i for all year, EUR/year, following Formula (6). The data in Table 4 show that the installation of infrared heaters provided the smallest energy and cost savings, while the most effective was the window replacement. The thermal insulation of walls produces effects with values between the other STMMs.

Table 4. Energy and savings in a cold season.

STMM	Energy Demand, Q_0, GJ/year	Energy Demand Q_i, GJ/Year	Energy Saving ΔQ_{cs}, GJ/Year	Savings K_i, EUR/Year
	Before Change	After Change		
Window replacement	40	20	20	280
Wall thermal insulation	40	22	18	252
Installation of infrared heaters	40	24	16	224

The data in Table 5 show the effect of two STMMs in a warm season. Energy needs before thermal modernization Q_0 and after Q_i, energy saving ΔQ_{cs} and savings K_i for the warm season were obtained. The installation of the two-flow air diffuser (TFAD) gives nearly as little savings in thermal energy and costs as the variable mode of the ventilation system with automation system.

Table 5. Energy and savings in a warm season.

STMM	Energy Needs, Q_0, GJ/Year	Energy Needs Q_i, GJ/Year	Energy Saving ΔQ_{ws} GJ/Year	Savings K_i, EUR/Year
	Before Change	After Change		
TFAD installation	9.233	6.487	2.746	109
Variable mode of the ventilation system (VAV)	9.233	6.122	3.111	124

Table 6 refers to the characteristics of the YTMMs, which are a combination of two seasonal thermal modernization measures. Their number is obtained according to the rule of combinatorics. Since the cold season includes three options, and the warm season includes two options, the number of YTMMs is six. The cost of energy in the cold and warm seasons were considered separately; however, in order to estimate the annual effect, these values must be considered aggregate. Energy demand before thermal modernization Q_0 and after Q_i, energy saving ΔQ_{cs} and savings K were obtained throughout the year. Thermomodernization measures related to partition insulation provide the greatest energy savings. At the same time, the greatest financial and energy savings were obtained from a combination of window replacement and the application of variable modes of the ventilation system.

Table 6. Annual energy and savings for YTMMs.

YTMMs Combinations of STMMs	Energy Demand, Q_0, GJ/Year	Energy Demand Q_i, GJ/Year	Energy Saving ΔQ_{cs}, GJ/Year	Savings K_i, EUR/Year
	Before Change	After Change		
Window replacement and TFAD installation	49.233	26.487	22.746	390
Window replacement and VAV	49.233	26.122	23.111	404
Wall thermal insulation and TFAD installation	49.233	28.487	20.746	362
Wall thermal insulation and VAV	49.233	28.122	21.111	376
Infrared heaters installation and TFAD installation	49.233	30.487	18.746	334
Infrared heaters installation and VAV	49.233	30.122	19.111	384

However, it is important to compare the investment costs, which may have a great influence on the total costs of the final choice of the modernization option. I_i is the investment cost of STMMs in the cold and warm seasons, and their sum is presented in Table 7, where, as well as economic indicators of combinations of STMMs for year-round operation, i.e., SPBT, NPVR and IRR are displayed. The highest investment costs for the warm season are related to the options where the walls are insulated, and it leads to the highest payback time of up to 6 years. The window replacement option is also characterized by high investment costs. Simultaneously, the differences in annual savings for these four options are not very diverse; however, the NPVR is negative for options involving window replacement. For the cold season, the option related to improving the U coefficient of the walls is more favorable in terms of both investment costs and payback time. Options for the modernization of heating and ventilation systems in the building are cheaper in terms of investment costs, and bring measurably large annual savings. Radiant heaters are nearly as energy efficient as TFAD devices. The payback time of these investments is very short, less than 1 year. However, examining the data presented in Table 7 for both seasons, it is clearly visible that the application of the system based on the two-flow air diffuser (TFAD) each time is beneficial.

Table 7. Economic indicators as an effect of thermal modernization for YTMMs.

YTMM	Investment Cost, I_i, EUR	Annual Saving K_i, EUR/Year	SPBT, Year	NPVR, EUR	IRR, %
Window replacement and TFAD installation	1904	390	4.9	−60	7.8
Window replacement and VAV	2006	404	5.0	−92	7.7
Wall thermal insulation and TFAD installation	2098	362	5.8	+1710	6.9
Wall thermal insulation and VAV	2200	376	5.8	+1778	6.8
Infrared heaters and TFAD installation	157	334	0.5	+1422	26
Infrared heaters installation and VAV	259	384	0.5	+1390	22

Energy costs are related to the costs of heat energy, electricity and refrigeration energy. The proper algorithm should be implemented in the building operation management system in order to provide reliable data for building electricity consumption management [75]. Since electricity in Ukraine is cheaper than thermal energy, it is obvious that the operating annual costs of electric heating and the use of electric infrared heaters will give the expected effect compared to a traditional heating system based on gas supply. Only questions of capital costs for installing the appropriate equipment arise. When estimating the cost of cooling energy, it is necessary to take into account the coefficient of performance of the refrigerating device. The cost of cooling energy is significantly higher than the cost of thermal energy. The cooling demand in the warm season is smaller than the heating demand in the cold season, but the costs of cooling are greater. Since the cost of the equipment of the corresponding ventilation systems should be taken into account, the question of the efficiency of these systems throughout the year is also of great interest.

In order to optimize both thermal engineering and cost-effectiveness across all solutions, optimal options for thermal modernization, as well as heating and ventilation systems, were identified. The year-round thermal modernization measures presented in Tables 6 and 7 are arranged vertically in order of increasing Simple Payback Time from minimum to maximum, and their cumulative effect is arranged horizontally, marked with a "+" sign. Additional measures for the new TMOs were highlighted (bolded) in Table 8. Various TMOs have been determined, namely:

- TMO I: Infrared heaters and TFAD installation;
- TMO II: Infrared heater installation and TFAD installation + **VAV;**
- TMO III: Infrared heater installation, TFAD installation and VAV + **Window replacement;**
- TMO IV (as TMO III): Infrared heater installation, TFAD installation, VAV and window replacement (no measures added);

- TMO V: Infrared heaters installation, TFAD installation, VAV and window replacement + **Wall thermal insulation**
- TMO VI (as TMO V): Infrared heater installation, TFAD installation, VAV, window replacement and wall thermal insulation (no measures added).

Table 8. Optimization of combinations of STMMs.

YTMM	Thermal Modernization Options (TMO)					
	I	II	III	IV	V	VI
Infrared heaters and TFAD installation	+	+	+	+	+	+
Infrared heater installation and **VAV**		+	+	+	+	+
Window replacement and TFAD installation			+	+	+	+
Window replacement and VAV				+	+	+
Wall thermal insulation and TFAD installation					+	+
Wall thermal insulation and VAV						+
Indicators						
Investment cost I, EUR	157	357	2163	2163	4163	4163
Annual savings K, EUR	334	458	738	738	990	990
SPBT, year	0.5	0.8	2.9	2.9	4.2	4.2
NPVR, EUR	1422	1806	1324	1324	516	516
IRR, %	26%	22%	11%	11%	9%	9%

Table 8 presents the results of the annual savings and investment profitability indices, which were calculated based on the aforementioned methods (see Section 2.1) and the optimization methodology described in [2,5].

The NPVR for the option with wall insulation had the lowest value of all optimization combinations due to the high investment costs. The IRR value for this investment was 9%, above the recommended value of 8%, which means that despite the high expenses, the investment is still profitable.

The economic indicators resulting from the TMO search (Table 8) show that the most profitable annual thermal modernization option is the aggregation of the installation of infrared heater and a two-flow air diffuser (TFAD) with variable mode of the ventilation system. The economic effect in terms of NPVR is the highest, due to energy saving of this optimal thermal modernization option over the normative 15 years.

This type of analysis is particularly significant nowadays when society is focused on environmentally friendly solutions, the prices of energy carriers change dramatically and decision processes require the consideration of many options in order to find the most optimal solution. The method described makes it easy to determine the benefits of the improvements applied for the entire year, taking into account both the cold and warm seasons. The method is based on popular and well-known indicators, and is easy to implement. This tool does not require a large amount of data, and allows for the elimination of unacceptable solutions that are not subject to detailed and more complex economic analyses.

The application of more complex methods [76], like whole-year energy simulations in the evaluation of building performance combined with complex economic evaluations of the investment, requires the involvement of additional energy simulation specialists and economic analysts, thus affecting the cost of such analyses and, ultimately, the proposed solution.

4. Conclusions

In many studies on lowering energy demand in buildings, thermal modernization measures are considered seasonally. For the cold season, these activities focus mainly on the building envelope, the heating and DHW systems, and heat sources, while in the warm season, the focus is on the ventilation system.

This work presents a new approach that allows users to determine, by the highest value of the NPVR, the economic efficiency of thermomodernization options as aggregated modernization activities, not only in the cold and warm seasons separately, but throughout the whole year. Our analysis eliminates thermomodernization options that are unacceptable or uneconomic, or not subjected to detailed sensitivity analyses.

This analysis was carried out in order to examine a new approach and to determine the optimal thermal modernization option based on a comprehensive consideration of the thermal modernization measures for the production building. Initially, seasonal thermal modernization measures for the cold and warm seasons were proposed separately. The options included wall insulation, window replacement, the installation of infrared heater, the installation of a two-flow air diffuser (TFAD) and VAV. A combination of measures for the warm and cold seasons, which create year-round thermal modernization measures, was proposed. Then, yearly cost savings and energy savings were calculated. The most profitable options were window replacement and the installation of VAV in the ventilation system by application of an integrated actuator, resulting in annual savings of 21.111 GJ of energy and 376 EUR. However, both radiant heaters and TFADs are nowhere near as investment cost-consuming as window replacement and wall thermal insulation, so an economic analysis was utilized. The economic analysis of the thermal modernization options followed the methods recommended by the United Nations Industrial Development Organization. Finally, the undertaken calculations led to the development of the optimal option in terms of the highest NPVR value. In the analyzed case, in order to decrease costs and energy usage, infrared heaters and the installation of a TFAD with a variable air volume ventilation system were recommended. Simultaneously, the Simple Payback Time for this option was less than one year. Additionally, the installation of a TFAD brings benefits in terms of economics and comfort for building users.

In future research, it is suggested that thermal modernization measures to reduce energy consumption for cooling buildings should be investigated. This is especially important in industrial facilities where heat gains in summer can be significant. Predicting energy consumption for different geographical zones and different building types under the proposed approach can also be further explored. Thus, it is necessary to take a year-round approach and define year-round indicators for evaluating thermal modernization investment in respective geographical zones.

Author Contributions: Conceptualization, O.V. and E.D.; methodology, I.A.; software, O.V. and I.A.; formal analysis, O.S.; investigation, I.S.; resources, I.A.; data curation, O.V.; writing—original draft preparation, N.S., O.V., E.D. and M.L.; writing—review and editing, E.D., M.L., O.V. and N.S.; supervision, O.V.; funding acquisition, O.V, E.D., M.L. and N.S. All authors have read and agreed to the published version of the manuscript.

Funding: This research received no external funding.

Data Availability Statement: The original contributions presented in the study are included in the article, further inquiries can be directed to the corresponding authors.

Conflicts of Interest: The authors declare no conflicts of interest.

References

1. Cao, Y.; Yang, J.; Li, J. Energy-Saving Research on Residential Gas Heating System in Cold Area Based on System Dynamics. *Int. J. Heat Technol.* **2020**, *38*, 457–462. [CrossRef]
2. Haq, H.; Valisuo, P.; Kumpulainen, L.; Tuomi, V. An Economic Study of Combined Heat and Power Plants in District Heat Production. *Clean. Eng. Technol.* **2020**, *1*, 100018. [CrossRef]
3. Yosifova, V.; Chikurtev, D.; Petrov, R. Research and Analysis of Modern Space Heating Technologies and Management for Industrial Buildings. *IOP Conf. Ser. Mater. Sci. Eng.* **2020**, *878*, 012010. [CrossRef]
4. Jouhara, H.; Khordehgah, N.; Almahmoud, S.; Delpech, B.; Chauhan, A.; Tassou, S.A. Waste Heat Recovery Technologies and Applications. *Therm. Sci. Eng. Prog.* **2018**, *6*, 268–289. [CrossRef]

5. Kaplun, V.; Shtepa, V.; Makarevych, S. Neural Network Modelling of Intelligent Energy Efficiency Control in Local Polygeneration Microgrid with Renewable Sources. In Proceedings of the 2020 IEEE KhPI Week on Advanced Technology (KhPIWeek), Kharkiv, Ukraine, 5–10 October 2020; IEEE: Kharkiv, Ukraine, 2020; pp. 98–102. [CrossRef]

6. Paraschiv, S.; Bărbuţă-Mişu, N.; Paraschiv, L.S. Technical and Economic Analysis of a Solar Air Heating System Integration in a Residential Building Wall to Increase Energy Efficiency by Solar Heat Gain and Thermal Insulation. *Energy Rep.* **2020**, *6*, 459–474. [CrossRef]

7. Ulewicz, M.; Zhelykh, V.; Furdas, Y.; Kozak, K. Assessment of the Economic Feasibility of Using Alternative Energy Sources in Ukraine. In Proceedings of the EcoComfort 2020. Lecture Notes in Civil Engineering; Springer: Cham, Switzerland, 2021; pp. 482–489. [CrossRef]

8. Lis, A. Renewable Energy Sources and Rationalisation of Energy Consumption in Buildings as a Way to Reduce Environmental Pollution. *Heat. Vent. Sanit.* **2020**, *6*, 332–339.

9. Auffenberg, F.; Snow, S.; Stein, S.; Rogers, A. A Comfort-Based Approach to Smart Heating and Air Conditioning. *ACM Trans. Intell. Syst. Technol.* **2018**, *9*, 1–20. [CrossRef]

10. Kapalo, P.; Vilčeková, S.; Mečiarová, Ľ.; Domnita, F.; Adamski, M. Influence of Indoor Climate on Employees in Office Buildings—A Case Study. *Sustainability* **2020**, *12*, 5569. [CrossRef]

11. Kapalo, P.; Sulewska, M.; Adamski, M. Examining the Interdependence of the Various Parameters of Indoor Air. In *Proceedings of the EcoComfort 2020. Lecture Notes in Civil Engineering*; Springer: Cham, Switzerland, 2021; pp. 150–157. [CrossRef]

12. Teleszewski, T.; Gładyszewska-Fiedoruk, K. Changes of Carbon Dioxide Concentrations in Classrooms: Simplified Model and Experimental Verification. *Polish J. Environ. Stud.* **2018**, *27*, 2397–2403. [CrossRef]

13. Orman, Ł.J.; Honus, S.; Jastrzębska, P. Investigation of Thermal Comfort in the Intelligent Building in Winter Conditions. *Rocz. Ochr. Sr.* **2023**, *25*, 45–54. [CrossRef]

14. Latosińska, J.; Gawdzik, J.; Honus, S.; Orman, Ł.J.; Radek, N. Waste for Building Material Production as a Method of Reducing Environmental Load and Energy Recovery. *Front. Energy Res.* **2023**, *11*, 1279337. [CrossRef]

15. Turski, M. Eco-Development Aspect in Modernization of Industrial System. *E3S Web Conf.* **2018**, *44*, 00181. [CrossRef]

16. Voznyak, O.; Spodyniuk, N.; Savchenko, O.; Dovbush, O.; Kasynets, M.; Datsko, O. Analysis of Premise Infrared Heating and Ventilation with an Exhaust Outlet and Flat Decking Air Flow. *Diagnostyka* **2022**, *23*, 2022207. [CrossRef]

17. Zender–Świercz, E. A Review of Heat Recovery in Ventilation. *Energies* **2021**, *14*, 1759. [CrossRef]

18. Ratajczak, K.; Michalak, K.; Narojczyk, M.; Amanowicz, Ł. Real Domestic Hot Water Consumption in Residential Buildings and Its Impact on Buildings' Energy Performance—Case Study in Poland. *Energies* **2021**, *14*, 5010. [CrossRef]

19. Starzec, M.; Kordana-Obuch, S.; Piotrowska, B. Evaluation of the Suitability of Using Artificial Neural Networks in Assessing the Effectiveness of Greywater Heat Exchangers. *Sustainability* **2024**, *16*, 2790. [CrossRef]

20. Ołtarzewska, A.; Krawczyk, D.A. Analysis of the Influence of Selected Factors on Heating Costs and Pollutant Emissions in a Cold Climate Based on the Example of a Service Building Located in Bialystok. *Energies* **2022**, *15*, 9111. [CrossRef]

21. Fedorczak-Cisak, M.; Knap, K.; Kowalska-Koczwara, A.; Pachla, F.; Pekarchuk, O. Energy and Cost Analysis of Adapting an Existing Building to 2017 Technical Requirements and Requirements for NZEB. *IOP Conf. Ser. Mater. Sci. Eng.* **2019**, *471*, 112094. [CrossRef]

22. Savchenko, O.; Lis, A. Economic Indicators of a Heating System of a Building in Ukraine and Poland. *Bud. Zoptymalizowanym Potencjale Energetycznym* **2020**, *10*, 89–94. [CrossRef]

23. Basińska, M.; Koczyk, H. Doświadczenia Po Przeprowadzonej Modernizacji Obiektu Do Standardu Pasywnego. *Czas. Tech. Bud.* **2009**, *106*, 3–10.

24. Orman, Ł.J.; Krawczyk, N.; Radek, N.; Honus, S.; Pietraszek, J.; Dębska, L.; Dudek, A.; Kalinowski, A. Comparative Analysis of Indoor Environmental Quality and Self-Reported Productivity in Intelligent and Traditional Buildings. *Energies* **2023**, *16*, 6663. [CrossRef]

25. Rosolski, S. New Technical Solutions in Upgrading Historic Building To Standard of Nearly Zero-Energy Building. Example of Tenement House in Poznań. *Space Form* **2021**, *2021*, 333–352. [CrossRef]

26. Hurnik, M.; Ferdyn-Grygierek, J.; Kaczmarczyk, J.; Koper, P. Thermal Diagnosis of Ventilation and Cooling Systems in a Sports Hall—A Case Study. *Buildings* **2023**, *13*, 1185. [CrossRef]

27. Popiolek, Z.; Kateusz, P. Comprehensive on Site Thermal Diagnostics of Buildings—Polish Practical Experience. *Arch. Civ. Eng.* **2017**, *2*, 125–132. [CrossRef]

28. Stypka, T.; Flaga-Maryańczyk, A. Możliwości Stosowania Zmodyfikowanej Metody AHP w Problemach Inżynierii Środowiska. *Ekon. Sr.* **2016**, *2*, 15. Available online: https://yadda.icm.edu.pl/baztech/element/bwmeta1.element.baztech-0309dd2c-5377-4 2d4-9510-d0332fe11301 (accessed on 3 January 2024).

29. Valancius, R.; Jurelionis, A.; Dorosevas, V. Method for Cost-Benefit Analysis of Improved Indoor Climate Conditions and Reduced Energy Consumption in Office Buildings. *Energies* **2013**, *6*, 4591–4606. [CrossRef]

30. Chinese, D.; Nardin, G.; Saro, O. Multi-Criteria Analysis for the Selection of Space Heating Systems in an Industrial Building. *Energy* **2011**, *36*, 556–565. [CrossRef]

31. Siudek, A.; Klepacka, A.M.; Florkowski, W.J.; Gradziuk, P. Renewable Energy Utilization in Rural Residential Housing: Economic and Environmental Facets. *Energies* **2020**, *13*, 6637. [CrossRef]

32. Lopes, J.; Oliveira, R.; Banaitiene, N.; Banaitis, A. A Staged Approach for Energy Retrofitting an Old Service Building: A Cost-Optimal Assessment. *Energies* **2021**, *14*, 6929. [CrossRef]
33. Biserni, C.; Valdiserri, P.; D'Orazio, D.; Garai, M. Energy Retrofitting Strategies and Economic Assessments: The Case Study of a Residential Complex Using Utility Bills. *Energies* **2018**, *11*, 2055. [CrossRef]
34. Mahdi, A.M. Energy Audit a Step to Effective Energy Management. *Int. J. Trend Res. Dev.* **2018**, *5*, 521–525.
35. Gokul, M.; Ponraj, S.; Rohan, S.; Venkatesh, M.; Prince Winston, D.; Praveen Kumar, B. Energy Audit for Industries and Institution. *Int. J. Adv. Res. Manag. Archit. Technol. Eng.* **2017**, *3*, 199–205.
36. Zarzeka-Raczkowska, E.; Raczkowski, A.; Pomorski, P. Technical and Economic Analysis of Thermal Modernisation Process of a Detached House. In *Proceedings of the Modern Materials, Installations and Constructions Technologies*; Pope John Paul II State School Heigher Education in Biała Podlaska: Biała Podlaska, Poland, 2013; pp. 22–27.
37. Szałański, P.; Kowalski, P.; Cepiński, W.; Kęskiewicz, P. The Effect of Lowering Indoor Air Temperature on the Reduction in Energy Consumption and CO_2 Emission in Multifamily Buildings in Poland. *Sustainability* **2023**, *15*, 12097. [CrossRef]
38. Krawczyk, D.A. The Optimum Variants of Warming up Walls, Roof, Windows Change and a Heating System Modernization in the Typical School according to Its Localization. *Instal* **2004**, *10*, 28–31.
39. Przesmycka, N.; Strojny, R.; Życzyńska, A. Modernisation of Hospital Buildings Built in the 20th Century in the Context of Architectural, Functional and Operational Problems. *Architectus* **2023**, *74*, 59–67. [CrossRef]
40. Krawczyk, D.A. Theoretical and Real Effect of the School's Thermal Modernization—A Case Study. *Energy Build.* **2014**, *81*, 30–37. [CrossRef]
41. Gładyszewska-Fiedoruk, K.; Krawczyk, D.A. The Possibilities of Energy Consumption Reduction and a Maintenance of Indoor Air Quality in Doctor's Offices Located in North-Eastern Poland. *Energy Build.* **2014**, *85*, 235–245. [CrossRef]
42. Lis, A.; Spodyniuk, N. The Quality of the Microclimate in Educational Buildings Subjected to Thermal Modernization. *E3S Web Conf.* **2019**, *100*, 00048. [CrossRef]
43. Bøhm, B. Production and Distribution of Domestic Hot Water in Selected Danish Apartment Buildings and Institutions. Analysis of Consumption, Energy Efficiency and the Significance for Energy Design Requirements of Buildings. *Energy Convers. Manag.* **2013**, *67*, 152–159. [CrossRef]
44. Hałacz, J.; Skotnicka-Siepsiak, A.; Neugebauer, M. Assessment of Reducing Pollutant Emissions in Selected Heating and Ventilation Systems in Single-Family Houses. *Energies* **2020**, *13*, 1224. [CrossRef]
45. Ferdyn-Grygierek, J.; Grygierek, K. HVAC Control Methods for Drastically Improved Hygrothermal Museum Microclimates in Warm Season. *Build. Environ.* **2019**, *149*, 90–99. [CrossRef]
46. Ratajczak, K.; Amanowicz, Ł.; Szczechowiak, E. Assessment of the Air Streams Mixing in Wall-Type Heat Recovery Units for Ventilation of Existing and Refurbishing Buildings toward Low Energy Buildings. *Energy Build.* **2020**, *227*, 110427. [CrossRef]
47. Zender-Świercz, E.; Piotrowski, J. Thermomodernization a Building and Its Impact on the Indoor Microclimate. *Struct. Environ.* **2013**, *5*, 37–40.
48. Tomczak, K.; Kinash, O. Assessment of the Validity of Investing in Energy-Efficient Single-Family Construction in Poland—Case Study. *Arch. Civ. Eng.* **2016**, *62*, 119–138. [CrossRef]
49. Voznyak, O.T.; Sukholova, I.Y.; Savchenko, O.O.; Dovbush, O.M. Termorenovatsiya Systemy Kondytsionuvannia Povitria Vyrobnychykh Prymishchen (Thermal Renewal of the Air Conditioning System of Industrial Premises). *Visnyk Odes'koyi Derzhavnoyi Akad. Budivn. Arhit.* **2017**, *68*, 114–120. (In Ukrainian)
50. Lei, L.; Shao, S. Prediction Model of the Large Commercial Building Cooling Loads Based on Rough Set and Deep Extreme Learning Machine. *J. Build. Eng.* **2023**, *80*, 107958. [CrossRef]
51. Lei, L.; Shao, S.; Liang, L. An Evolutionary Deep Learning Model Based on EWKM, Random Forest Algorithm, SSA and BiLSTM for Building Energy Consumption Prediction. *Energy* **2024**, *288*, 129795. [CrossRef]
52. The United Nations Industrial Development Organization (UNIDO). Available online: https://www.unido.it/about.php (accessed on 3 January 2024).
53. Spodyniuk, N.; Voznyak, O.; Savchenko, O.; Sukholova, I.; Kasynets, M. Optimization of Heating Efficiency of Buildings above Underground Coal Mines by Infrared Heaters. *Nauk. Visnyk Natsionalnoho Hirnychoho Universytetu* **2022**, *3*, 100–106. [CrossRef]
54. Knapik, M. Analiza Zakresu Termomodernizacji Oraz Wykorzystania Energii Odnawialnych Dla Budynku Mieszkalnego w Celu Przekształcenia Go w Obiekt Nisko-Energochłonny, Politechnika Krakowska, Kraków. 2019. Available online: https://repozytorium.biblos.pk.edu.pl/resources/41737 (accessed on 1 February 2024).
55. Available online: https://www.britannica.com/place/Ukraine (accessed on 1 February 2024).
56. *DBN V.2.6-31:2021 Standard*; Thermal Insulation and Energy Efficiency of Buildings. SE National Research Institute of Building Construction (NDIBK), 2021. Available online: https://online.budstandart.com/ua/catalog/doc-page.html?id_doc=98037 (accessed on 1 February 2024).
57. Sarbu, I.; Tokar, A. Numerical Modelling of High-Temperature Radiant Panel Heating System for an Industrial Hall. *Int. J. Adv. Appl. Sci.* **2018**, *5*, 1–9. [CrossRef]
58. Yosifova, V.; Chikurtev, D. Development of Module System for Intelligent Control of Infrared Heating. *AIP Conf. Proc.* **2022**, *2449*, 020006. [CrossRef]
59. Kebede, T.; Esakki, B. Effect of Various Infrared Heaters on Sintering Behavior of UHMWPE in Selective Inhibition Sintering Process. *Int. J. Eng. Adv. Technol.* **2019**, *8*, 725–731. [CrossRef]

60. Samek, L.; De Maeyer-Worobiec, A.; Spolnik, Z.; Kontozova, V. The Impact of Electric Overhead Radiant Heating on the Indoor Environment of Historic Churches. *J. Cult. Herit.* **2007**, *8*, 361–369. [CrossRef]

61. Vidyarthi, S.K.; El Mashad, H.M.; Khir, R.; Zhang, R.; Tiwari, R.; Pan, Z. Evaluation of Selected Electric Infrared Emitters for Tomato Peeling. *Biosyst. Eng.* **2019**, *184*, 90–100. [CrossRef]

62. Yadav, G.; Gupta, N.; Sood, M.; Anjum, N.; Chib, A. Infrared Heating and Its Application in Food Processing. *Pharma Innov. J.* **2020**, *9*, 142–151.

63. Kordana, S. The Identification of Key Factors Determining the Sustainability of Stormwater Systems. *E3S Web Conf.* **2018**, *45*, 00033. [CrossRef]

64. Cepiński, W.; Szałański, P. Increasing the Efficiency of Split Type Air Conditioners/Heat Pumps by Using Ventilating Exhaust Air. *E3S Web Conf.* **2019**, *100*, 00006. [CrossRef]

65. Voznyak, O.; Myroniuk, K.; Sukholova, I.; Kapalo, P. The Impact of Air Flows on the Environment. In *Proceedings of the CEE 2019. Lecture Notes in Civil Engineering*; Springer: Cham, Switzerland, 2020; Volume 47, pp. 534–540. [CrossRef]

66. Voznyak, O.; Spodyniuk, N.; Sukholova, I.; Savchenko, O.; Kasynets, M.; Datsko, O. Diagnosis of Three Types Damages to the Ventilation System. *Diagnostyka* **2022**, *23*, 2022102. [CrossRef]

67. Voznyak, O.; Savchenko, O.; Spodyniuk, N.; Sukholova, I.; Kasynets, M.; Dovbush, O. Improving of Ventilation Efficiency at Air Distribution by the Swirled Air Jets. *Pollack Period.* **2022**, *17*, 123–127. [CrossRef]

68. Kordana, S.; Słyś, D. Decision Criteria for the Development of Stormwater Management Systems in Poland. *Resources* **2020**, *9*, 20. [CrossRef]

69. Wojtacha-Rychter, K.; Smoliński, A. Study of the Hazard of Endogenous Fires in Coal Mines—A Chemometric Approach. *Energies* **2018**, *11*, 3047. [CrossRef]

70. Osypenko, V.; Kaplun, V. Inverse Conversion of Transition Matrices Method for Polygeneration Microgrid Dynamic Electricity Cost Prediction. In Proceedings of the 2022 IEEE 17th International Conference on Computer Sciences and Information Technologies (CSIT), Lviv, Ukraine, 10–12 November 2022; pp. 568–571. [CrossRef]

71. Zhang, H.; Li, K.; Zhao, L.; Jia, L.; Kaita, M.; Wan, F. Modeling of a Wellhead Heating Methodology With Heat Pipes in Coal Mines. *ASME J. Energy Resour. Technol.* **2021**, *143*, 010907. [CrossRef]

72. Basok, B.; Davydenko, B.; Pavlenko, A.M. Numerical Network Modeling of Heat and Moisture Transfer through Capillary-Porous Building Materials. *Materials* **2021**, *14*, 1819. [CrossRef] [PubMed]

73. Gorobets, V.G.; Trokhaniak, V.I.; Rogovskii, I.L.; Titova, L.L.; Lendiel, T.I.; Dudnyk, A.O.; Masiuk, M.Y. The Numerical Simulation of Hydrodynamics and Mass Transfer Processes for Ventilating System Effective Location. *INMATEH Agric. Eng.* **2018**, *56*, 185–192.

74. Trokhaniak, V.I.; Rogovskii, I.L.; Titova, L.L.; Popyk, P.S.; Bannyi, O.O.; Luzan, P.H. Computational Fluid Dynamics Investigation of Heat-Exchangers for Various Air-Cooling Systems in Poultry Houses. *Bull. Karaganda Univ. Phys. Ser.* **2020**, *1*, 125–134. [CrossRef]

75. Lei, L.; Wu, B.; Fang, X.; Chen, L.; Wu, H.; Liu, W. A Dynamic Anomaly Detection Method of Building Energy Consumption Based on Data Mining Technology. *Energy* **2023**, *263*, 125575. [CrossRef]

76. Kalina, J.; Pohl, W. Technical and Economic Analysis of a Multicarrier Building Energy Hub Concept with Heating Loads at Different Temperature Levels. *Energy* **2024**, *288*, 129882. [CrossRef]

Article

Comparative Analysis of Subjective Indoor Environment Assessment in Actual and Simulated Conditions

Łukasz Jan Orman [1,*], Natalia Siwczuk [1], Norbert Radek [2], Stanislav Honus [3], Jerzy Zbigniew Piotrowski [1] and Luiza Dębska [1]

1   Faculty of Environmental Engineering, Geodesy and Renewable Energy, Kielce University of Technology, al. Tysiaclecia P.P. 7, 25-314 Kielce, Poland; nkrawczyk@tu.kielce.pl (N.S.); piotrowski@tu.kielce.pl (J.Z.P.); ldebska@tu.kielce.pl (L.D.)
2   Faculty of Mechatronics and Mechanical Engineering, Kielce University of Technology, al. Tysiaclecia P.P. 7, 25-314 Kielce, Poland; norrad@tu.kielce.pl
3   Faculty of Mechanical Engineering, VSB-Technical University of Ostrava, 17. Listopadu 2172/15, 708 00 Ostrava, Czech Republic; stanislav.honus@vsb.cz
*   Correspondence: orman@tu.kielce.pl

Abstract: This paper experimentally analyses an indoor environment assessment of a large group of respondents regarding their subjective perception of overall comfort, indoor air quality and humidity. The questionnaire survey was applied as a testing method together with measurements of the physical parameters conducted with a microclimate meter. Two types of environment were analysed: educational rooms and the climate chamber. The comparative analysis of the sensations experienced within them indicates that they generate quite similar responses; however, some discrepancies have been identified. The overall comfort of the climate chamber was typically assessed as being higher than that of the educational rooms at the same air temperature. The most favourable air temperature in the climate chamber was ca. 20.7 °C, while in the educational rooms it was ca. 22.3 °C. The most preferable conditions in the climate chamber occurred at a thermal sensation vote of −0.4 ("pleasantly slightly cool"), while in the educational rooms it occurred at +0.2 ("neutral/pleasantly slightly warm"). Quite strong correlations between overall comfort and indoor air quality as well as between humidity assessment and humidity preference votes were observed, which did not seem to depend on the type of environment. These findings are important because results from the simulated conditions are often used in the analyses of actual living/working environments.

Keywords: indoor environment; microclimate; questionnaire study

Citation: Orman, Ł.J.; Siwczuk, N.; Radek, N.; Honus, S.; Piotrowski, J.Z.; Dębska, L. Comparative Analysis of Subjective Indoor Environment Assessment in Actual and Simulated Conditions. *Energies* **2024**, *17*, 656. https://doi.org/10.3390/en17030656

Academic Editors: Katarzyna Ratajczak and Łukasz Amanowicz

Received: 30 December 2023
Revised: 24 January 2024
Accepted: 26 January 2024
Published: 30 January 2024

1. Introduction

The building sector is considered to be a significant consumer of energy. The operation of buildings—including services such as heating, ventilation and air conditioning (HVAC systems)—accounts for 30% of global total energy consumption [1]. Moreover, nowadays people tend to spend up to ca. 90% of their time indoors [2,3] and expect satisfactory indoor environmental conditions that would ensure the overall comfort and well-being of room users. However, this might require energy input for proper air treatment processes. Finding a solution to providing acceptable indoor environments for the people while keeping the building's energy demand low is an issue of utmost importance.

An indoor environment assessment is performed in order to determine human subjective responses. These responses depend on a number of physical parameters within rooms such as ambient and radiation temperature, air movement, relative humidity, carbon dioxide concentration, etc. [4], as well as outdoor air parameters [5]. However, there are also variables that can affect differences in the perception of individuals present in the same room. People from different backgrounds and locations can experience indoor environments differently [6,7]. Montazami et al. [8] revealed a relation between children's socio-economic background and their perception of thermal comfort in classrooms. The age of the respondents also plays a role. According to [9], older people have different comfort needs in relation to their younger counterparts, but the differences are not so big. In [10], it was reported that younger people (below 25 y.o.) expressed comfort at a slightly but not significantly higher temperature than older ones. Indraganti and Rao [11] observed a considerable but poor correlation between age and thermal sensation as well as overall comfort. On the other hand, an important factor affecting subjective human responses is gender. It proved to be a significant aspect in the thermal comfort study [7] with female respondents having a narrower neutral temperature range and adjusting their clothing level more swiftly as temperature changes. The study [11] revealed that women showed slightly higher thermal sensation, with a preference for a warmer environment. At the same time their thermal acceptance was higher than that in men. Statistically significant gender differences were confirmed in [10], regarding comfort temperature, thermal acceptability as well as regarding the use of controls such as windows. Karjalainen [12] reported that women were less satisfied with thermal environments than men and felt both cold and hot more often than men. According to [13], thermal comfort temperature was significantly higher for women than for men (overall 24.0 °C and 23.2 °C, respectively). A study conducted by Parsons [14] for identical clothing and activity confirmed the differences in thermal responses for both genders. The author concluded that women tend to be cooler than men in cool conditions. Moreover, Luo et al. [15] pointed out that pregnancy and menopause status could influence thermal sensations in rooms. Additionally, individual preferences can have an impact on the subjective assessment. It needs to be noted, however, that many of the above-mentioned factors might be of less importance to the final results when studies are conducted with a large number of participants.

The studies of indoor environment assessment are conducted either in actual rooms of residential, office or public utility buildings (e.g., field—based studies) or in simulated conditions of climate chambers.

Fang et al. [16] carried out field-based experiments into thermal comfort in classrooms of the City University of Hong Kong. The comfort range was found to be within the temperature values of 21.56–26.75 °C. A study [17] of thermal indoor environments in Italian university classrooms showed that the students expressed "very satisfactory" thermal responses for the air temperature values of 20.9–26.4 °C. However, in a later paper [18] the same authors reported different thermal sensation votes for various classrooms. Krawczyk et al. [19] focused on comfort perceptions expressed in the questionnaires accompanied by temperature and humidity measurements at Polish and Spanish universities. It was reported that the optimal temperature values depended on the location (for Polish students it was 21.7–22.3 °C, while for the Spanish students it was 23.3–24.8 °C). Similarly, Sakellaris et al. [20] confirmed that relations between indoor environment and users' comfort depend upon the socio-cultural context, as well as personal and building features. Their digital questionnaire survey covered over 7400 workers in almost 170 office buildings in eight European countries. Kim et al. [21] carried out field-based tests in a university building in the United Arab Emirates. With regards to the overall indoor environmental quality, over half of the respondents felt uncomfortable and the remaining ones felt neutral. These unfavourable sensations resulted mostly from cold and stuffy air conditions as well as draught. Altomonte et al. [22] indicated that there are some crucial physical factors that influence overall comfort (namely lighting conditions, temperature, sound and air quality), but other (non-physical) factors can play a role, too. Moreover, Göçer et al. [23] reported that indoor and outdoor noise, restricted access to daylight as well as to the outside view, building/work aesthetics and personal control over building systems might have an impact on the occupants' satisfaction. The overall satisfaction can also be influenced by indoor air quality, which is usually associated with the carbon dioxide concentration. Thus, guidelines regarding limiting values of CO_2 are proposed [24,25]. Vilcekova et al. [26] studied indoor air quality in Slovak primary school classrooms. It was reported that some 53% of the students rated the air quality as acceptable while almost 73% rated it as stuffy. Furthermore, they regarded poor air quality as the most significant problem. Aguilar et al. [27] indicated the presence of poor air quality in Spanish university classrooms, where the carbon dioxide level exceeded the allowable limit. However, Kim et al. [21] reported that even at a low CO_2 concentration, the respondents assessed the indoor air quality as low (with 63% of the respondents feeling uncomfortable). A lack of operable windows might have been a source of this problem. Similarly, inadequate ventilation proved to be a problem (to be addressed using different techniques [28–31]); however, it is not so prevalent and/or serious, especially in modern intelligent buildings [32] or even in public utility buildings located in a tropical climate [33]. Despite the wealth of research results supporting the claim that air quality is related to the carbon dioxide concentration, Clements et al. [34] reported poor user satisfaction despite low carbon dioxide concentration, which might have been caused by the absence of natural light and the presence of noise, rather than a consequence of air quality degradation. Similarly, people who studied at an elevated temperature of 30 °C considered the air quality to be worse in relation to a thermal environment of 22 °C, as reported in [35].

The tests in simulated conditions are typically conducted for much smaller numbers of participants. Geng et al. [36] carried out research in a 35 m² room at Tsinghua University in Beijing (China) with an adjusted indoor atmosphere, in which 21 people participated. The biggest overall dissatisfaction was observed at the air temperature of 16.2 °C (about 70% were dissatisfied), while the air temperature of 24 °C proved to be most favourable. The study also showed that air quality satisfaction was as a result of the air temperature (with the biggest satisfaction observed for the lowest air temperature of 16.2 °C and the biggest dissatisfaction observed for 22 °C and 24 °C), which supports the claims presented in [34,35] regarding the fact that CO_2 levels might not be the only source of poor-quality sensations.

Climate chamber experiments enable experiments to be carried out in a broad range of indoor environmental parameters. Yang et al. [37] conducted tests in a climate chamber in China and collected 440 thermal responses from eighty students (22 thermal conditions were created in the chamber with an ambient temperature that varied from 26 to 32 °C and a relative humidity that varied from 40 to 90%). The authors proved the inaccuracy of the selected thermal comfort model. Ji et al. [38] performed thermal comfort tests in a climate chamber in order to determine how the previous thermal environment may influence the present sensations. The authors claimed that humans' evaluations may be considered a combination of both past and present feelings. Zhang et al. [39] studied the sensations of 60 people in the climate chamber at temperatures of 20–32 °C and humidity from 50% to 70% in order to determine if there is any difference between the responses of the respondents from buildings equipped with a centralised air conditioning system (half of the group) and with a split-type air conditioner (the other half of the group). The results showed that there were differences in the mean skin temperature values, but no significant differences were observed for other physiological responses. Similarly, both groups reported the same neutral temperatures (almost 27 °C). Soebarto et al. [40] compared the thermal responses of 22 older and 20 younger people. Four test conditions were set in the environmental chamber. The authors found no considerable differences between the thermal sensations and the comfort of older vs. younger subjects. Chen et al. [41] used a climatic chamber for investigations of the thermal responses of 26 elderly Nordic people. It was reported that the neutral temperature of these elderly respondents from Northern Europe was 26 °C, while their preferred temperature was 26.5 °C. Jian et al. [42] studied the thermal sensations of 29 people from China in a climate chamber. The authors reported that the sensations changed even after the ambient temperature became stable, which is an important finding that should be considered during the analysis of climate chamber test results. Ahmad et al. [43] conducted tests in a climate chamber with temperatures in the range of 19–29 °C. An air temperature of 23 °C proved to be most satisfactory. The authors found a linear correlation between overall thermal comfort and thermal sensations. Dong et al. [44] analysed the thermal responses of 22-year-old migrants in China in the climate chamber.

The air temperature varied from 18 to 33 °C, while relative humidity varied from 42 to 52%. The subjective thermal sensation vote ranged from −2.8 (at 18 °C) to +2.7 (at 33 °C). The most favourable value of 0.1 was obtained at 26 °C. In the study, thermal comfort was also tested as well as skin temperature as a marker of thermal adaptation. In [45], climate chamber tests on 20 walking men were presented. The operative temperature ranged from 23 to 29 °C, while the air velocity ranged from 0.05 to 0.82 m/s. Increased velocity values led to improved thermal comfort in the simulated summer conditions. Very recently, Upadhyay et al. [46] examined the influence of relative humidity and air velocity on thermal sensation, comfort and preferences in a sub-tropical climate. The tests were carried out in a controlled climate chamber. Sixteen people took part in it, while each underwent one hundred and forty experimental sets covering the temperature range of 20–40 °C, relative humidity of 30–70% and air velocity of 0.25–2.0 m/s. The comfort temperature was found to vary from 25.7 to 32.9 °C. On the other hand, Yang et al. [47] aimed to determine thermal responses at high temperature values (29–40 °C), resembling those encountered in a spinning workshop. Forty students participated in the tests. They were exposed to over a hundred state points in a controlled climate chamber. Temperatures above 38 °C were considered to be unacceptable.

Indoor environment studies are conducted either in actual rooms of residential, office or public utility buildings or in simulated conditions of climate chambers. However, people only live in "real" buildings. Thus, the question arises if data obtained in climate chambers can be used in practice (for example, for the design of heating and air conditioning systems, or for developing reliable models of thermal comfort). There is a question if building standards, which were developed based on climate chamber data, can provide accurate results regarding people's sensations in actual rooms of various buildings. This issue was raised by de Dear [48] who claimed that climate chambers have certain constraints in relation to field-based studies such as sample size, demographics, measurement procedures, questionnaires, clothing resistance, etc. Thus, the test results obtained in climate chamber experiments might have limited application to the development of guidelines and the possibility of reaching a conclusion about the nature of indoor environment assessment by the respondents in actual rooms. What is more, psychological aspects related to the fact that respondents are situated in a closed space of a chamber might also play a role. Consequently, there might be differences between the results obtained in the actual and simulated conditions. Moreover, Sakellaris et al. [20] confirmed that room users' perceptions depend upon their socio-cultural context, as well as personal and building features. Thus, in the present study all the experimental conditions are the same for both environments tested. The current paper is focused on the comparison of human responses regarding the overall comfort, air quality and humidity assessment in these two different environments with the same experimental conditions (regarding measurement methodology, demographics, etc.). Typically, studies are conducted either in actual rooms or in simulated conditions of climate chambers. A direct comparison of the data for these two environments might be practically impossible or at least very challenging due to differences in the study groups and experimental conditions. The present study is free from these limitations. It provides experimental data of a uniform nature for both environments. Such a study has not been found in the literature and the present paper aims to bridge this research gap.

2. Materials and Methods

This study took place in the field-based environment of educational rooms (lecture/classrooms) located in five university buildings in Kielce (central Poland), as well as in the climate chamber, which is also situated in one of the buildings. Figure 1a presents two out of the five buildings, where the tests occurred, while Figure 1b shows the climate chamber (located in the laboratory hall behind the buildings shown in Figure 1a).

(a)   (b)

Figure 1. Locations of the measurements: (**a**) building "Energis" (in the front) and building "A" (in the back) of Kielce University of Technology; (**b**) climate chamber in the laboratory hall of building "A".

The experimental procedure consisted of conducting direct measurements of the physical parameters within each educational room and the climate chamber. The following values were recorded with the use of a microclimate meter: air and globe temperature, air speed, relative humidity and CO_2. At the same time, anonymous questionnaires containing questions about the subjective assessment of sensations regarding the indoor environment were handed over to the occupants who completed them within a few minutes.

The microclimate meter recorded the environment parameters at the rate of one second. However, an analysis of the data was conducted for the values of these parameters as they were at the moment of completing the questionnaires by the respondents. The changes of these parameters within this short period of time were well within the error bands of the measuring system. Table 1 shows the basic technical data of the testing device (a system consisting of the microclimate meter with the probes).

Table 1. Details of the measuring device (according to the manufacturer's data [49]).

No	Parameter	Measuring Range	Measuring Accuracy
1	Air temperature	−20–70 °C	±0.3 °C
2	Air speed	0–5 m/s	±(0.03 m/s + 4% of the obtained value)
3	CO_2 level	0–10,000 ppm	±50 ppm + 3% of the obtained value

The measuring system consisting of the microclimate meter with adequate probes was mounted on the tripod and placed as close as possible to the area where the occupants were seated in the educational rooms (most often in the centre of these rooms) and in the centre of the climate chamber. Figure 2 presents the microclimate meter situated in a lecture room of the "Energis" building.

During this study, the respondents filled in anonymous questionnaires containing questions on their subjective assessment of the environmental conditions within the rooms. In total, 1302 questionnaire forms were collected in 92 educational rooms of the 5 university buildings. The tests in the climate chamber produced 480 questionnaires covering 60 datasets (with 10 values of air temperature and 3 values of relative humidity for 2 types of clothing: summer and winter).

The number of people in each educational room ranged from 10 to 56. The surface area of the rooms varied from 24.5 m^2 to 461.6 m^2, while their cubature varied from 68.5 m^3 to 1430.9 m^3. The details of the geometry of the educational rooms are provided in Table S1 of the Supplementary Material, together with time and weather details at the moment of conducting the measurements. The area of the climate chamber equaled 4.05 m^2, while its cubature was 9.3 m^3. The participants in both of the environmental conditions were students, the average age and BMI (body mass index defined as a ratio of weight in kilograms divided by height in meters squared) of the respondents in the field-based tests were 22.4 y.o. and 23.3 kg/m^2, while in the climate chamber they were 25.8 y.o. and 23.4 kg/m^2, respectively. The average clothing thermal insulation was 0.59 clo in the case of the field-based tests and 0.65 clo in the climate chamber tests. All the respondents were Polish and came from the same cultural background. Thus, the physical parameters were similar and despite a large sample size, there should be no significant differences between the respondents participating in this study in both indoor environments (educational rooms and the climate chamber). The total share of women and men was very similar in the educational rooms and amounted to 49% of women and 51% of men, while in the climate chamber experiments it was 62% and 38%, respectively.

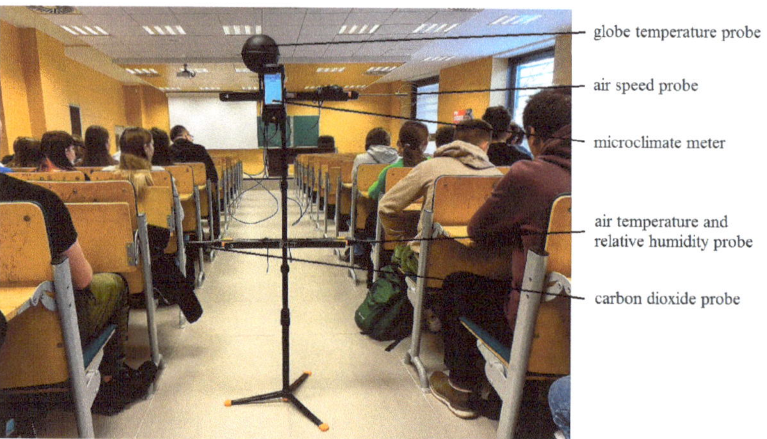

Figure 2. The microclimate meter with the probes.

The selection of the participants was based on their consent to complete the questionnaires and thus take an active part in this study. Random sampling was impossible due to the low total number of students. The study group covered the students typically in their twenties. The sample group in each case was a regular student group, consequently a diverse and proper representation of the total population of this age group was considered to be achieved. The minimal number of people in the group was set at 10 persons, while the maximum amounted to 56. Naturally, the increase in the number of students in each group leads to improved sampling validity, but this is limited these days due to low enrollment. The test results for each group formed a set of two types of data: objective properties of the indoor environment and subjective assessment of the features of this environment (expressed as average values of the answers provided in the questionnaires). Consequently, possible cause–effect relations can be analysed.

The design of the questionnaire form was influenced by the standards observed in [50,51] as well as journal papers [10,52,53].

3. Results and Discussion

During this study, the physical parameters of the indoor environment were recorded simultaneously to the process of completing the questionnaire forms. Air temperature ranged from 20.0 to 29.7 °C in the educational rooms and from 19.0 to 28.3 °C in the climate chamber. The respective ranges of relative humidity in both environments were 19.7–65.8% and 18.6–74.3%.

The subjective assessment considers all the conditions within rooms that affect human beings is commonly referred to as "occupants' satisfaction" [21,23], "overall comfort sensation" [54], "overall indoor environmental comfort" [55] or "overall comfort" (which seems to be most commonly used in the literature, e.g., in [20]). In the present study, the last term will be used in the analysis.

The assessment of the overall comfort of room users was carried out by the respondents providing an answer to the question "How do you rate your overall comfort?" The answers to choose from varied from "very good" (+2) through "good" (+1), "neutral" (0), "bad" (−1) to "very bad" (−2). Figure 3 presents the distribution of the votes (as the frequency count) for the climate chamber and for the educational rooms. The data regarding the educational rooms have been analysed by the authors in [56] separately for the intelligent and traditional buildings.

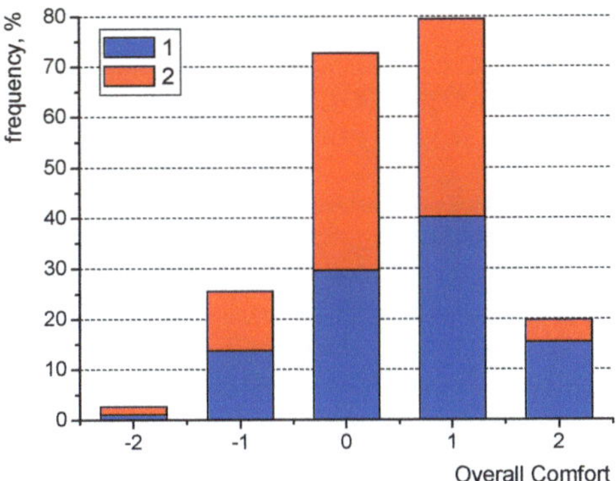

Figure 3. Distribution of Overall Comfort votes in the climate chamber (1) and educational rooms (2).

The majority of the respondents (in both environments) felt fine and marked the "neutral" or "good/very good" answers. The proportion of "good" and "bad" answers were very similar in both room types. The largest difference was observed for the response "+2"—the students felt very good in the climate chamber in comparison with the educational rooms, which might be attributed to some other factors such as stress during the classes. In both cases, the share of very negative responses was marginal. In fact, a similar percentage share of the negative sensations ("−1" and "−2") was recorded for both environments, which might disagree with the findings in [23] that restricted access to daylight as well as to the outside view might have an impact on the occupants' satisfaction.

The overall comfort might depend on a number of factors; however, indoor air temperature seems to be one of the most important parameters. Figure 4 presents the relation between the average values of the overall comfort (OC) for each room in the educational buildings and in the climate chamber, and the air temperature (T).

The most favourable range of air temperature for both environments seems to be from 20.5 to 23 °C, which is in agreement with data presented in [19], where the comfort temperatures for the Polish students ranged from 21.7 to 22.3 °C, as well as in agreement with [43], where the temperature of 23 °C was reported as most satisfactory. However, data for Chinese people of a similar age presented in [26] showed that a higher air temperature value of 24 °C proved to be most favourable. On the other hand, elderly Nordic people expressed an even higher preferred temperature of 26.5 °C [41].

An important finding is that the overall comfort for the climate chamber was typically higher than that for the educational rooms at the same air temperature values. This might be related to the influence of other factors such as the lack of stress in the chamber as opposed to students' participation that is required in a lecture/classrooms.

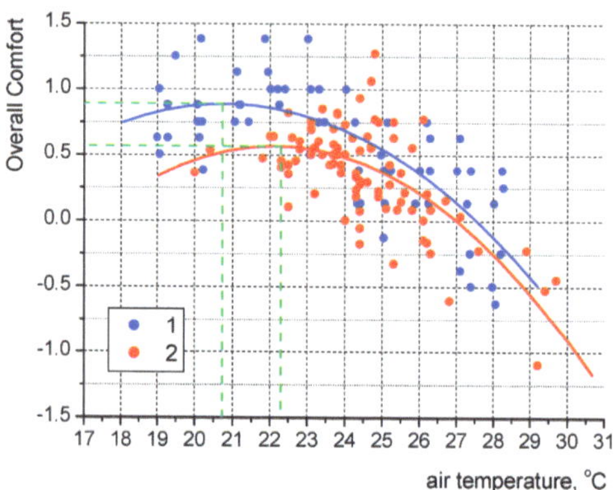

Figure 4. Mean Overall Comfort vs. air temperature—data for 60 sessions in the climate chamber (1) and 92 rooms in the educational buildings (2); blue and red lines: polynomial fits for (1) and (2), respectively; green lines: indicators of maximal OC values.

The most satisfactory air temperature value in the climate chamber experiments was ca. 20.7 °C, while in the educational rooms it was ca. 22.3 °C. This shift can be explained by the presence of increased air movement (often caused by draughts) in the lecture/classrooms, especially in the summer season due to open windows.

The obtained polynomial correlations for the data recorded in the climate chamber (subscript "1") and the educational rooms (subscript "2") are as follows:

$$OC_1 = -0.0194T^2 + 0.8083T - 7.5126 \left(R^2 = 0.57 \right) \tag{1}$$

$$OC_2 = -0.0237T^2 + 1.0494T - 11.045 \left(R^2 = 0.47 \right) \tag{2}$$

Air temperature might indeed have an impact on the overall comfort of room users; however—due to the highly subjective perception of temperature and the individual thermal preferences—it might be vital to study not only the influence of the temperature, but the correlation between the subjective assessment of thermal sensations and overall comfort.

The question in the questionnaire related to thermal sensations was "How do you assess your current thermal sensation". The respondents could choose from the following: "too hot" (+3), "too warm" (+2), "pleasantly warm" (+1), "neutral" (0), "pleasantly cool" (−1), "too cool" (−2), "too cold" (−3). Figure 5 presents the relation between average values of thermal sensations (Thermal Sensation Vote—TSV) and overall comfort (OC) for each of the 92 rooms of the university buildings and 60 sessions in the climate chamber.

The obtained polynomial correlations for the data recorded in the climatic chamber (subscript "1") and the educational rooms (subscript "2") are as follows:

$$OC_1 = -0.1836TSV^2 - 0.1607TSV + 0.8695 \left(R^2 = 0.56 \right) \tag{3}$$

$$OC_2 = -0.2155TSV^2 - 0.0363TSV + 0.5938 \left(R^2 = 0.60 \right) \tag{4}$$

As opposed to the findings of Ahmad et al. [43], who found a linear correlation between overall thermal comfort and thermal sensations in the climate chamber experiments, the curves in the present study are polynomials (similar to the findings of Majewski et al. [57]). The second order polynomial trend line is much more understandable due to the fact that the optimal/most favourable set of parameters for the respondents must exist. These most preferable conditions seem to be at TSV = −0.4: "pleasantly slightly cool" (for the climate chamber) and TSV = +0.2: "neutral/pleasantly slightly warm" (for the educational rooms).

This shift—as already mentioned in Figure 4—can be explained by elevated air movement in the lecture/classrooms during their normal operation. At the same time, it needs to be added that OC data for the climate chamber tests are typically higher than that for the educational rooms. This phenomenon can be attributed to other unfavourable factors present in the lecture/classrooms such as increased stress levels.

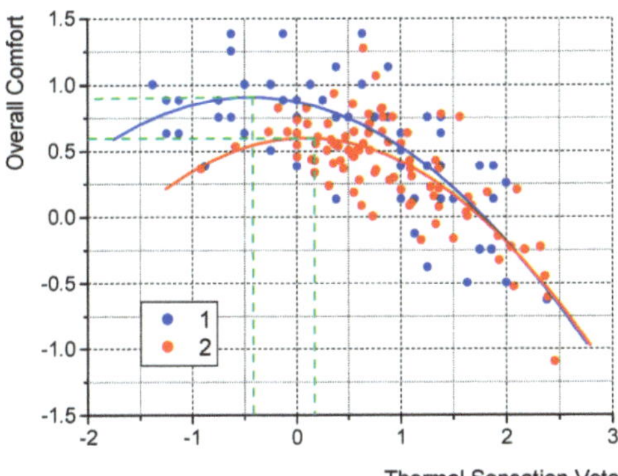

Figure 5. Mean Overall Comfort vs. Thermal Sensation Vote—data for 60 sessions in the climatic chamber (1) and 92 rooms in the educational buildings (2); blue and red lines: polynomial fits for (1) and (2), respectively; green lines: indicators of maximal OC values.

Overall comfort might also be influenced by indoor air quality. The respondents were asked to assess the quality of air during this study. They marked the following answers in the questionnaire regarding this feature: "v. good" (+2), "good" (+1), "neither good nor bad" (0), "bad" (−1), "v. bad" (−2). Figure 6 presents the dependence between the mean values of the subjective assessment of air quality (Indoor Air Quality Vote—IAQV) in the rooms and the overall comfort experienced by the people in these rooms (also as mean values of the answers provided in the questionnaires). The data regarding the educational rooms have been analysed by the authors in [56]; however, they have the addition of a distinction regarding the type of building.

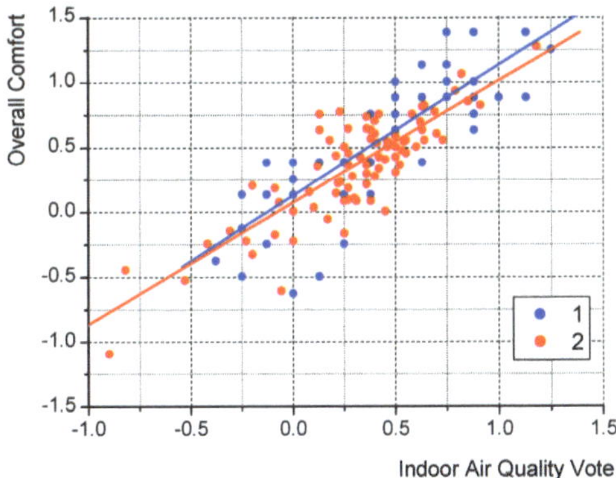

Figure 6. Mean Overall Comfort vs. Indoor Air Quality Vote—data for 60 sessions in the climatic chamber (1) and 92 rooms in the educational buildings (2); blue and red lines: linear fits for (1) and (2), respectively.

The obtained linear correlations for the data recorded in the climate chamber (subscript "1") and the educational rooms (subscript "2") are quite strong and the equations take the following form:

$$OC_1 = 1.0082 IAQV + 0.1245 \left(R^2 = 0.67\right) \tag{5}$$

$$OC_2 = 0.9359 IAQV + 0.0709 \left(R^2 = 0.69\right) \tag{6}$$

As can be seen in the figure, both red and blue trend lines are close to each other, which proves that the correlation between overall comfort and indoor air quality might not depend on the type of environment (actual vs. simulated). It needs to be emphasised that the red trendline (representing a large dataset of 1302 questionnaires from the educational rooms) provides an almost ideal representation of this phenomenon: if the air quality is good (IAQV = 1), overall comfort is also fine (OC = 1). If the air quality is perceived as neutral (IAQV = 0), the mean value of overall comfort is also close to zero. Data for the climate chamber seem to show a little divergence, which might be caused by the smaller sample size. However, the character of the changes is the same.

A dependence of perceived air quality vs. carbon dioxide concentration was also tested by the authors, but no correlation was found (the regression coefficients for both environments' data were close to zero). This disagrees with data presented in [27]; however, a number of other researchers (e.g., [28,34,36]) claim that the perception of room users might be shaped by other factors (such as the presence of odours [36] or the absence of natural light [34]).

Overall comfort should also depend on relative humidity; however, this parameter is often neglected in the analyses found in the literature. In the present study, the respondents were asked to assess the humidity within rooms based on the 5-point scale: "too humid" (+2), "quite humid" (+1), "pleasantly" (0), "quite dry" (−1), "too dry" (−2). Figure 7 presents the distribution of the Humidity Assessment Votes—HAV (as the frequency count), for the climate chamber and for the educational rooms.

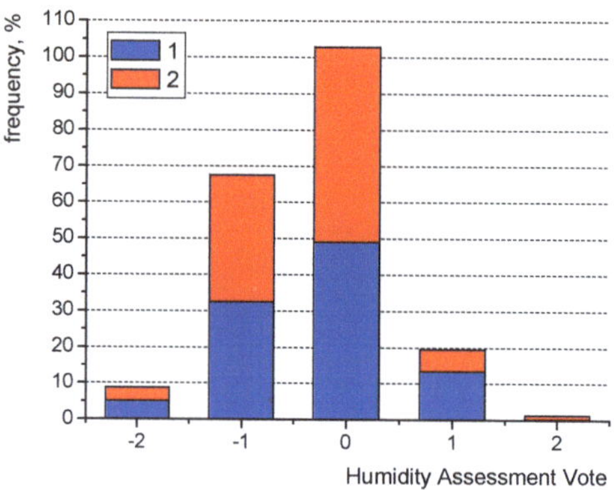

Figure 7. Distribution of Humidity Assessment Votes in the climatic chamber (1) and educational rooms (2).

The figure clearly shows that the relative humidity was found to be satisfactory (answer "0") by about half of the respondents both in the climate chamber and in the educational rooms. Similarly, about the same share of the respondents (ca. 33%) in both environments also considered the indoor air to be too dry ("−1"). This might suggest that human perceptions regarding humidity in both of the analysed environments are very similar.

The next question dealt with individual preferences (called Humidity Preference Vote—HPV) and read "How would you like to change the air humidity?". The possible answers were "more humid" (+1), "no change" (0) and "drier" (−1). Figure 8 shows the distribution of the votes (as the frequency count) for the climate chamber and for the educational rooms.

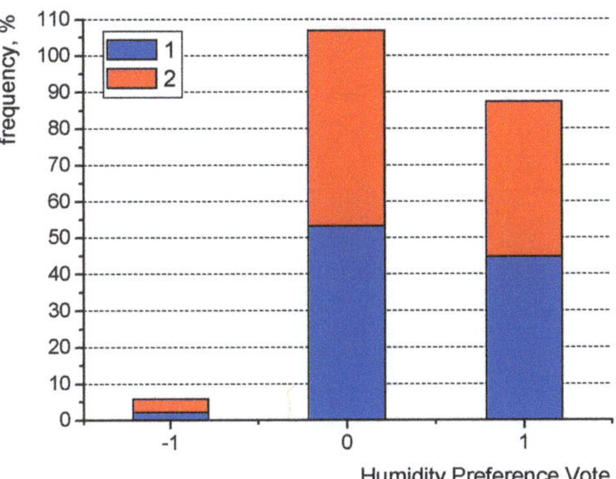

Figure 8. Distribution of Humidity Preference Votes in the climatic chamber (1) and educational rooms (2).

As can be seen, the largest share of the respondents did not want any change with regards to air humidity for both of the considered environments. However, many opted for a more humid condition. Both figures (Figures 7 and 8) seem to show the same character of human response for significantly different environment types (lecture/classrooms in university buildings and the climate chamber), but for a similar range of relative humidity variations. Confirmation of the same character of human perception can be found in Figure 9, where the relation between the mean Humidity Assessment Vote for each room has been shown against the mean Humidity Preference Vote.

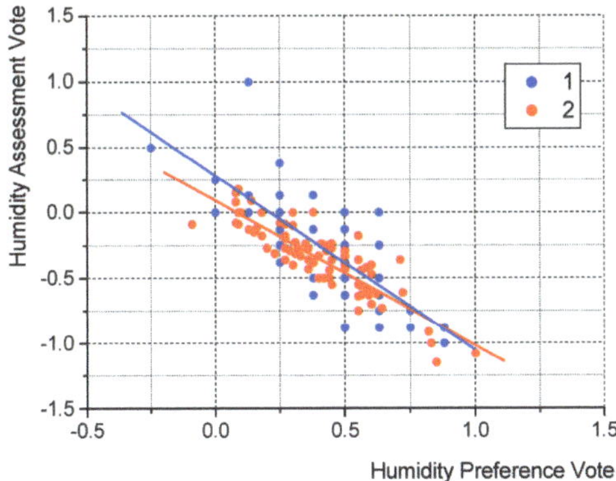

Figure 9. Mean Humidity Assessment Vote vs. Humidity Preference Vote—data for 60 sessions in the climatic chamber (1) and 92 rooms in the educational buildings (2); blue and red lines: linear fits for (1) and (2), respectively.

As could be anticipated if the respondents felt that the air was too dry (marked as "−1" in the questionnaire regarding humidity assessment HAV), they would also prefer the air to be more humid ("+1" in the questionnaire regarding HPV). Figure 9 shows this clearly with both the red and blue trend lines present at point (1.0; −1.0). Similarly, if the respondents felt fine (HAV = 0), they would not welcome any change to the air humidity and would keep it unchanged (HPV = 0). This is clearly visible for the educational buildings (as in Figure 6), but the data for the climate chamber show some deviation. This might also be explained by a large number of people occupying the educational rooms (1302 questionnaires collected there), whereas the data for the climate chamber may simply be less accurate due to the sample size. This assumption is backed by a lower value for the coefficient of determination for climate chamber data (as given below). However, this graph seems to properly describe the relative interconnection between the considered humidity sensations. The linear correlation equations for the data from the climatic chamber (subscript "1") and the educational rooms (subscript "2") are as follows:

$$\text{HAV}_1 = -1.3355\text{HPV} + 0.2845 \left(R^2 = 0.60 \right) \tag{7}$$

$$\text{HAV}_2 = -1.1084\text{HPV} + 0.0928 \left(R^2 = 0.77 \right) \tag{8}$$

The perception of air humidity (AHV) is undoubtedly influenced by the values of relative humidity in the closed spaces. However, the dependence turned out not to be quite as straightforward as has been presented below in Figure 10. Both the environments were tested in similar humidity conditions: the relative humidity values varied from 19.7 to 65.8% in the educational buildings, while in the climate chamber values varied from 19.0 to 73.4%.

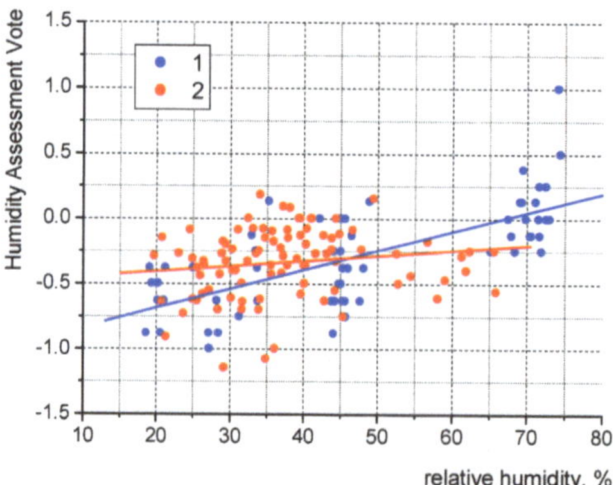

Figure 10. Mean Humidity Assessment Vote vs. relative humidity (rh)—data for 60 sessions in the climatic chamber (1) and 92 rooms in the educational buildings (2); blue and red lines: linear fits for (1) and (2), respectively.

The rise in relative humidity typically led to elevated HAV values; however, the clear trend can only be noticed in the case of the climate chamber tests. In the case of the educational rooms, the dependence is not so obvious. It might be linked with a large number of other parameters of the indoor air of the rooms, which were not considered in the analyses. The coefficient of determination of the linear correlation for the data from the educational rooms was only 0.02, thus the equation is not presented; however, this coefficient amounted to 0.50 in the case of the climatic chamber experiments with the following equation of the linear fit (the blue line in Figure 10):

$$\text{HAV}_1 = 0.0146\text{rh} - 0.9806 \tag{9}$$

The lack of correlation regarding the educational buildings (where relative humidity did not exceed 66%) might be in agreement with data presented by Kong et al. [58], who reported almost no or very marginal interconnection between mean humidity sensation vote and air humidity in the range of 20–70% in their tests carried out in China. The strong impact of humidity can be observed beyond the value of 70% [58,59], especially at high indoor air temperatures. In the present study, such values were recorded in the climate chamber and the influence of these data points on the trend line can be seen in Figure 10.

As discussed earlier, the overall comfort of room users was quite well correlated with thermal and air quality assessment (Figures 5 and 6). It needs to be verified if the same holds true for indoor air humidity. Figure 11 presents a relation between overall comfort vs. Humidity Assessment Vote of the respondents—the data points represent the mean values of the questionnaire surveys in both analysed environments, namely the climate chamber (represented as "1" in the figure) and the educational rooms (represented as "2").

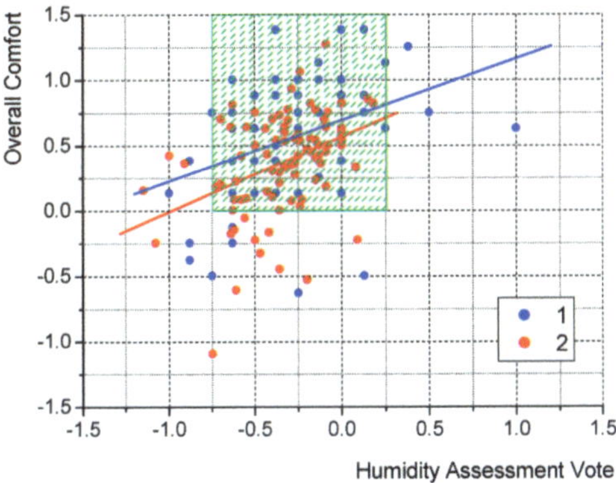

Figure 11. Mean Overall Comfort vs. Humidity Assessment Vote—data for 60 sessions in the climatic chamber (1) and 92 rooms in the educational buildings (2); blue and red lines: linear fits for (1) and (2), respectively; green box: the area of positive overall comfort.

It seems that the character of the interconnection between overall comfort and humidity assessment is very similar for both environments (actual and simulated). The linear fittings of the data show the same trend (although the coefficients of determination in both cases are low and amount to $R^2 = 0.148$ for the climate chamber experiments and $R^2 = 0.156$ in the case of the educational rooms) and the majority of the data points for both environments fall in the range of $-0.75 < HAV < +0.25$ when the respondents experienced positive overall comfort ($OC > 0$). This further supports the claim that the sensations in the educational rooms and the climatic chamber are of the same nature.

4. Conclusions

This study was conducted in university rooms and a climate chamber. Based on the analysis of 1302 questionnaires collected in lecture/classrooms and 480 questionnaires from the climate chamber as well as microclimate meter data analysis, the following conclusions can be drawn:

1. Overall comfort in the climate chamber was typically higher than that in the educational rooms at the same air temperature values. It might be related to the impact of other factors such as the lack of stress in the climate chamber as opposed to students' presence in lecture/classrooms.

2. The most favourable air temperature value in the climate chamber experiments was ca. 20.7 °C, while in the educational rooms it was ca. 22.3 °C. This shift can be explained by the presence of increased air movement (often caused by draughts) in the lecture/classrooms, especially in the summer season possibly due to open windows.

3. The most preferable conditions in the climate chamber occurred at a thermal sensation vote of TSV = −0.4 ("pleasantly slightly cool"), while in the educational rooms it was TSV = +0.2 ("neutral/pleasantly slightly warm"). This shift can also be explained by elevated air movement in the lecture/classrooms during their normal operation.

4. Overall comfort data for the climate chamber tests were typically higher than that for the educational rooms for the same thermal sensation votes. As mentioned above, this phenomenon can be attributed to other unfavourable factors present in the lecture/classrooms such as increased stress levels.

5. Quite strong correlations between overall comfort and indoor air quality as well as between humidity assessment and humidity preference votes were observed, which did not seem to depend on the type of environment.

This study took place in central Poland (Europe), which is characterised by a moderate climate with four distinct seasons (spring: March/June, summer: June/September, autumn: September/December and winter: December/March). The participants of this study were Polish students mostly in their twenties, thus the conclusions might be limited to this age group based in an Eastern/Central European location.

Supplementary Materials: The following supporting information can be downloaded at: https://www.mdpi.com/article/10.3390/en17030656/s1, Table S1: Geometrical details of the educational rooms together with time and outside air temperature data.

Author Contributions: Conceptualization, Ł.J.O.; methodology, Ł.J.O. and N.S.; software, N.R., S.H. and J.Z.P.; investigation, N.S.; resources, Ł.J.O. and N.R.; data curation, N.S., N.R., S.H., J.Z.P. and L.D.; writing—original draft preparation, Ł.J.O., N.S., N.R. and S.H.; writing—review and editing, Ł.J.O., N.S., N.R., S.H., J.Z.P. and L.D.; funding acquisition, Ł.J.O., N.R. and S.H. All authors have read and agreed to the published version of the manuscript.

Funding: The work in this paper was supported by the project: "REFRESH—Research Excellence For REgion Sustainability and High-tech Industries, (VP2), (Reg. No.: CZ.10.03.01/00/22_003/0000048) co-funded by the European Union".

Data Availability Statement: Data available upon request from the corresponding author.

Conflicts of Interest: The authors declare no conflicts of interest.

References

1. International Energy Agency, Tracking Buildings. Available online: https://www.iea.org/energy-system/buildings (accessed on 9 December 2023).
2. Brasche, S.; Bischof, W. Daily time spent indoors in German homes—Baseline data for the assessment of indoor exposure of German occupants. *Int. J. Hyg. Environ. Health* **2005**, *208*, 247–253. [CrossRef]
3. Klepeis, N.E.; Nelson, W.C.; Ott, W.R.; Robinson, J.P.; Tsang, A.M.; Switzer, P.; Behar, J.V.; Hern, S.C.; Engelmann, W.H. The National Human Activity Pattern Survey (NHAPS): A resource for assessing exposure to environmental pollutants. *J. Expo. Anal. Environ. Epidemiol.* **2001**, *11*, 231–252. [CrossRef] [PubMed]
4. Ratajczak, K.; Amanowicz, Ł.; Pałaszyńska, K.; Pawlak, F.; Sinacka, J. Recent Achievements in Research on Thermal Comfort and Ventilation in the Aspect of Providing People with Appropriate Conditions in Different Types of Buildings—Semi-Systematic Review. *Energies* **2023**, *16*, 6254. [CrossRef]
5. Basińska, M.; Ratajczak, K.; Michałkiewicz, M.; Fuć, P.; Siedlecki, M. The Way of Usage and Location in a Big City Agglomeration as Impact Factors of the Nurseries Indoor Air Quality. *Energies* **2021**, *14*, 7534. [CrossRef]
6. Cao, B.; Luo, M.; Li, M.; Zhu, Y. Too cold or too warm? A winter thermal comfort study in different climate zones in China. *Energy Build.* **2016**, *133*, 469–477. [CrossRef]
7. Hwang, R.-L.; Lin, T.-P.; Kuo, N.-J. Field experiments on thermal comfort in campus classrooms in Taiwan. *Energy Build.* **2006**, *38*, 53–62. [CrossRef]
8. Montazami, A.; Gaterell, M.; Nicol, F.; Lumley, M.; Thoua, C. Impact of social background and behaviour on children's thermal comfort. *Build. Environ.* **2017**, *122*, 422–434. [CrossRef]
9. Van Hoof, J.; Schellen, L.; Soebarto, V.; Wong, J.K.W.; Kazak, J.K. Ten questions concerning thermal comfort and ageing. *Build. Environ.* **2017**, *120*, 123–133. [CrossRef]
10. Indraganti, M.; Ooka, R.; Rijal, H.B. Thermal comfort in offices in India: Behavioral adaptation and the effect of age and gender. *Energy Build.* **2015**, *103*, 284–295. [CrossRef]
11. Indraganti, M.; Rao, C.D. Effect of age, gender, economic group and tenure on thermal comfort: A field study in residential buildings in hot and dry climate with seasonal variations. *Energy Build.* **2010**, *42*, 273–281. [CrossRef]
12. Karjalainen, S. Gender differences in thermal comfort and use of thermostats in everyday thermal environments. *Build. Environ.* **2007**, *42*, 1594–1603. [CrossRef]

13. Maykot, J.K.; Rupp, R.F.; Ghisi, E. A field study about gender and thermal comfort temperatures in office buildings. *Energy Build.* **2018**, *178*, 254–264. [CrossRef]

14. Parsons, K.C. The effects of gender, acclimation state, the opportunity to adjust clothing and physical disability on requirements for thermal comfort. *Energy Build.* **2002**, *34*, 593–599. [CrossRef]

15. Luo, M.; Wang, W.; Ke, K.; Cao, B.; Zhai, Y.; Zhou, X. Human metabolic rate and thermal comfort in buildings: The problem and challenge. *Build. Environ.* **2018**, *131*, 44–52. [CrossRef]

16. Fang, Z.; Zhang, S.; Cheng, Y.; Fong, A.M.L.; Oladokun, M.O.; Lin, Z.; Wua, H. Field study on adaptive thermal comfort in typical air conditioned classrooms. *Build. Environ.* **2018**, *133*, 73–82. [CrossRef]

17. Buratti, C.; Ricciardi, P. Adaptive analysis of thermal comfort in university classrooms: Correlation between experimental data and mathematical models. *Build. Environ.* **2009**, *44*, 674–687. [CrossRef]

18. Ricciardi, P.; Buratti, C. Environmental quality of university classrooms: Subjective and objective evaluation of the thermal, acoustic, and lighting comfort conditions. *Build. Environ.* **2018**, *127*, 23–36. [CrossRef]

19. Krawczyk, D.A.; Gładyszewska-Fiedoruk, K.; Rodero, A. The analysis of microclimate parameters in the classrooms located in different climate zones. *App. Therm. Eng.* **2017**, *113*, 1088–1096. [CrossRef]

20. Sakellaris, I.A.; Saraga, D.E.; Mandin, C.; Roda, C.; Fossati, S.; De Kluizenaar, Y.; Carrer, P.; Dimitroulopoulou, S.; Mihucz, V.G.; Szigeti, T.; et al. Perceived Indoor Environment and Occupants' Comfort in European "Modern" Office Buildings: The OFFICAIR Study. *Int. J. Environ. Res. Public Health* **2016**, *13*, 444. [CrossRef]

21. Kim, Y.K.; Abdou, Y.; Abdou, A.; Altan, H. Indoor Environmental Quality Assessment and Occupant Satisfaction: A Post-Occupancy Evaluation of a UAE University Office Building. *Buildings* **2022**, *12*, 986. [CrossRef]

22. Altomonte, S.; Allen, J.; Bluyssen, P.M.; Brager, G.; Heschong, L.; Loder, A.; Schiavon, S.; Veitch, J.A.; Wang, L.; Wargocki, P. Ten questions concerning well-being in the built environment. *Build. Environ.* **2020**, *180*, 106949. [CrossRef]

23. Göçer, Ö.; Candido, C.; Thomas, L.; Göçer, K. Differences in Occupants' Satisfaction and Perceived Productivity in High- and Low-Performance Offices. *Buildings* **2019**, *9*, 199. [CrossRef]

24. *EN 16798-1:2019*; Energy Performance of Buildings—Ventilation for Buildings—Part 1: Indoor Environmental Input Parameters for Design and Assessment of Energy Performance of Buildings Addressing Indoor Air Quality, Thermal Environment, Lighting and Acoustics—Module M1-6. European Committee for Standardization: Brussels, Belgium, 2019.

25. Borowski, M.; Zwolińska, K.; Czerwiński, M. An Experimental Study of Thermal Comfort and Indoor Air Quality—A Case Study of a Hotel Building. *Energies* **2022**, *15*, 2026. [CrossRef]

26. Vilcekova, S.; Meciarova, L.; Burdova, E.K.; Katunska, J.; Kosicanova, D.; Doroudiani, S. Indoor environmental quality of classrooms and occupants' comfort in a special education school in Slovak Republic. *Build. Environ.* **2017**, *120*, 29–40. [CrossRef]

27. Aguilar, A.J.; de la Hoz-Torres, M.L.; Ruiz, D.P.; Martínez-Aires, M.D. Monitoring and Assessment of Indoor Environmental Conditions in Educational Building Using Building Information Modelling Methodology. *Int. J. Environ. Res. Public Health* **2022**, *19*, 13756. [CrossRef]

28. Basińska, M.; Michałkiewicz, M.; Ratajczak, K. Effect of Air Purifier Use in the Classrooms on Indoor Air Quality—Case Study. *Atmosphere* **2021**, *12*, 1606. [CrossRef]

29. Zender-Swiercz, E. Analysis of the impact of the parameters of outside air on the condition of indoor air. *Int. J. Environ. Sci. Technol.* **2017**, *14*, 1583–1590. [CrossRef]

30. Zender-Swiercz, E. Improving the indoor air quality using the individual air supply system. *Int. J. Environ. Sci. Technol.* **2018**, *15*, 689–696. [CrossRef]

31. Amanowicz, Ł.; Ratajczak, K.; Dudkiewicz, E. Recent Advancements in Ventilation Systems Used to Decrease Energy Consumption in Buildings—Literature Review. *Energies* **2023**, *16*, 1853. [CrossRef]

32. Rolando, D.; Mazzotti Pallard, W.; Molinari, M. Long-Term Evaluation of Comfort, Indoor Air Quality and Energy Performance in Buildings: The Case of the KTH Live-In Lab Testbeds. *Energies* **2022**, *15*, 4955. [CrossRef]

33. Aflaki, A.; Esfandiari, M.; Jarrahi, A. Multi-Criteria Evaluation of a Library's Indoor Environmental Quality in the Tropics. *Buildings* **2023**, *13*, 1233. [CrossRef]

34. Clements, N.; Zhang, R.; Jamrozik, A.; Campanella, C.; Bauer, B. The Spatial and Temporal Variability of the Indoor Environmental Quality during Three Simulated Office Studies at a Living Lab. *Buildings* **2019**, *9*, 62. [CrossRef]

35. Lan, L.; Wargocki, P.; Wyon, D.P.; Lian, Z. Effects of thermal discomfort in an office on perceived air quality, SBS symptoms, physiological responses, and human performance. *Indoor Air* **2011**, *21*, 376–390. [CrossRef]

36. Geng, Y.; Ji, W.; Lin, B.; Zhu, Y. The impact of thermal environment on occupant IEQ perception and productivity. *Build. Environ.* **2017**, *121*, 158–167. [CrossRef]

37. Yang, Y.; Li, B.; Liu, H.; Tan, M.; Yao, R. A study of adaptive thermal comfort in a well-controlled climate chamber. *App. Therm. Eng.* **2015**, *76*, 283–291. [CrossRef]

38. Ji, W.; Cao, B.; Luo, M.; Zhu, Y. Influence of short-term thermal experience on thermal comfort evaluations: A climate chamber experiment. *Build. Environ.* **2017**, *114*, 246–256. [CrossRef]

39. Zhang, Z.; Zhang, Y.; Khan, A. Thermal comfort of people from two types of air-conditioned buildings—Evidences from chamber experiments. *Build. Environ.* **2019**, *162*, 106287. [CrossRef]

40. Soebarto, V.; Zhang, H.; Schiavon, S. A thermal comfort environmental chamber study of older and younger people. *Build. Environ.* **2019**, *155*, 1–14. [CrossRef]

41. Chen, M.; Farahani, A.V.; Kilpeläinen, S.; Kosonen, R.; Younes, J.; Ghaddar, N.; Ghali, K.; Melikov, A.K. Thermal comfort chamber study of Nordic elderly people with local cooling devices in warm conditions. *Build. Environ.* **2023**, *235*, 110213. [CrossRef]
42. Jian, Y.; Liu, S.; Bian, M.; Liu, Z.; Liu, S. Climate Chamber Experiment Study on the Association of Turning off Air Conditioning with Human Thermal Sensation and Skin Temperature. *Buildings* **2022**, *12*, 472. [CrossRef]
43. Ahmad, R.I.; Norfadzilah, J.; Raemy, M.Z. Human Responses to the Thermal Comfort in Air-Conditioned Building: A Climate Chamber Study. *Int. J. Integr. Eng.* **2022**, *14*, 287–295. [CrossRef]
44. Dong, Y.; Shi, Y.; Liu, Y.; Rupp, R.F.; Toftum, J. Perceptive and physiological adaptation of migrants with different thermal experiences: A long-term climate chamber experiment. *Build. Environ.* **2022**, *211*, 108727. [CrossRef]
45. Jia, X.; Wang, J.; Zhu, Y.; Ji, W.; Cao, B. Climate chamber study on thermal comfort of walking passengers with elevated ambient air velocity. *Build. Environ.* **2022**, *218*, 109100. [CrossRef]
46. Upadhyay, K.; Elangovan, R.; Subudhi, S. Establishing thermal comfort baseline in a sub-tropical region through a controlled climate chamber study. *Adv. Build. Energy Res.* **2023**. [CrossRef]
47. Yang, R.; Yang, L.; Wei, J. Research on the thermal environment in a climate chamber with different high-temperature combinations. *Adv. Build. Energy Res.* **2023**. [CrossRef]
48. De Dear, R. Thermal comfort in practice. *Indoor Air* **2004**, *14* (Suppl. S7), 32–39. [CrossRef] [PubMed]
49. Available online: www.testo.com (accessed on 10 December 2023).
50. *EN 15251:2007*; Indoor Environmental Input Parameters for Design and Assessment of Energy Performance of Buildings Addressing Indoor Air Quality, Thermal Environment, Lighting and Acoustics. CEN: Brussels, Belgium, 2007.
51. *ISO 28802:2012*; Ergonomics of the Physical Environment—Assessment of Environments by Means of an Environmental Survey Involving Physical Measurements of the Environment and Subjective Responses of People. European Committee for Standardization: Brussels, Belgium, 2012.
52. Singh, M.K.; Kumar, S.; Ooka, R.; Rijal, H.B.; Gupta, G.; Kumar, A. Status of thermal comfort in naturally ventilated classrooms during the summer season in the composite climate of India. *Build. Environ.* **2018**, *128*, 287–304. [CrossRef]
53. Nematchoua, M.K.; Tchinda, R.; Orosa, J.A. Thermal comfort and energy consumption in modern versus traditional buildings in Cameroon: A questionnaire-based statistical study. *Appl. Energy* **2014**, *114*, 687–699. [CrossRef]
54. Kosmopoulos, P.; Galanos, D.; Anastaselos, D.; Papadopoulos, A. An Assessment of the Overall Comfort Sensation in Workplaces. *Int. J. Vent.* **2012**, *10*, 311–322. [CrossRef]
55. Morandi, F.; Pittana, I.; Cappelletti, F.; Gasparella, A.; Tzempelikos, A. Assessing Overall Indoor Environmental Comfort and Satisfaction: Evaluation of a Questionnaire Proposal by Means of Statistical Analysis of Responses. In Proceedings of the IAQ 2020: Indoor Environmental Quality Performance Approaches, Athens, Greece, 4–6 May 2022.
56. Orman, Ł.J.; Krawczyk, N.; Radek, N.; Honus, S.; Pietraszek, J.; Dębska, L.; Dudek, A.; Kalinowski, A. Comparative Analysis of Indoor Environmental Quality and Self-Reported Productivity in Intelligent and Traditional Buildings. *Energies* **2023**, *16*, 6663. [CrossRef]
57. Majewski, G.; Orman, Ł.J.; Telejko, M.; Radek, N.; Pietraszek, J.; Dudek, A. Assessment of Thermal Comfort in the Intelligent Buildings in View of Providing High Quality Indoor Environment. *Energies* **2020**, *13*, 1973. [CrossRef]
58. Kong, D.; Liu, H.; Wu, Y.; Li, B.; Wei, S.; Yuan, M. Effects of indoor humidity on building occupants' thermal comfort and evidence in terms of climate adaptation. *Build. Environ.* **2019**, *155*, 298–307. [CrossRef]
59. Buonocore, C.; De Vecchi, R.; Scalco, V.; Lamberts, R. Influence of relative air humidity and movement on human thermal perception in classrooms in a hot and humid climate. *Build. Environ.* **2018**, *146*, 98–106. [CrossRef]

Article

Radiators Adjustment in Multi-Family Residential Buildings—An Analysis Based on Data from Heat Meters

Karol Bandurski *, **Andrzej Górka and Halina Koczyk**

Faculty of Environmental Engineering and Energy, Institute of Environmental Engineering and Building Installations, Poznań University of Technology, Berdychowo 4, 60-965 Poznań, Poland; andrzej.gorka@put.poznan.pl (A.G.); halina.koczyk@put.poznan.pl (H.K.)
* Correspondence: karol.bandurski@put.poznan.pl

Abstract: Energy is consumed in buildings through the use of various types of energy systems, which are controlled by the occupants via provided interfaces. The quality of this control should be verified to improve the efficiency of the systems and for the comfort of the occupants. In the case of residential buildings, due to privacy reasons, it is problematic to directly monitor human–building interactions using sensors installed in dwellings. However, data from increasingly common smart meters are easily available. In this paper, the potential use of data from heat meters is explored for the analysis of occupant interactions with space-heating (SH) systems. A pilot study is conducted based on a one-year set of daily data from 101 dwellings. First, the identification of an indoor temperature and a strategy for thermostatic radiator valve (TRV) adjustments for all the investigated dwellings is presented. Second, the performed analysis suggests that 96% of the households did not use the automatic adjustment function of the TRVs since adjustments using the on–off mode were the most common, which could be empirical evidence for Kempton's theory on mental models of home heating controls. The reasons for this could be the weakness of the TRV as an SH interface and the technical specificity of the analyzed SH (its supply temperature). The preliminary investigation confirms the potential of the proposed methodology, but further research is needed.

Keywords: heating; smart meters; TRV; occupant behavior; residential building thermostats

Citation: Bandurski, K.; Górka, A.; Koczyk, H. Radiators Adjustment in Multi-Family Residential Buildings—An Analysis Based on Data from Heat Meters. *Energies* **2023**, *16*, 7485. https://doi.org/10.3390/en16227485

Academic Editor: Francesco Minichiello

Received: 12 September 2023
Revised: 27 October 2023
Accepted: 31 October 2023
Published: 8 November 2023

1. Introduction

Decreasing the energy used in residential buildings while maintaining high-level comfort is one of the major challenges that face the ongoing energy conservation transition, as declared by the European Union [1]. In working towards this goal, it is important to understand where and why energy is consumed by the users in the buildings [2]. Occupants use energy to achieve comfort, and they can continue to do so by taking advantage of HVAC (heating, ventilation, and air conditioning) with the provided building system interface. Research on how to achieve comfort [3], as well as the interactions with the system interface [4,5], is important because the integration of these two processes is crucial for the performance of the building. If the interface is not adapted to the requirements of the occupants, then achieving comfort will be difficult for them. The consequences of this would be a loss of trust in the control system [6], unexpected use of the building's systems, and/or unnecessary use of energy. Examples of these processes have been collected by O'Brien and Gunay [7] or Sarran et al. [8].

Basic residential building interfaces include windows, movable shades, thermostats, and light switches [5]. A simple Scopus search (date: 26 October 2023, search within: "Article title, Abstract, Keywords: "occupant behaviour" AND "model" AND "windows"; "occupant behaviour" AND "model" AND "blinds"; "occupant behaviour" AND "model" AND "thermostats"; "occupant behaviour" AND "model" AND "lighting") shows that the most common research relates to the control of windows (254 papers) and lighting

(152 papers), and much less frequently thermostats (56 papers) and shading (40 papers), which indicates that there is still a lack of recognition of these processes. It might seem that thermostat control, especially in terms of heating, can be simplified to controlling the temperature that is maintained in the building, which affects transmission heat loss. However, the way in which a water heating system is controlled significantly affects, for example, the distribution of heat loss in the system, which may increase the indoor temperature in an uncontrolled way and, in turn, may cause occupants to open a window. There is, thus, a relationship between the building's systems and the building envelope. In addition, the control of the heating system also influences the peak loads or the operation of the auxiliary equipment of the system. Therefore, an analysis of the use of heating systems is of great importance for future designs that reduce energy loss and peak loads.

A general overview of factors that influence the ways in which space heating (SH) is adjusted was prepared by Wei et al. [9]; such heating systems can be classified as air- or water-based. In general, the former heating type is more common in North America, and the latter is more common in Europe. Both forms can be controlled by smart thermostats, the usage of which is most described in the literature for SH interfaces due to remote data collections performed by their manufacturers [10–15]. However, in, for example, Poland, the Czech Republic, and Bulgaria [16], but also Germany and the UK, the most common interfaces for SH are thermostatic radiator valves (TRVs); to illustrate this point, it is known that 45% of households in Poland have a TRV [17]. Lomas et al. [18], in their critical review, explained that the energy-saving effect of thermostats on energy consumption for SH in residential buildings is not well documented. Their studies demonstrate that the energy efficiency of the zonal temperature control is moderately documented. They consider the depth of research on the savings that result from TRV installations and the superiority of smart thermostats over standard thermostats to be of a low rate. In a subsequent paper, Lomas et al. confirmed the contribution of TRVs towards a small amount of energy savings because of SH zonal control [19]; this study used the latest generation of TRVs, which contain electronics with remote control functionality. Another point of view in the evaluation of TRVs is provided by Cholewa et al. [20]. They described six years of heat consumption measurements for the SH requirements of, among others, three multi-family buildings, where conventional radiator valves were replaced with TRVs. Their measurements refer to the three years of energy consumption before and the three years after the valve replacement. The average saved energy, due to their use and the system's hydraulic balancing, was 17%.

To conclude the above findings, there is still a relatively low amount of research focusing on human interactions with heating controls. One of the most popular heating controls is the TRV, the energy efficiency of which is questionable. Therefore, there is a need to better understand its usage by occupants and to assess its usability for energy-efficient building operations.

In regard to the TRV–occupant interactions, the literature is limited to surveys [21–23] or measurement research based on only a selection of households [19,24,25], which involve a lot of additional sensors inside the dwellings. The authors have found only one expectation: a monitoring campaign conducted in the UK, where 47 flats were observed using smart heating control systems, which were implemented in buildings based on funding support from a European project [25]; however, the presented results refer to a specific group of users: elderly residents in a care home. From the point of view of continuous control of the energy efficiency of the operation of heating systems, these are not suitable methods of monitoring. The first method is subjective [26] and is limited to only collecting information once, as later repetition may be tedious for respondents. The second method requires an advanced monitoring system installed in the dwellings. The collected data are very informative but could be seen as personal data, which would not be acceptable for most residents. In the case of smart heating controls, acquiring the funds for such upgrades is problematic. Therefore, a methodology for analyzing the operational quality of the heating systems with TRVs or others based on water heaters that involve readily available data, such as heat meters, would be valuable. Such approaches will be objective and will not disturb household privacy. Moreover, the collected data will not contain any direct information about the indoor environment of the dwellings. To date, methods for heat meter data analysis have been developed solely for studying the overall energy performance of a building [27–29].

The questions for this research are: (1) What information regarding the occupant control of the SH can be extracted from heat meters? and (2) Is it possible, based on this information, to learn more about people's use of TRV? Do they use TRV in accordance with its design? This paper proposes a methodology for determining parameters characterizing occupant–SH system interactions based on radiator model fitting to heat meter data. Therefore, the privacy zone of the occupants is not disturbed. Preliminary results are presented using data from a multi-family housing estate. Additionally, the methodology includes an algorithm for separating the heat consumption of the SH and the domestic hot water (DHW) from the dwelling heat meter data. Correlation between the dwelling characteristics and the observed TRV usage models is investigated.

2. Methodology

2.1. Research Object

Data for the analysis was collected from a multi-family housing estate in Poland, in a humid continental climate with mild summers and rainfall all year round, Dfb according to Köppen-Geiger climate classification [30]. The building estate was constructed at the beginning of the 21st century (Figure 1). It consists of 4 buildings with 108 apartments with a total area of 6752 m². The six-story buildings, the first story of which is a basement and the final two are two-story apartments, are represented in Figure 2. Table 1 summarizes the parameters of the dwelling types.

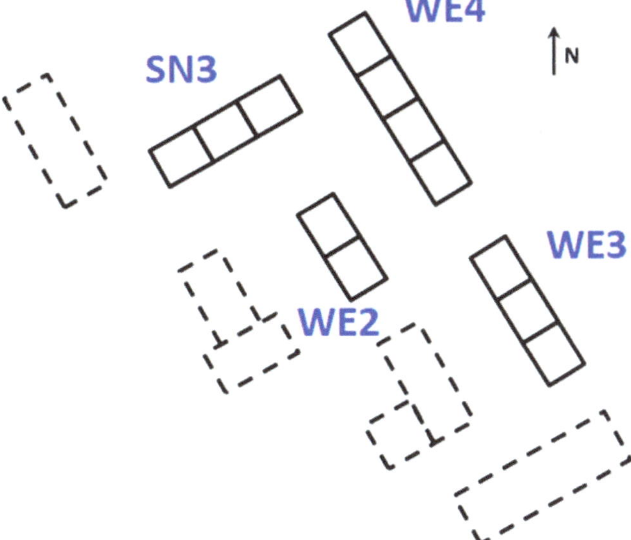

Figure 1. Plan of the studied housing estate. Solid lines represent the surveyed buildings and their staircases, and the dashed lines correspond to the buildings surrounding the housing estate. Adapted with permission from ref. [31].

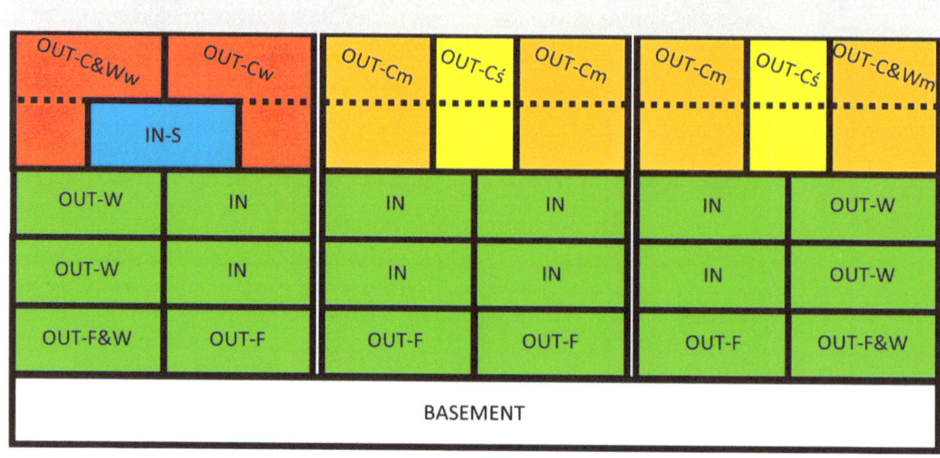

Figure 2. Types of dwellings in the analyzed buildings. Colors represent dwellings with identical floor areas. Two-story apartments are located on the top floor (except IN-S). The figure shows the layout of the three staircase buildings, for which the two-staircase building has an A and a B staircase, and the four-staircase building has an A staircase and three B staircases. Adapted with permission from ref. [31].

Table 1. Characteristics of the surveyed dwellings (see Figure 2 for the type of dwelling notation). Adapted with permission from ref. [31].

Types of Dwellings	$A_f\,[\mathrm{m^2}]$	$H_{tr}\,[\mathrm{W/K}]$	$C_m\,[\mathrm{kJ/K}]$	$A_{win}\,[\mathrm{m^2}]$	$\dot{Q}_{D,SH,max}\,[\mathrm{W}]$
IN	53	22.54	29.98	9.04	1180
IN-S	36	17.27	16.65	6.78	880
OUT-W	53	31.81	29.88	9.87	1510
OUT-Cm	83	38.56	46.44	13.34	2410
OUT-Cs	79	42.20	35.88	15.84	2250
OUT-Cw	104	49.75	55.58	16.73	2990
OUT-F	53	30.76	29.98	9.04	1570
OUT-CWm	83	56.96	46.70	15.54	3060
OUT-CWw	104	68.30	55.37	18.93	3630
OUT-FW	53	40.03	29.88	9.87	1920

2.2. Residential Thermal Stations System, Heat Network, and Gas Boilers

Heating for the housing estate is provided by the local gas boilers located in the WE3 building. An underground heating network (HN) supplies the heated water from the gas boilers to WE2, WE4, and SN3 buildings. The heating network supplies the water temperature ($t_{\mathrm{HN},sup}$) to the residential thermal stations (RTSs), which all year round equals approximately 70 °C. The RTS are single-function residential thermal stations in which the DHW is prepared or the water is passed through to supply the SH. This solution reduces the distribution heat losses due to the same pipes being used for the SH and DHW distributions.

However, it enforces the maintained high supply temperature all year round to enable the preparation of the DHW.

2.3. Data

Measurements were performed from March 2015 to February 2016. Figure 3 shows a scheme depicting the research facility with the variables measured, estimated, and determined through a data analysis described in this paper.

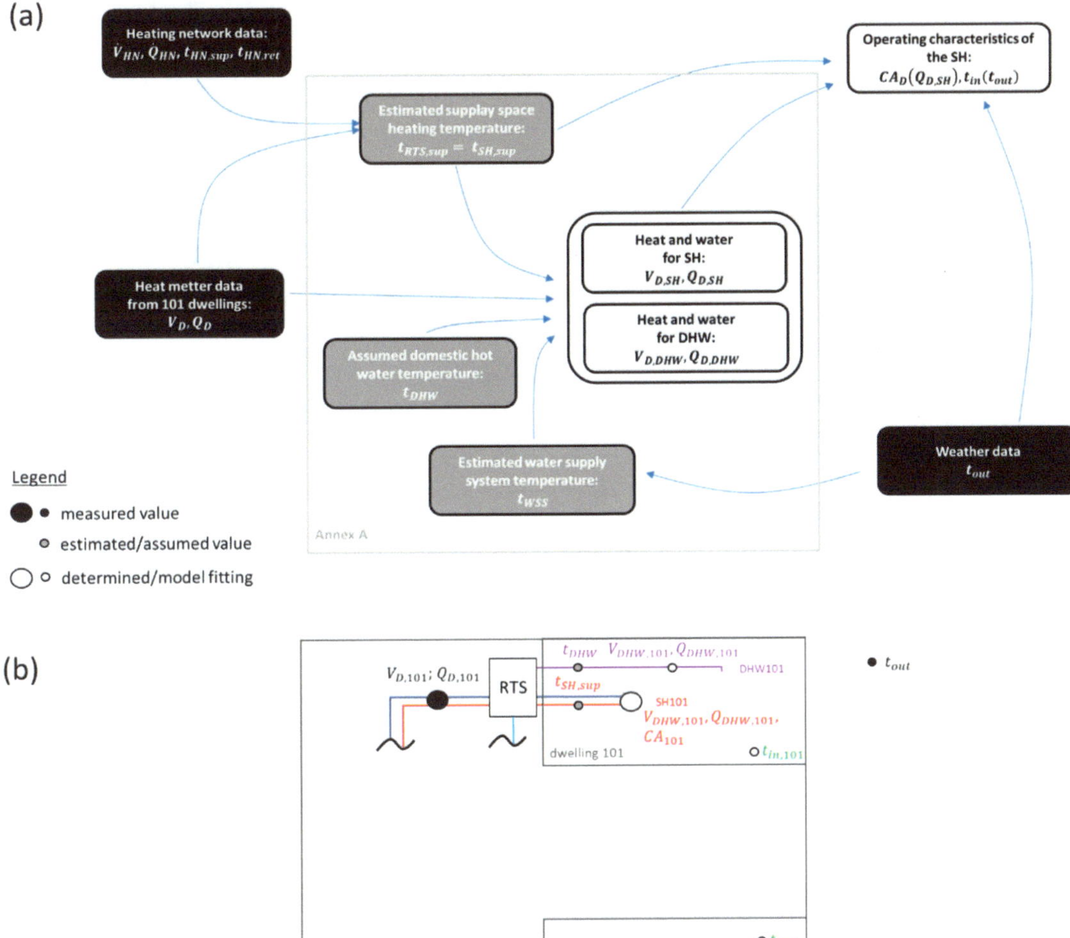

Figure 3. Scheme of (**a**) the performed analysis and (**b**) the research facility. Indication of the variables measured, estimated, and determined by a model fitting.

2.3.1. Heat Consumption and Heating Water Flow

Data from the dwellings was collected via Kamstrup MULTICAL302 heat meters, which were installed in 101 apartments. The remaining four apartments were not included in the analysis. The heat meter memory contained 460 daily records of the heat consumption and the heating water consumption of the dwelling. The memory also contained records with hourly and monthly data, but they were not included in the analysis. It is noteworthy that this type of heat meter, which is widespread, has a standard memory resolution (0.01 GJ), which makes daily analysis difficult and hourly analysis impossible. The problem is noticeable in the results (Section 3): models for the dwelling with very low heat consumption represent a sharp decrease in accuracy. The low resolution arises from the limited internal memory of the meters. However, it can be modified during production (subject to customer orders). This dataset is part of the ASHRAE Global Occupant Behavior Database [32,33], which is an open database.

Data on the gas boiler operations was obtained from the operator (Veolia Energia Poznań SA, Poznań, Poland). The data are provided in 15-min timesteps and include the supplied and return water temperature, the instantaneous thermal power of the boilers, the instantaneous heating water flow, the total heat transferred to the heat network, and the total volume of the heating water that flowed through the HN.

2.3.2. Outdoor Air Temperature

Outdoor air temperatures for the analysis were measured at a weather station located 3 km from the buildings under the study (Figure 4). Figure 5 compares outdoor temperature distribution in the city, taking measurements from 2015 onwards. It can be seen that, in 2015 and 2016, this does not differ significantly in terms of average, median, 25th, and 75th centile values from other years. The only noticeable difference is that one of the warmest winters was observed in 2015; however, this season is not included in the analysis (see Figure 4).

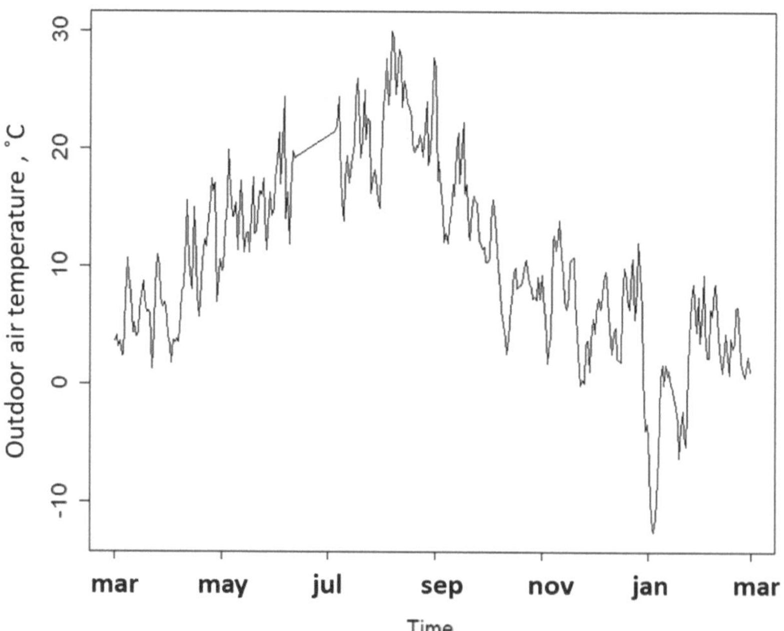

Figure 4. Outdoor air temperature. The end of June represents a period in which the measuring system failed. The straight lines provide an interpolation of the missing data.

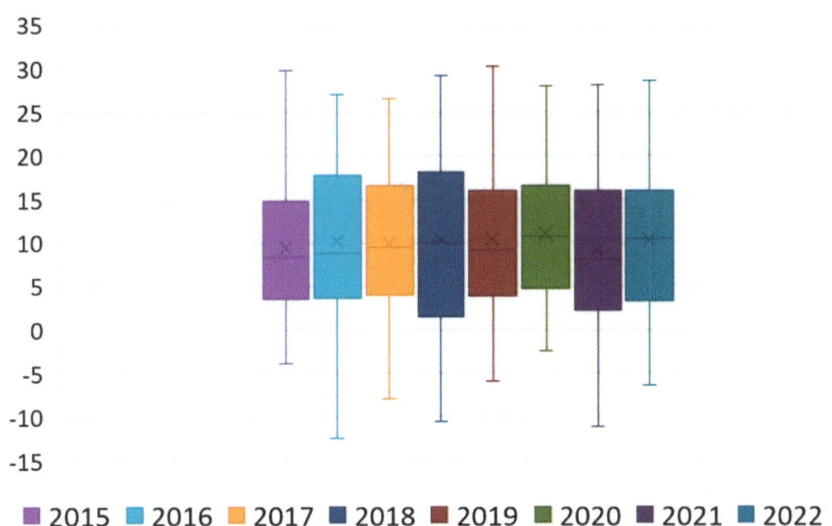

Figure 5. Yearly outdoor air temperature distribution, based on mean daily data, in the city where heat consumption data were collected.

2.4. SH Usage Model

For the analysis, only SH data were used. The methodology of the separation of heat and heating water consumption for the SH and the DHW from heat meter readings is described in Supplementary Material S1.

The power of the radiator, but also that of the SH, can be described by the following model [34–37]:

$$\dot{Q}_{D,SH} = \dot{m}_{D,SH} \cdot c_w \cdot (t_{SH,sup} - t_{in}) \cdot \left\{ 1 - \left[\frac{\dot{m}_{D,SH} \cdot c_w}{\dot{m}_{D,SH} \cdot c_w + n \cdot CA_D \cdot (t_{SH,sup} - t_{in})^n} \right]^{1/n} \right\} \quad (1)$$

Based on this research, it is possible to determine the values of the following variables found in Equation (1):

- $\dot{m}_{D,SH}$, the mean daily mass flow rate of the heating water in a given dwelling that is determined on the basis of the following expression:

$$\dot{m}_{D,SH} = \frac{V_{D,SH}}{24 \cdot 3600} \rho_w \quad (2)$$

in which the daily heating water consumption ($V_{D,SH}$) was measured by a heat meter, and the water densities (ρ_w) were calculated according to ref. [38] for the mean water temperature in the HN $(t_{HN,sup} + t_{HN,ret})/2$,

- c_w, the water-specific heat capacity was calculated according to ref. [38] for the mean water temperature in the HN,
- $t_{SH,sup}$, the mean daily supply water temperature of SH; its determination is described in Supplementary Material S1, Equation (S1.1),
- n, the radiator type exponent is assumed to be 0.3, which is the same value for all radiators based on the analysis of the technical documentation and the information from the manufacturer.

The values of t_{in} and CA_D are unknown. The next section proposes the possible methods for controlling the heating system; a set of corresponding assumptions for these values are included.

2.5. Assumption for Occupant–SH Interaction

Two variables in Equation (1) determine how the SH is adjusted. These are:

- t_{in}, the temperature in the dwellings,
- CA_D, the product of the radiator heating area (A_D) in the dwelling and the coefficient of the heat transfer intensity between the radiators and the indoor environment.

The following proposed values for these variables are based on occupant behavior assumptions, which will later be verified based on the collected data.

2.5.1. Indoor Temperature in the Dwelling t_{in}

The temperature in the dwellings was not measured, but data can be found in the literature regarding the temperature maintained in dwellings as a function of the outdoor temperature [39–46]. The occupant behavior assumptions for this analysis were prepared based on the data collected in multi-family buildings in Estonia [47], whose building cultural and climatic context is similar to that of Poland. Five models for the indoor temperature adjustment as a function of the outdoor temperature are proposed (Figure 6):

$$t_{var,high} = \max(-0.05t_{out} + 24.57, 0.33t_{out} + 20.17) \tag{3a}$$

$$t_{con,high} = \max(24, 0.33t_{out} + 20.17) \tag{3b}$$

$$t_{con,medium} = \max(22, 0.33t_{out} + 18.67) \tag{3c}$$

$$t_{con,low} = \max(20, 0.33t_{out} + 17.17) \tag{3d}$$

$$t_{var,low} = \max(0.05t_{out} + 19.57, 0.33t_{out} + 17.17) \tag{3e}$$

Models with a *var* subscript assume a change in indoor temperature during the heating season, while models with a *con* subscript assume that (during the heating season) indoor temperature is relatively constant.

2.5.2. Radiators Operation Adjustment CA_D

The way that the TRV operates provides:

- hydraulic balancing of the system by the installers,
- opening and closing of the heating water flow through radiators via manual adjustment by the occupants,
- automated maintenance of the thermal comfort conditions in a room set by the occupants.

The final function is the basic energy-saving feature of the TRV. It operates thanks to an internal mechanism that smoothly regulates the heating water flow through the radiator as a function of the room temperature. In the most common variant, the TRV mechanism is based on the thermal expansion phenomenon of the substance; it does not require an external power supply. For a given setting of the TRV, the power of the radiator is adjusted in a manner that maintains the set temperature of the room. Usually, the TRV does not have precise temperature value settings, but only those over the range from 1 to 5, since the resulting internal temperature that the TRV maintains also depends on the location of the heater in the room.

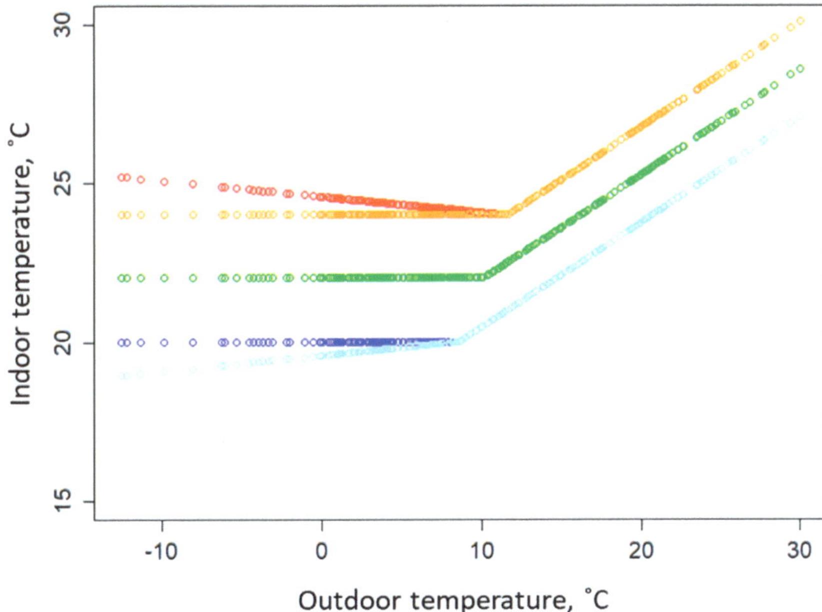

Figure 6. Analyzed models for the indoor temperature adjustment as a function of outdoor temperature: $t_{var,high}$ (red, Equation (3a)), $t_{con,high}$ (orange, Equation (3b)), $t_{con,medium}$ (green, Equation (3c)), $t_{con,low}$ (blue, Equation (3d)), and $t_{var,low}$ (light blue, Equation (3e)).

Each dwelling is equipped with radiators with known maximum power and type; also, the design temperature difference between the radiator and room is known. The following expression was used to calculate the maximum product of the radiator heating area and the coefficient of the heat transfer intensity ($CA_{D,max}$) base on these data:

$$CA_{D,max} = \frac{\dot{Q}_{D,SH,max}}{\Delta t_{SH-in,log}^{1+n}} \tag{4}$$

There are two opposite ways to control the heating system:

- Automated, in which the radiator/SH power control is based on the automated control of flow through the TRV. Hence, the heating area of the radiators and the number of active radiators are not changed. Only power output from the radiator is regulated by its temperature. In this approach, the heating area of the system is fixed so that:

$$CA_{D,con} = const \tag{5}$$

- Manual, in which the power control of the radiators/SH is based on manually opening and closing the given radiators. This is performed without the use of the TRV function that adjusts the flow through the radiator to the set temperature and the actual temperature in the room. Then, the control of the radiator/SH power is based on an adjustment of the heating surface of the radiators, which can be simplified as:

$$CA_{D,var} = CA_{D,max}\frac{Q_{D,SH}}{max(Q_{D,SH})} \tag{6}$$

Equations (5) and (6) define the two proposed intensity adjustment models for the SH operation. Additionally, the heating system can be only partially used in a given dwelling, even with the smallest loads, in which case the maximum heating area of the radiators is smaller than the available one ($A_{D,max}$). It is also possible that the heat transfer coefficient between the radiator and the surroundings, C, will be more intense compared to standard operations due to the simultaneous ventilation (by opening windows) and space heating. On this basis, the following assumptions for regulating the intensity of the radiator operation have been proposed:

- Automated:

$$CA_{D,con,v.high} = 1.50\, CA_{D,max} \tag{7a}$$

$$CA_{D,con,high} = 1.25\, CA_{D,max} \tag{7b}$$

$$CA_{D,con,medium} = CA_{D,max} \tag{7c}$$

$$CA_{D,con,low} = 0.75\, CA_{D,max} \tag{7d}$$

$$CA_{D,con,v.low} = 0.5\, CA_{D,max} \tag{7e}$$

- Manual:

$$CA_{D,var,v.high} = 1.50\, CA_{D,var} \tag{7f}$$

$$CA_{D,var,high} = 1.25\, CA_{D,var} \tag{7g}$$

$$CA_{D,var,medium} = CA_{D,var} \tag{7h}$$

$$CA_{D,var,low} = 0.75\, CA_{D,var} \tag{7i}$$

$$CA_{D,var,v.low} = 0.5\, CA_{D,var} \tag{7j}$$

All the multipliers (from 0.5 to 1.5) refer to the change in the heating area (A_D) or change in the coefficient of the intensity of heat exchange between the radiator and the surrounding environment (C_D). The heating area should not be higher than the available heating area in the dwelling. Therefore, multipliers above 1 suggest an intensification of C_D, e.g., by intense ventilation of the heated spaces.

2.6. Method of Estimation of the Occupant–SH Interaction Characteristic

To assess which of the control methods is most likely to be applied in a given dwelling, calculations based on Equation (1) and all combinations of the above-described assumptions are compared with the measured heat consumption given from the following:

$$\dot{Q}_{D,SH,meas} = \frac{Q_{D,SH}}{24 \times 3600} \tag{8}$$

Here, the root mean square errors (RMSEs) are used as indicators that assess the fitting models with a given occupant–SH interaction assumption, i.e.,:

$$RMSE = \frac{\sqrt{\left(\dot{Q}_{D,SH,meas} - \dot{Q}_{D,SH}\right)^2}}{\dot{Q}_{D,SH,meas}} \tag{9}$$

3. Results

3.1. Model for All the Buildings

First, a model for the whole housing estate is investigated by summing the heat and heating water consumption from all the metered dwellings. The best-fit models are based on assumptions $CA_{D,var,medium} + t_{con,medium}$, which represent the middle of the considered values. The RMSE of the model predictions is 7% (Figure 7). The following models have a similar accuracy: $CA_{D,var,medium} + t_{con,high}$ (RMSE = 8%) and $CA_{D,var,medium} + t_{var,high}$ (RMSE = 9%).

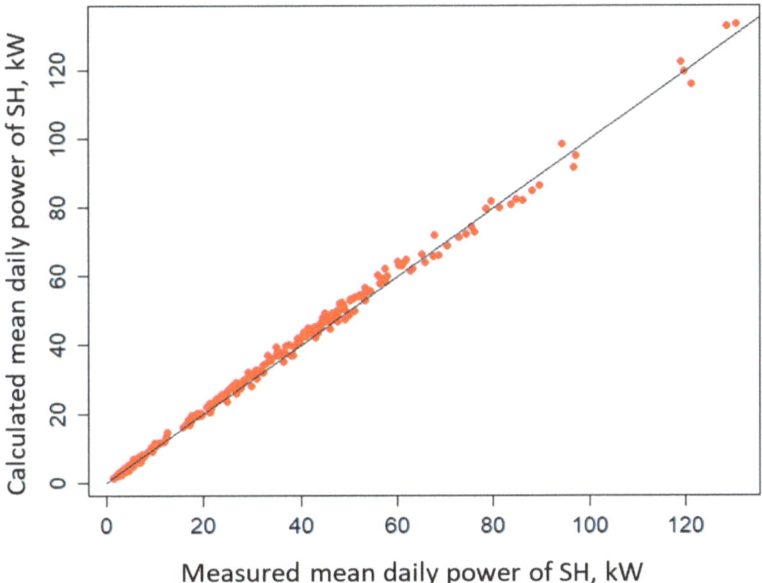

Figure 7. Measured values of the mean daily heat power delivered to the metered apartments of the housing estate compared with those predicted by the best-fit model ($CA_{D,var,medium} + t_{con,medium}$).

3.2. Dwellings Models

Table 2 shows several models of a given type found for the 101 investigated dwellings. The graphical comparison of the model results with the measurements for each dwelling is presented in Supplementary Material S2. The heating system control in the largest number of dwellings is described by the models $CA_{D,var,medium} + t_{con,medium}$ and $CA_{D,var,medium} + t_{con,high}$; the 17 dwellings are fitted for each.

The fitting of the models is shown as a histogram in Figure 8. Here, it can be seen that more than half of the models have an RMSE of less than 30%, and 80% of the models have an RMSE of less than 50%. We take RMSE < 25% as the range of the most reliable models (the red line in Figure 8). The accuracy of the models is correlated exponentially with the dwelling's heat consumption for the SH (Figure 9). This means that, for households with low heat consumption, low heat meter resolution causes a high RMSE (see Section 2.3.1).

Among the occupant–SH interaction models investigated, only 4% of the dwellings were fitted with models with a control based on automated flow adjustment by the TRV, and none of these have RMSE values lower than 25%. The models of the remaining dwellings suggest that the SH operation control is mainly based on the control of the operation time and the number of radiators (models $CA_{D,var,...}$): 69% (81% for RMSE < 25%) of the dwellings controlled the radiator area over the whole range of design heating areas ($CA_{D,var,medium}$), and in 28% (17% for RMSE < 25%) of the dwellings CA_D is larger than the designed one. As noted in Section 2.5.2, this could be due to a larger coefficient C, which represents a more intense heat transfer between the radiator and the indoor environment compared with standard conditions, e.g., those caused by simultaneous ventilation and heating of the dwellings.

Table 2. Number of dwellings for which a model with given assumptions is fitted. The value in brackets is the number of models for which RMSE < 25% is provided.

	$t_{var,high}$	$t_{con,high}$	$t_{con,medium}$	$t_{con,low}$	$t_{var,low}$	**Total**
$CA_{D,var,v.high}$	0	0	0	3(0)	0	3(0)
$CA_{D,var,high}$	2(0)	2(2)	8(1)	4(2)	8(3)	24(8)
$CA_{D,var,medium}$	11(7)	17(12)	17(11)	11(4)	13(5)	69(39)
$CA_{D,var,low}$	0	0	0	0	1(1)	1(1)
$CA_{D,var,v.low}$	0	0	0	0	0	0
$CA_{D,con,v.high}$	0	1(0)	0	0	0	1(0)
$CA_{D,con,high}$	0	0	0	0	0	0
$CA_{D,con,medium}$	0	0	0	0	0	0
$CA_{D,con,low}$	0	0	0	0	1(0)	1(0)
$CA_{D,con,v.low}$	0	2(0)	0	0	0	2(0)
Total	13(7)	22(14)	25(12)	18(6)	23(9)	101(48)

Indoor temperature adjustment in a dwelling is definitely more varied. The most widely used is the mean internal temperature profile, $t_{con,medium}$, which has been adjusted for the whole housing estate, employed in 25% (25% for RMSE < 25%) of the dwellings. The most rarely used profile is the one in which, for the duration of the heating period, there is an increase in the internal temperature compared to the transitional season ($t_{var,high}$); this applies to 13% (14% for RMSE < 25%) of the apartments.

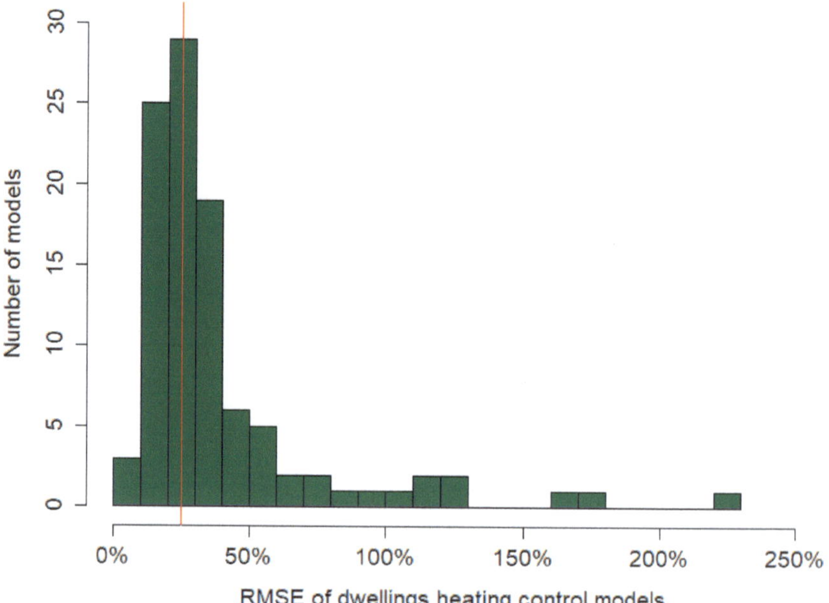

Figure 8. Histogram of the RMSE values for the dwelling models. The red line indicates the position of 25% RMSE, below which represents the most reliable models.

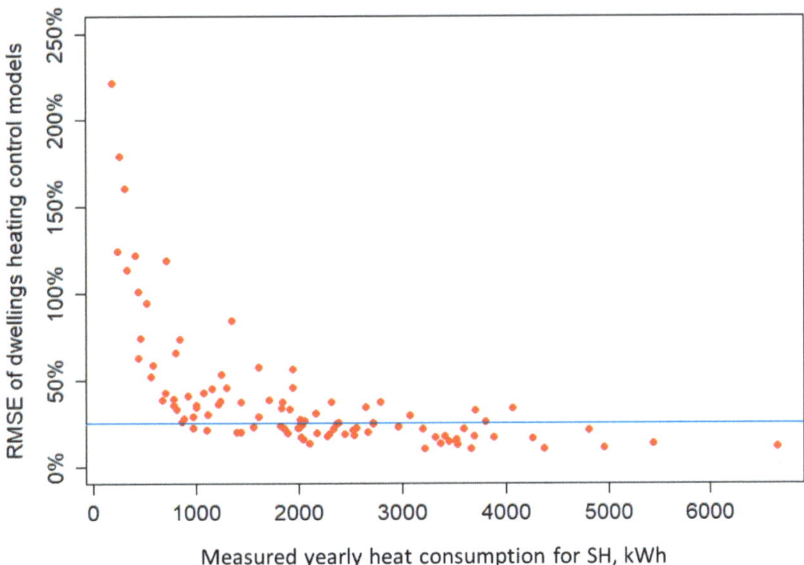

Figure 9. RMSE values of the dwelling models against the yearly heat consumption of the dwellings. The blue line indicates the position of 25% RMSE, below which represents the most reliable models.

3.3. Models of Occupant–SH Interaction and Dwelling Energy Characteristics

The χ^2 coefficient is used to check if there is a correlation between the two fitting models and the floor on which the dwelling is located, the heat transmittance of the dwelling, and the block to which the dwelling belongs. Only one of the six cases is statistically confirmed to correlate, i.e., the p-value of the test is less than 0.1. The correlation between the model of the internal temperature and the block where the apartment is located has the value $p = 0.06$. Table 3 summarizes the number of indoor temperature models in each block. It can be seen here that only within building WE3 are the proportion of the models with higher internal temperatures ($t_{var,high}$ and $t_{con,high}$) greater than those with lower ones ($t_{var,low}$ and $t_{con,low}$). In this building, there is a gas boiler room in the basement, so, on one hand, it is an object in which the heat losses from the boiler room heat a part of dwellings and, on the other hand, $t_{SH,sup}$ is higher than in the other buildings. From the results presented, it seems more important that the heat losses from the boiler room cause higher temperatures in some dwellings. Introducing a significantly lower $t_{SH,sup}$ in the model for those dwellings causes the temperature in the apartments to be lower than in reality. In other words, the effect of specifying an underestimated $t_{SH,sup}$ in the model of these dwellings would produce an overrepresentation of the dwellings with low indoor temperatures rather than higher temperatures.

Table 3. Number of dwellings for which the indoor temperature model is fitted by building type. The boiler is located in WE3, so this is where the highest supply temperature of the RTS resides.

	$t_{var,high}$	$t_{con,high}$	$t_{con,medium}$	$t_{con,low}$	$t_{var,low}$	Total
WE2	1	2	5	4	5	17
WE3	5	8	4	7	1	25
SN3	6	3	5	3	8	25
WE4	1	9	11	4	9	34

3.4. Models of Occupant–SH Interaction and Heat Consumption

The method by which the SH is adjusted may influence the heat consumption of the dwelling. However, the coefficient of the heat transmittance of the dwelling and, in the case of multi-family buildings, the heat exchange between the adjacent dwellings are also meaningful. Figure 10 shows the annual heat consumption of the dwellings described by each assumption. Looking at the median and mean values of the SH operation intensity, the models with higher values have a lower heat consumption per floor area, but these do not have statistically significant differences according to the ANOVA test, i.e., $p > 0.1$. In the case of the indoor temperature models, the effect on heat consumption is more pronounced; it is statistically significant according to the ANOVA test, i.e., $p < 0.1$. However, it is difficult to explain from the physical point of view because the highest heat consumption occurs for dwellings with mean indoor temperature values.

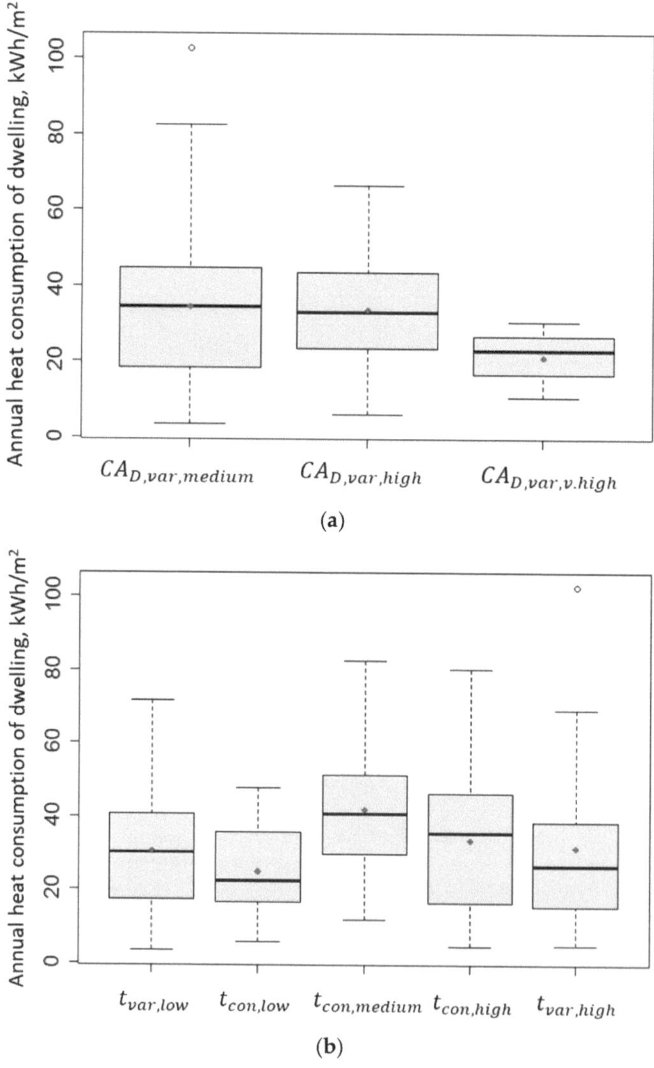

Figure 10. Annual heat consumption of the dwellings for SH purposes based on models with (**a**) SH operation intensity and (**b**) indoor temperatures.

Figures 11 and 12a show the length of the heating season for the models. Only the relationship between the indoor temperature model and the length of the heating season is statistically significant (ANOVA test, $p < 0.1$). On the one hand, the statistical significance vanishes (ANOVA test, $p > 0.1$) after limiting the set of models to the most reliable ones (RMSE < 25%). In contrast, examining only the models at $t_{con,...}$ and their medians (Figure 12b), the relationship is easier to interpret physically: the longer heating seasons are observed for the dwellings in which higher indoor temperatures are maintained. Next, the $t_{var,...}$ models imply a more specific use of the SH, which requires more research.

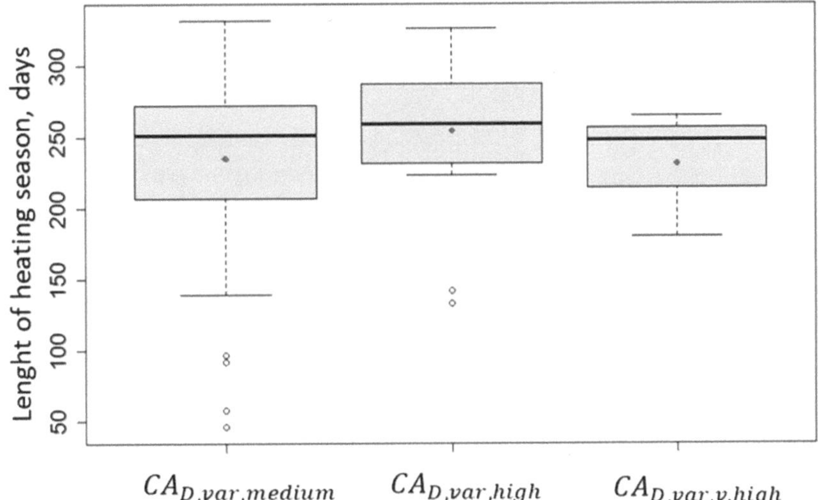

Figure 11. Length of the heating season from the models of the SH operation intensity.

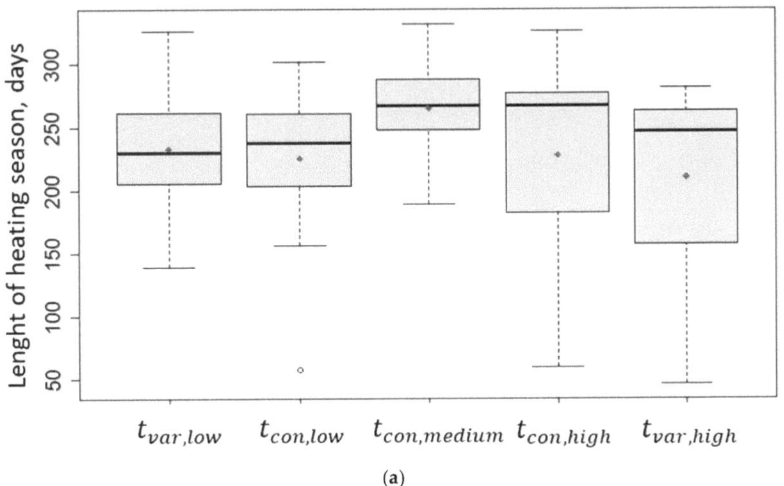

(a)

Figure 12. *Cont.*

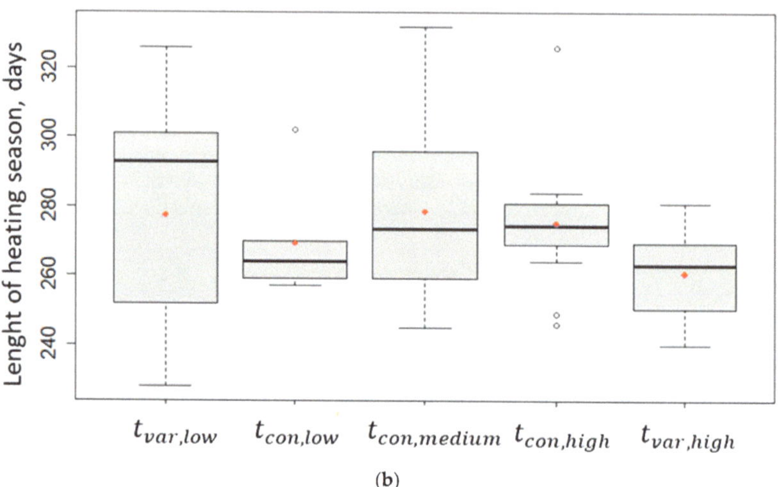

(**b**)

Figure 12. Length of the heating season from the indoor temperature models: (**a**) all the models and (**b**) the most reliable models (where RMSE < 25%).

4. Discussion

4.1. Sources of Uncertainty

The first issue to consider is how reliable the presented method is. As noted in the Introduction, the results provided here are preliminary. For a complete analysis of this approach, an experiment is needed in which, in addition to data from the heat meters, there will be data on the room temperature, the radiator temperature, and the window openness. It would also be valuable to have better-prepared heat meters with a higher resolution (see Section 2.3.1). It will then be possible to interpret precisely how the values obtained from this approach relate to the room conditions and the radiator operation, for example, whether t_{in} in Equation (1) is correlated with the room operative temperature or whether this correlation is strongly disturbed by radiator location. This is a challenge for all such indirect analyses that relate to an energy signature [27].

Second, it is important to note the specifics of the system under study. This is a constant-temperature heat network due to using the RTS, with direct coupling between the heat source and the radiators. The year-round high value of $t_{SH,sup}$ makes the operation of the TRV, when maintaining a fixed set point position (i.e., during the automatic flow control), less stable. For a given position, the TRV has a defined characteristic relating to the dependence of the opening degree and the temperature of the integrated sensor (i.e., the control stem). The flow rate and, thus, the radiator power at a constant, $t_{SH,sup}$, is a function of the opening degree of the TRV and the pressure generated by the pump in the network. If the pump operates in constant-pressure mode (which is how it is set up in this installation), then the power of the radiator can be assumed to depend on t_{in}, regardless of the heating load of the room. Next, the decrease in the power occurs as a result of an increase in t_{in} and only after another decrease will t_{in} begin to increase. In the case of the variable $t_{SH,sup}$, as a function of the external temperature, using the heating (TRVs have a hysteresis curve that stabilizes the operation slightly: on an increase in t_{in}, a given value corresponds to a lower flow than the same value during a decrease of t_{in} [48]) curve, the operation of the TRV would be more stable. Then, the heater power will also be regulated by the change in $t_{SH,sup}$ and, thus, correlates with the room heat load. Due to the constant $t_{SH,sup}$, a manual control enables higher comfort compared with the automatic TRV operation. Thirdly, in the case of the flow systems, the pressure system in the pipes and the proper balancing of the overall system are important [20,48]. In this aspect, the system was not evaluated.

At the end of this Section, it is noteworthy that the model that best describes the total consumption of all the metered dwellings is consistent with expectations. The following models describe it:

- $CA_{D,var,medium}$, in which the control is performed by changing the heating area or the operating time of the radiators, with maximum utilization of the available heating area, $\sum A_{D,max}$, in peak consumption. This is without increasing the heat transfer coefficient, C, between the radiator and the room.
- $t_{con,medium}$ representing the average internal temperature values observed in similar buildings during direct measurements [47].

4.2. The Actual Use of TRVs and Its Implications

As the Introduction notes, using the TRVs does not guarantee energy savings [18]. Although some datasets in the literature support this possibility [20], it is unclear whether energy savings can be achieved by balancing the system with a TRV or by automated control of the room temperature. However, a common problem with TRVs is their acceptance by users. As Kempton [49] noted over 30 years ago, the user perception and the use of thermostats can be divided in two ways. Kempton formulated them as a feedback theory and a valve theory. The first method of perception is consistent with the technical properties of the thermostats and accounts for the mechanism of their operation. The second method treats the thermostats as devices that directly serve to activate a heat (or cooling) source, perhaps with a specific power, but not self-regulating as a function of a set temperature. According to Kempton, 25–50% of Americans follow the valve theory. It can be assumed that after so much time has passed since this research, this state has changed in favor of the feedback theory as a result of public education. However, is it a matter of the education of society or more a matter of perception and the needs of individual people, for which the age of technology is less important? Occupant behaviors consistent with valve theory were also observed by Galvin [50] in thermostat data from Germany collected in 2012 and Aragon et al. in thermostat data from elderly residents collected in the UK in 2019/2020 [25]. A different picture is shown by Karjalainen's [22] research on how thermostats (including TRVs) were used in office buildings and homes in Finland, which was conducted about 15 years ago. They show that more than 60% of people in the living room adjust thermostats less than once a month and 80% less than once a week, but valve theory would suggest a more frequent adjustment of the settings. The response to feeling cold is usually to dress warmer (for more than 50% of the cases), only around 20% first increase the thermostat setting. This is more of a savings strategy, as Karjalainen reported no major problems with the technology: the Finns are highly rated, at 4.5/5 on average, for the accessibility of these thermostats and the ease of changing their settings. Only less than 20% felt that the room temperature response to the change in settings was slow or very slow.

The results presented in this paper are more supportive of Kempton's observations, with almost all the dwellings best described by the model that assumes manual control of the radiator area: increasing heat loads are covered by either running more (area) radiators or operating them for more extended periods. Flow control, which simultaneously causes lower flow rates and a higher decrease in heating water temperature as it flows through the system, is not only invisible in the surveyed buildings but also in the vast majority of dwellings. The discrepancy between the control methods declared by Karjalainen's questionnaires [22] and those resulting from the present analysis, on the assumption that both studies are reliable, can be explained by the occupancy of the dwellings. According to research [51], in Poland, the degree of overpopulation of apartments is four times higher than in Finland; in the studied housing estates, the population density was at least 22 m^2 per person. This situation causes the internal heat gains to have a much greater share in the energy balance of the dwelling and significantly affects the dynamics of this balance on the heat load side. This favors the dynamics in the use of the TRV and windows (natural ventilation) to obtain the effect faster without waiting for the automatic response of the TRV. A similar explanation could be valid for research by Galvin [50], which was performed on data from low-energy houses, so internal heat gains also have a greater share in the energy balance than in the case of standard buildings. As a result, the occupants cover their needs by not using TRV controllability.

From an energy efficiency standpoint, a failure to utilize the TRV automatic control functionality of the system is a problem. Such a situation is associated with higher flows in the system, i.e., a larger pump energy consumption, alongside a high heating medium temperature that leads to higher distribution losses. Moreover, radiators reach higher temperatures that cause higher ventilation heat losses when windows are open and higher heat losses by more intensive heat transfer to internal surfaces of the building envelope. This heat does not warm the space but directly dissipates to the outdoor environment.

4.3. Presented Approach Application

Assuming the robustness of the presented model, it could deliver crucial information about occupant-heating system interactions. These data could act as a basis for feedback provided to occupants, which will be the first step to educating or helping them toward more efficient use of their heating system. The next steps could be in two directions. One is the use of data to share the heat consumption bill between dwellings in one building. So far, methods based on dwelling heat meters or radiator heat allocators neglect heat transfer between dwellings that could be estimated based on the presented model, as one of the model outputs is the indoor temperature. The second possibility is to use information from the model to control the dwelling heating systems more individually, but such an approach needs to be developed with a deeper understanding of the user expectations regarding the heating system.

5. Conclusions and Future Work

The method for analyzing the daily data from heat meters, which includes the volume of heat delivered to the apartment and the volume of heating water that flows through the residential system, is presented. Based on these data, the paper answers two research questions:

(1) What information regarding the occupant control of the SH can be extracted from heat meters?

A novel approach for heat meter data analysis is presented. The employed model for the SH system, which is fitted to gather a relevant dataset, enables us to estimate indoor temperatures maintained in the dwelling, the intensity of the use of the radiators, and the way that the TRV is used. The analysis resulted in the description of the dwellings using the SH model parameters and examined the correlation of the model parameters with the characteristics of the dwellings. Two correlations were observed. The first correlation was between the estimated internal dwelling temperature and the length of the heating period of the dwelling. The median length of the heating period was lower for dwellings in which the lower internal temperatures were predicted (Figure 12b). The second correlation relates to the estimation by the model of the internal dwelling temperature and the building name to which particular apartments belong (Table 3). Observed higher estimated internal temperatures in building WE3 could be caused by large heat gains from the boiler room located in that building, which disturbed the SH usage in some of the dwellings of this building.

(2) Is it possible, based on this information, to learn more about people's use of TRV?

The model of the SH system used was fitted to gather a dataset that enables an estimation of the intensity of the use of the radiators and the way that the TRV is operated. The research indicated that the data from the vast majority of the dwellings (i.e., 96%) do not match the model of the automated control TRV, which means that users employed the TRV according to valve theory [49], i.e., without account for the important functionality of these devices, which self-adjusts as a function of the set indoor temperature.

The additional output of the presented research, given in Supplementary Material S1, is a method for separation of the registered heat consumption on the SH and the DHW consumption, accounting for the water system supply temperature during the year as well as the variation of the DHW consumption during the week (i.e., weekdays, Saturdays, and Sundays).

The presented method needs further improvement by a more comprehensive research project. There are two possible paths. The first is to employ laboratory space [37] or one-room measurements and observe the heat and water flow of a single radiator. In such experiments, the ventilation air temperature and its flow could be modified as radiator surroundings (e.g., furniture or location of ventilation air inlet) as well as water pressure and temperature in the heating system. The results could be used to compare the estimated CA_D and t_{in} to real values or ventilation and surround modifications. The second approach is to conduct the presented analysis in a building where smart TRV is implemented, e.g., ref. [25], and verify if the recorded use of the TRV could be found in heat meter data for the whole dwelling/building.

The improved model will be an easy-to-use tool to analyze the use of SH, which can be employed to control (or tariff) these systems or to advise users. The results of the study on the use of the TRV suggest a need to rethink the radiator control for multi-family buildings in thermally renovated and newly constructed buildings.

Supplementary Materials: The following supporting information can be downloaded at: https://www.mdpi.com/article/10.3390/en16227485/s1, Supplementary Material S1: estimation of water supply system temperature; estimation of supply temperature for residential thermal station and space heating; algorithm for separation of heat consumption for space heating and domestic hot water. Supplementary Material S2: the graphical comparison of the model results with the measurements for each dwelling.

Author Contributions: Conceptualization, K.B.; methodology, K.B.; investigation, K.B.; resources, K.B.; data curation, K.B.; writing—original draft preparation, K.B.; writing—review and editing, K.B. and A.G. and H.K.; visualization, K.B.; supervision, A.G. and H.K. All authors have read and agreed to the published version of the manuscript.

Funding: This research received no external funding. The APC was funded by Poznań University of Technology grant number SBAD PB 0981 and SBAD MK 0980.

Institutional Review Board Statement: Not applicable.

Informed Consent Statement: Not applicable.

Data Availability Statement: Part of data (heat meters data) are available from the ASHRAE Global Occupant Behavior Database: https://ashraeobdatabase.com (accessed on 26 October 2023).

Acknowledgments: The authors would like to thank our students, the manager of the investigated building, and the service staff for supporting this research. Special thanks also to Veolia Energia Poznań SA for data collected from their heating network and Kamstrup Sp. z o.o. for support in data collection.

Conflicts of Interest: The authors declare no conflict of interest.

Nomenclature

Abbreviations

DHW	domestic hot water
HN	heat network
RTS	residential thermal station
SH	space heating
TRV	thermostatic radiator valve
WSS	water system supply

Dates

$d_{V,0}$	start of general vacation
$d_{V,1}, d_{V,2}$	start and end date of school vacations
$d_{SH,0}, d_{SH,1}$	estimated start and end of the heating season

Temperatures

t_{DHW}	DHW temperature at the tapping point, °C
t_{in}	daily mean indoor temperature of the apartment, °C
$t_{HN,ret}$	daily mean return temperature of HN, °C
$t_{HN,sup}$	daily mean supply temperature of HN, °C
t_{out}	daily mean outdoor temperature, °C
$t_{RTS,sup}$	daily mean supply temperature of RTS, °C
$t_{SH,sup}$	daily mean supply temperature of SH, °C
$\Delta t_{SH-in,log}$	daily mean logarithmic temperature difference between the radiator temperature and the indoor temperature of the dwelling, °C
t_{WSS}	daily mean WSS temperature on RTS supply, °C

Energy, Power, and Flow

$\dot{m}_{D,SH}$	mean daily mass flow of heating water in SH of the dwelling, kg/s
Q_D	daily heat consumption of the dwelling, J
$Q_{D,DHW,wd}$, $Q_{D,DHW,st}$, $Q_{D,DHW,sn}$	daily heat consumption of the dwelling for DHW on weekdays, Saturdays, and Sundays, respectively, J
$Q_{D,SH}$	daily heat consumption of the dwelling for SH, J
$\dot{Q}_D$	mean daily heating power of RTS and radiators in the dwelling, W
$\dot{Q}_{D,SH}$	mean daily heating power of SH (sum of radiators power) in the dwelling, W
$\dot{Q}_{D,SH,max}$	maximum heating power of SH (sum of radiators power) in the dwelling, W
$\dot{Q}_{D,SH,meas}$	measured mean daily heating power of SH (sum of radiators power) in the dwelling, W
$\dot{Q}_{HN}$	mean daily heating power of HN, W
$\dot{Q}_{HN,loss,sup}$	mean daily heat loss from the supply pipe of HN, W
V_D	daily heating water flow through the dwelling, m³
$V_{D,DHW,wd}$, $V_{D,DHW,st}$, $V_{D,DHW,sn}$	daily heating water flow in dwelling RTS for DHW, on weekdays, Saturdays, and Sundays, respectively, m³
$V_{D,SH}$	daily heating water flow in the dwelling for SH, m³
$\dot{V}_{HN}$	mean daily volume flow of heating water in HN, m³/s

Other Physical Variables

A_D	heating area of the radiator, m²
$A_{D,max}$	total heating area of the radiators in the dwelling, m²
A_f	floor area of dwelling, m²
A_{win}	window area, m²
C	coefficient of the intensity of heat exchange between the radiator and the surrounding environment, W/(m²K)
C_m	thermal capacity of dwelling walls, kJ/K
c_w	specific heat of water, J/(kgK)
H_{tr}	heat transfer coefficient of the dwelling envelope, W/K
k_{Htr}	correction factor $\dot{Q}_D$, which accounts for dwellings and staircases without heat meters, unitless
$k_{q(twss)}$	coefficient concerning the influence of t_{WSS} on the change in heat consumption for DHW, unitless
$k_{v(twss, tRTS,sup)}$	coefficient concerning the influence of t_{WSS} and $t_{RTS,sup}$ on the change in heating water flow for DHW, unitless
n	exponent of the thermal characteristics of the radiator, unitless
ρ_w	water density, kg/m³

References

1. European Commission. *Communication from the Commission to the European Parliament, the Council, the European Economic and Social Committee and the Committee of the Regions a Renovation Wave for Europe—Greening Our Buildings, Creating Jobs, Improving Lives;* European Commission: Brussels, Belgium, 2020.
2. Janda, K.B. Buildings don't use energy: People do. *Archit. Sci. Rev.* **2011**, *54*, 15–22. [CrossRef]
3. Vellei, M.; de Dear, R.; Inard, C.; Jay, O. Dynamic thermal perception: A review and agenda for future experimental research. *Build. Environ.* **2021**, *205*, 108269. [CrossRef]
4. O'Brien, W.; Day, J.; Brackley, C. How Building Interfaces Affect Occupant Experience. *ASHRAE J.* **2021**, *63*, 28–32,34–35.
5. Day, J.K.; McIlvennie, C.; Brackley, C.; Tarantini, M.; Piselli, C.; Hahn, J.; O'Brien, W.; Rajus, V.S.; De Simone, M.; Kjærgaard, M.B.; et al. A review of select human-building interfaces and their relationship to human behavior, energy use and occupant comfort. *Build. Environ.* **2020**, *178*, 106920. [CrossRef]
6. Karjalainen, S. Should it be automatic or manual—The occupant's perspective on the design of domestic control systems. *Energy Build.* **2013**, *65*, 119–126. [CrossRef]
7. O'Brien, W.; Gunay, H.B. The contextual factors contributing to occupants' adaptive comfort behaviors in offices—A review and proposed modeling framework. *Build. Environ.* **2014**, *77*, 77–87. [CrossRef]
8. Sarran, L.; Brackley, C.; Day, J.K.; Bandurski, K.; André, M.; Spigliantini, G.; Roetzel, A.; Gauthier, S.; Stopps, H.; Agee, P.; et al. Untold Stories from the Field: A Novel Platform for Collecting Practical Learnings on Human-Building Interactions. In Proceedings of the IAQ 2020: Indoor Environmental Quality Performance Approaches, Transitioning from IAQ to IEQ, Athens, Greece, 4–6 May 2022.
9. Wei, S.; Jones, R.; De Wilde, P. Driving factors for occupant-controlled space heating in residential buildings. *Energy Build.* **2014**, *70*, 36–44. [CrossRef]
10. Vellei, M.; O'Brien, W.; Martinez, S.; Le Dréau, J. Some evidence of a time-varying thermal perception. *Indoor Built Environ.* **2022**, *31*, 788–806. [CrossRef]
11. Stopps, H.; Touchie, M.F. Residential smart thermostat use: An exploration of thermostat programming, environmental attitudes, and the influence of smart controls on energy savings. *Energy Build.* **2021**, *238*, 110834. [CrossRef]
12. Hosseinihaghighi, S.; Panchabikesan, K.; Dabirian, S.; Webster, J.; Ouf, M.; Eicker, U. Discovering, processing and consolidating housing stock and smart thermostat data in support of energy end-use mapping and housing retrofit program planning. *Sustain. Cities Soc.* **2022**, *78*, 103640. [CrossRef]
13. Sarran, L.; Gunay, H.B.; O'Brien, W.; Hviid, C.A.; Rode, C. A data-driven study of thermostat overrides during demand response events. *Energy Policy* **2021**, *153*, 112290. [CrossRef]
14. Huchuk, B.; O'Brien, W.; Sanner, S. A longitudinal study of thermostat behaviors based on climate, seasonal, and energy price considerations using connected thermostat data. *Build. Environ.* **2018**, *139*, 199–210. [CrossRef]
15. Woo Ham, S.; Karava, P.; Bilionis, I.; Braun, J. A data-driven model for building energy normalization to enable eco-feedback in multi-family residential buildings with smart and connected technology. *J. Build. Perform. Simul.* **2021**, *14*, 343–365. [CrossRef]
16. Csoknyai, T.; Hrabovszky-Horváth, S.; Georgiev, Z.; Jovanovic-Popovic, M.; Stankovic, B.; Villatoro, O.; Szendrő, G. Building stock characteristics and energy performance of residential buildings in Eastern-European countries. *Energy Build.* **2016**, *132*, 39–52. [CrossRef]
17. Główny Urząd Statystyczny. Zużycie Energii w Gospodarstwach Domowych w 2018 roku n.d. Available online: https://stat.gov.pl/obszary-tematyczne/srodowisko-energia/energia/zuzycie-energii-w-gospodarstwach-domowych-w-2018-roku,2,4.html (accessed on 26 October 2023).
18. Lomas, K.J.; Oliveira, S.; Warren, P.; Haines, V.J.; Chatterton, T.; Beizaee, A.; Prestwood, E.; Gething, B. Do domestic heating controls save energy? A review of the evidence. *Renew. Sustain. Energy Rev.* **2018**, *93*, 52–75. [CrossRef]
19. Lomas, K.J.; Allinson, D.; Watson, S.; Beizaee, A.; Haines, V.J.; Li, M. Energy savings from domestic zonal heating controls: Robust evidence from a controlled field trial. *Energy Build.* **2022**, *254*, 111572. [CrossRef]
20. Cholewa, T.; Balen, I.; Siuta-Olcha, A. On the influence of local and zonal hydraulic balancing of heating system on energy savings in existing buildings—Long term experimental research. *Energy Build.* **2018**, *179*, 156–164. [CrossRef]
21. Revell, K.M.A.; Stanton, N.A. Mental model interface design: Putting users in control of home heating. *Build. Res. Inf.* **2017**, *46*, 251–271. [CrossRef]
22. Karjalainen, S. Thermal comfort and use of thermostats in Finnish homes and offices. *Build. Environ.* **2009**, *44*, 1237–1245. [CrossRef]
23. Bruce-Konuah, A.; Jones, R.V.; Fuertes, A.; Messi, L.; Giretti, A. The role of thermostatic radiator valves for the control of space heating in UK social-rented households. *Energy Build.* **2018**, *173*, 206–220. [CrossRef]
24. Fabi, V.; Andersen, R.V.; Corgnati, S.P. Influence of occupant's heating set-point preferences on indoor environmental quality and heating demand in residential buildings. *HVAC R Res.* **2013**, *19*, 635–645.
25. Aragon, V.; James, P.; Gauthier, S. Revisiting Home Heat Control Theories through a UK Care Home Field Trial. *Energies* **2022**, *15*, 4990. [CrossRef]
26. Gauthier, S.; Shipworth, D. The gap between what is said and what is done: A method for distinguishing reported and observed responses to thermal discomfort. In Proceedings of the 8th Windsor Conference: Counting the Cost of Comfort in a Changing World, Windsor, UK, 10–13 April 2014.

27. Rasmussen, C.; Bacher, P.; Calì, D.; Nielsen, H.A.; Madsen, H. Method for Scalable and Automatised Thermal Building Performance Documentation and Screening. *Energies* **2020**, *13*, 3866. [CrossRef]
28. Tureczek, A.M.; Nielsen, P.S.; Madsen, H.; Brun, A. Clustering district heat exchange stations using smart meter consumption data. *Energy Build.* **2019**, *182*, 144–158. [CrossRef]
29. Aragon, V.; James, P.A.B.; Gauthier, S. The influence of weather on heat demand profiles in UK social housing tower blocks. *Build. Environ.* **2022**, *219*, 109101. [CrossRef]
30. Beck, H.E.; Zimmermann, N.E.; Mcvicar, T.R.; Vergopolan, N.; Berg, A.; Wood, E.F. Data Descriptor: Present and future Köppen-Geiger climate classification maps at 1-km resolution. *Sci. Data* **2018**, *5*, 1–12. [CrossRef] [PubMed]
31. Bandurski, K. Occupant Impact on the Energy Balance of Residential Buildings—Investigation and Modelling. Doctoral Dissertation, Poznan University of Technology, Poznan, Poland, 2021.
32. ASHRAE Global Occupant Behavior Database n.d. Available online: https://ashraeobdatabase.com/#/ (accessed on 27 January 2022).
33. Dong, B.; Liu, Y.; Mu, W.; Jiang, Z.; Pandey, P.; Hong, T.; Olesen, B.; Lawrence, T.; O'Neil, Z.; Andrews, C.; et al. A Global Building Occupant Behavior Database. *Sci. Data* **2022**, *9*, 369. [CrossRef] [PubMed]
34. Mielnicki, J.S. *Centralne Ogrzewanie: Regulacja i Eksploatacja*; Arkady: Warszawa, Poland, 1985.
35. Muniak, D. Charakterystyki cieplne grzejników w instalacjach centralnego ogrzewania. *Część I. Ciepłownictwo Ogrzew. Went.* **2013**, *44*, 241–245.
36. Kwiatkowski, J.; Cholewa, L. *Centralne Ogrzewanie Pomoce Projektanta*; Arkady: Warszawa, Poland, 1980.
37. Dzierzgowski, M. Verification and Improving the Heat Transfer Model in Radiators in the Wide Change Operating Parameters. *Energies* **2021**, *14*, 6543. [CrossRef]
38. Wojtkowiak, J.; Oleśkowicz-Popiel, C. *Eksperymenty w Wymianie Ciepła*; Wydawnictwo Politechniki Poznańskiej: Poznań, Poland, 2004.
39. Nicol, F. Temperature and adaptive comfort in heated, cooled and free-running dwellings. *Build. Res. Inf.* **2017**, *45*, 730–744. [CrossRef]
40. Boerstra, A.C.; Van Hoof, J.; Van Weele, A.M. A new hybrid thermal comfort guideline for the Netherlands: Background and development. *Archit. Sci. Rev.* **2015**, *58*, 24–34. [CrossRef]
41. Nicol, J.F.; Humphreys, M.A. Adaptive thermal comfort and sustainable thermal standards for buildings. *Energy Build.* **2002**, *34*, 563–572. [CrossRef]
42. Peeters, L.; de Dear, R.; Hensen, J.; D'haeseleer, W. Thermal comfort in residential buildings: Comfort values and scales for building energy simulation. *Appl. Energy* **2009**, *86*, 772–780. [CrossRef]
43. Carlucci, S.; Bai, L.; de Dear, R.; Yang, L. Review of adaptive thermal comfort models in built environmental regulatory documents. *Build. Environ.* **2018**, *137*, 73–89. [CrossRef]
44. Coley, D.; Herrera, M.; Fosas, D.; Liu, C.; Vellei, M. Probabilistic adaptive thermal comfort for resilient design. *Build. Environ.* **2017**, *123*, 109–118. [CrossRef]
45. Kane, T.; Firth, S.K.; Hassan, T.M.; Dimitriou, V. Heating behaviour in English homes: An assessment of indirect calculation methods. *Energy Build.* **2017**, *148*, 89–105. [CrossRef]
46. Kelly, S.; Shipworth, M.; Shipworth, D.; Gentry, M.; Wright, A.; Pollitt, M.; Crawford-Brown, D.; Lomas, K. Predicting the diversity of internal temperatures from the English residential sector using panel methods. *Appl. Energy* **2013**, *102*, 601–621. [CrossRef]
47. Ilomets, S.; Kalamees, T.; Vinha, J. Indoor hygrothermal loads for the deterministic and stochastic design of the building envelope for dwellings in cold climates. *J. Build. Phys.* **2017**, *41*, 547–577. [CrossRef]
48. EN 215:2019 Thermostatic Radiator Valves—Requirements and Test Methods n.d. Available online: https://standards.cencenelec.eu/dyn/www/f?p=CEN:110:0::::FSP_PROJECT,FSP_ORG_ID:65664,6112&cs=1E82804F8C28EDA99815971F676E02023 (accessed on 3 February 2022).
49. Kempton, W. Two Theories of Home Heat Control. *Cogn. Sci.* **1986**, *10*, 75–90. [CrossRef]
50. Galvin, R. Targeting "behavers" rather than behaviours: A "subject-oriented" approach for reducing space heating rebound effects in low energy dwellings. *Energy Build.* **2013**, *67*, 596–607. [CrossRef]
51. Eurostat. Dane Statystyczne Dotyczące Mieszkalnictwa—Statistics Explained n.d. Available online: https://ec.europa.eu/eurostat/statistics-explained/index.php?title=Archive:Housing_statistics/pl&oldid=498749#Jako.C5.9B.C4.87_mieszka.C5.84 (accessed on 4 February 2022).

 *energies*

Article

Seasonal Air Quality in Bedrooms with Natural, Mechanical or Hybrid Ventilation Systems and Varied Window Opening Behavior-Field Measurement Results

Magdalena Baborska-Narożny [1] **and Maria Kostka** [2,*]

1   Faculty of Architecture, Wrocław University of Science and Technology, ul. Bolesława Prusa 53/55, 50-317 Wrocław, Poland
2   Department of Air Conditioning, Heating, Gas Engineering and Air Protection, Wrocław University of Science and Technology, Norwida St. 4/6, 50-373 Wrocław, Poland
*   Correspondence: maria.kostka@pwr.edu.pl

Abstract: The article presents the results of measurements of temperature, relative humidity and CO_2 concentration in six single-family houses' bedrooms located in Poland, in Wrocław and vicinity, during two climatic seasons: summer–autumn and winter. Two buildings with natural ventilation (NV) were tested, three with mechanical ventilation with heat recovery (MV) and one with hybrid ventilation (HV)—mixed mode natural and mechanical. The behavior of residents regarding opening windows was analyzed and the influence of the changing internal and external conditions on their active reactions was examined. The analysis confirms and adds to the global discourse on the key impact of user behavior on securing healthy indoor air quality in housing, regardless of ventilation system or building energy standard. A disconnect exists between the observed window opening practices and typical design principles, assuming adjustment to a given ventilation system or changing weather conditions. The observations showed that in both analyzed seasons it was possible to obtain a good quality internal environment, in terms of CO_2 level, regardless of the ventilation system used in the building. However, unfavorable results were observed for one bedroom, in which the inhabitants do not adapt their behavior to local technical conditions. Taking into account the level of relative humidity (RH), much higher values were observed in the NV bedrooms in both analyzed periods. The obtained results were divided into IAQ classes in accordance with the EN 16798-1. The recorded values of the internal temperature confirm the significant influence of the location of the room in the building and the actions taken by the residents.

Keywords: bedroom ventilation; indoor air quality; window opening behavior; hybrid ventilation; MVHR; natural ventilation; housing

Citation: Baborska-Narożny, M.; Kostka, M. Seasonal Air Quality in Bedrooms with Natural, Mechanical or Hybrid Ventilation Systems and Varied Window Opening Behavior-Field Measurement Results. *Energies* **2022**, *15*, 9328. https://doi.org/10.3390/en15249328

Academic Editors: Katarzyna Ratajczak and Łukasz Amanowicz

Received: 10 November 2022
Accepted: 7 December 2022
Published: 9 December 2022

1. Introduction

According to the reports of the Statistics Poland [1], more than 350,000 single-family houses were commissioned for use in Poland over the years 2018–2021. As the mean household size is approx. 2.6 people, it can be assumed that nearly 1 million new bedrooms have been added to single-family housing. According to the Organization for Economic Co-operation and Development (OECD) data, the average sleep time in the 30 member countries is 8 h 24 min [2]. It is relevant to understand the factors that enhance or hinder the safety and comfort of the bedroom environment, where people spend almost 35% of their lives. In recent years, there have been several dozen publications on the bedroom internal environment. Research shows that the effective exchange of indoor air is crucial for sleep quality [3–5]. Sekhar, Akimoto et al. [6,7] provide a rich source of knowledge about relevant standards and research evidence. The authors summarized the findings on the basic bedroom air parameters, i.e., temperature, relative humidity, air exchange rate and carbon dioxide concentration. The research reveals a wide variety of internal conditions worldwide, which, apart from independent factors such as weather, external pollutants (e.g., noise) [8–11] or building characteristics, are also influenced by the residents and their strategies of cooperation with the building and its equipment [12–15]. Our previous research also suggests that the influence of residents is fundamental to the shaping of internal conditions and energy consumption in housing [16–18]. Canha et al. [19] focused on a review of field studies seeking to understand the bedroom environment, and concluded it was essential to provide further evidence from a "wider range of settings (including different countries)" ([19], p. 17). This prompted us to conduct research focused on the conditions of a temperate transitional climate, which is characteristic of Poland. On the one hand, this climate is characterized by the occurrence of cold and hot periods, negatively affecting the internal environment and forcing active methods of its maintenance (heating, cooling). On the other hand, it is distinguished by long periods of mild conditions, allowing for the passive functioning of buildings. Batog and Badura [20] demonstrated exceeded CO_2 concentrations in bedrooms in Poland by analyzing socialist blocks of flats that rely on natural ventilation, which were often inherently weakened by the lack of trickle vents. Our study looks at newly built energy efficient houses, equipped with three types of ventilation systems. Firstly, naturally ventilated homes, i.e., mainstream typology until recently for the Polish residential sector. Secondly, homes relying on whole house mechanical ventilation systems with heat recovery (MVHR), and lastly, those equipped with hybrid ventilation, allowing for a combination of passive and active methods of air exchange.

2. Materials and Methods

2.1. Case Study Characteristics

The reported data was collected as a part of a bigger ongoing research project focused on the influence of user behavior on thermal comfort and energy consumption in energy efficient new-built houses. Within the project, annual data is collected for 10 case studies of voluntarily-participating households. The data includes monitoring of internal environment conditions, focusing on thermal comfort, energy consumption and occupant feedback. Data collection began in the summer of 2021. Key selection criteria for inclusion into the study sample were: occupancy longer than 2 years at the beginning of the study (handover prior to 2019), house energy efficiency standards exceeding targets mandatory at the time of their design, and inclusion of systems supporting low energy goals such as mechanical ventilation with heat recovery, heat pumps or PVs. Location in the vicinity of Wrocław for all case-studies was preferable due to planned repeated on-site visits and similar climatic conditions. All the recruited houses are detached or semi-detached, built in Wrocław and vicinity, up to 40 km away from the city center. All were constructed between 2012 and 2017. The analyses presented in the article were based on the results of measurements collected in the main bedrooms of 6 naturally, mechanically and hybrid ventilated houses. The characteristic parameters of the monitored bedrooms are presented in Table 1.

Table 1. Monitored bedrooms characteristics.

House	NV1	NV2	MV1	MV2	MV3	HV1
Ventilation type *	NV	NV	MV	MV	MV	HV
Floor area, m^2/High, m	19.4/2.62	23/2.80	12/2.70	16/2.70	12/2.70	18.2/2.73
Volume, m^3	50.8	64.4	32.4	43.2	32.4	33.9
No. of occupants	2	2/3	2/3	2	2/1	2/3
Thermal mass	high	high	Low	low	low	medium
Floor	ground floor	ground floor	1st floor	1st floor	1st floor	1st floor
Window orientation	E	W/N	N	S/E	N	W
Noise exposure	medium	low	low	low	low	very low

* NV—natural ventilation, MV—natural ventilation with heat recovery, HV—hybrid ventilation (natural and mechanical change-over system).

For three buildings, the load bearing wall material is ceramic bricks; for four buildings, it is cross laminated timber, which leads to varied thermal mass (Table 1). Two of the analyzed houses rely on natural ventilation, and the rest are equipped with mechanical ventilation with heat recovery. Of the latter group, one house has a hybrid system, where the occupants can choose whether to use MV or to switch to NV. All the houses have floor heating systems and heat pumps. Two bedrooms are on ground floor level and four on the first floor, of which one is adjacent to a pitched roof.

All the bedrooms have openable and easily accessible windows. In terms of external noise, i.e., a factor potentially limiting night-time windows opening, all the houses are located in quiet neighborhoods, with NV1 relatively most exposed to potential noise from a nearby road and HV1 least exposed to noise. In terms of external air quality, four houses are located in Wrocław, and thus are exposed to poor air quality, mostly in the heating season [21]. HV1 and NV1 are located within villages with some buildings in the vicinity relying on solid fuel for heating and hot water. There are times of day when windows need to be closed to prevent polluted air from entering the house; however this is typically not during the analyzed hours of the night. NV2 is located min. 2 km away from any solid fuel heating sources and close to a wooded area, suggesting it has the lowest air pollution in the studied sample.

2.2. Methods

The analysis was performed for two periods covering the warm season (summer–autumn), when free running mode was allowed (22 August–31 October 2021) and winter (9 January–15 March 2022), when active heating season was required. The weather data was obtained from the weather station in the center of Wrocław [22]. Temperature, relative humidity and carbon dioxide concentration loggers have been installed in each bedroom. HOBO MX1102A and Comet Vision U3430 meters with parameters listed in Table 2 were used. The parameters of the measuring devices are presented in Table 2.

Table 2. CO_2, temperature and relative humidity (RH) monitoring equipment used in the study.

	Data Logger HOBO MX 1102A	**Data Logger Comet Vision U3430**
Measuring range	Temperature: 0–50 °C RH: 1–90% CO_2: 0 ppm–5000 ppm	Temperature: −20–60 °C RH: 0–100% CO_2: 0 ppm–5000 ppm
Accuracy	Temperature: ±0.21 °C from 0 °C to 50 °C RH: ±2% from 20% to 80% typical to a maximum of ±4.5% including hysteresis at 25 °C; below 20% and above 80% ± 6% typical CO_2: ±50 ppm ± 5% of reading at 25 °C, less than 90% RH non-condensing and 1013 mbar	Temperature: ±0.4 °C RH: ±1.8% CO_2: ±(50 ppm + 3% from reading) at 25 °C and 1013 hPa
Resolution	Temperature: 0.024 °C at 25 °C RH: 0.01% CO_2: 1 ppm	Temperature: 0.1 °C RH: 0.1% CO_2: 1 ppm
Sampling interval	15 min	15 min

After a walk-through accompanied by the residents, magnetic reed switches were installed in the windows indicated as those used to ventilate the bedrooms. Magnetic reed switches developed and manufactured by Efento were used for the tests. In five houses, reed switches were installed directly in the bedrooms. In one house, a reed switch was installed in the corridor in the immediate vicinity of the bedroom, as according to the residents, the bedroom window is always kept closed overnight. The reed switches record information on the opening of the windows in a 5 min time step (information on the opening status of the window in successive 5 min periods, not the actual duration of the opening). Table 3 summarizes the basic information on the location of the measuring equipment and the operating mode of the ventilation system in the analyzed measurement periods, and Figure 1 shows their location in buildings. All sensors were located at the height 0.6–0.7 m above floor level.

Table 3. Measuring equipment and operating mode of the ventilation system in buildings.

House	Type of Data Logger	Magnetic Reed Switch Location	Ventilation Operation Mode	
			Warm Season Summer–Autumn 22 August–31 October 2021	Heating Season Winter 9 January–15 March 2022
NV1	HOBO MX1102A	corridor	NV	NV
NV2	Comet Vision U3430	bedroom	NV	NV
MV1	Comet Vision U3430	bedroom	MV	MV
MV2	Comet Vision U3430	bedroom	MV	MV
MV3	Comet Vision U3430	bedroom	MV	MV
HV1	HOBO MX1102A	bedroom	HV-NV	HV-MV

Figure 1. Location of measuring equipment: (**a**) NV1; (**b**) NV2; (**c**) HV1; (**d**) MV1; (**e**) MV2; (**f**) MV3.

For the analyzed periods, the bedroom occupancy hours were assumed to be between 11 p.m. and 7 a.m. The registered air parameters were compared for the selected hours. In the next step, four periods covering three consecutive days were selected, for which a detailed profile of individual parameters was presented against the windows opening behavior.

The level of carbon dioxide concentration and the relative humidity (RH) in rooms were compared to the EN 16798-1 standard [23]. In terms of the level of CO_2, it classifies bedrooms in four IAQ categories depending on the internal increase in concentration in relation to the atmospheric air. The internal increase in value does not exceed 380 ppm for Category I and 950 ppm for Category IV. Assuming the average CO_2 concentration in the outside air is 400 ppm, it gives the final achieved concentration in the range of 780–1350 ppm. The EN 16798-1 standard does not classify residential buildings as requiring air humidification or dehumidification, however, for the purposes of this article, the measured values were compared with categories of rooms with controlled relative humidity. The highest Category I includes rooms with a RH from 30% to 50%, Category II from 25% to 65%, and the lowest Category III from 20% to 70%.

3. Analysis and Results

3.1. Carbion Dioxide Concentration

Figure 2 shows the concentration of carbon dioxide in bedrooms recorded in two measurement periods, between 11 p.m. and 7 a.m. During the warm period, three buildings were ventilated naturally and three were mechanically ventilated. In the cold period, two buildings were ventilated naturally and four mechanically (Table 3).

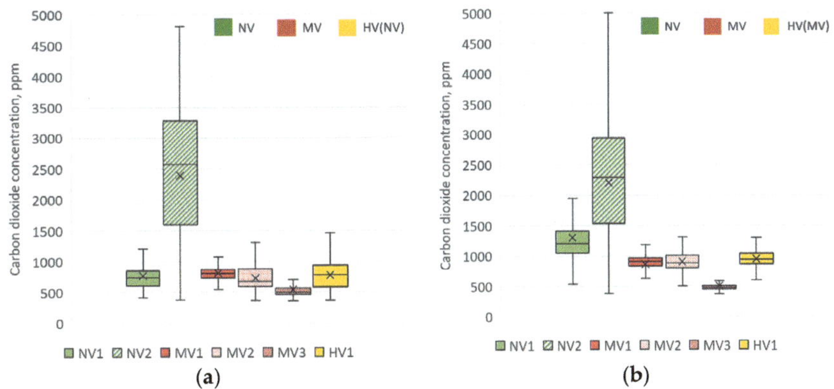

Figure 2. Bedroom carbon dioxide concentration, hours 11 p.m.–7 a.m.: (**a**) summer–autumn; (**b**) winter.

In the warm season, apart from the NV2 building, low levels of CO_2 concentration in the rooms were maintained. Average values for NV1, MV1, MV2, HV1 ranged from 763 ppm to 805 ppm, and for MV3 533 ppm. Maintaining the level I category according to EN 16798-1 was achieved 62% (NV1), 40% (MV1), 65% (MV2), 93% (MV3) and 51% (HV1) of the time. The periods of exceeding the level of IV category were much shorter—2% (MV1) and 3% (NV1). In MV2 and HV1, the period of exceeding class IV was <0.3% of the time, and in MV3, no exceeding was recorded. In the cold season, in mechanically ventilated buildings, CO_2 concentrations were lower than in the naturally ventilated NV1 building, but a general upward trend was observed in all buildings except MV3. Average values for MV1, MV2 and HV1 ranged from 861 ppm to 942 ppm, for NV1 it was 1299 ppm, and for MV3 it was 497 ppm. Class I maintenance was observed for 8% (NV1), 20% (MV1), 22% (MV2), 98% (MV3) and 10% (HV1) of the time. For all bedrooms with the MV system, the period of exceeding the IV category was $\leq 1\%$, while for the NV1 building it was 28%.

The maximum values of CO_2 concentration were recorded in the NV2 house. Regardless of the season, it remained at a very high level. The average value for the warm season was 2384 ppm, and for the cold season was 2197 ppm. Meeting the conditions of category I was achieved for 18% of measurements in the warm season and 12% of measurements in the winter. Exceeding the IV category was observed for 79% of the time, both in the warm and cold period. Several measuring points reached the value of 5000 ppm, which is the maximum recorded by the measuring device.

3.2. Temperature

Figure 3 shows the temperature in bedrooms recorded in two measurement periods, between 11 p.m. and 7 a.m.

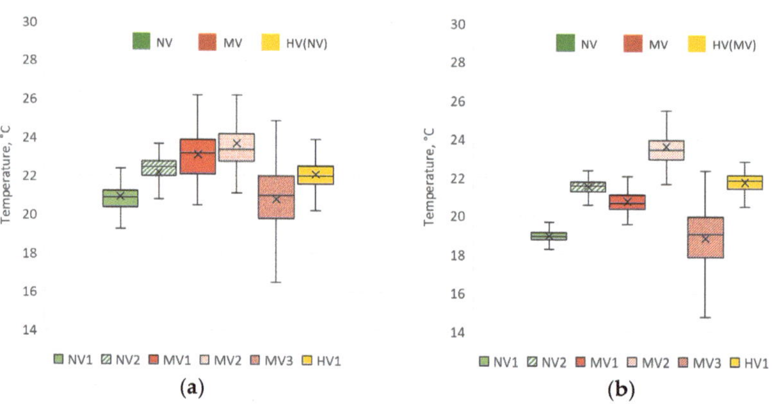

Figure 3. Bedroom temperature, hours 11 p.m.–7 a.m.: (**a**) summer–autumn; (**b**) winter.

The lowest average air temperatures in the bedrooms were recorded in the NV1 and MV3 buildings and are, respectively, 21.0 °C and 20.8 °C for the warm season and 18.8 °C and 18.9 °C for the winter season. The largest temperature fluctuations occurred in the MV3 building, which resulted from the frequent opening of windows by residents, discussed later in the study. The low average temperature in the NV1 bedroom has different contributing causes for each of two periods analyzed, with the exception of its ground floor location, previously linked with overall more stable and lower internal temperatures [16] than on higher floors. Otherwise, for the heating season, it was the NV1 residents' preference for lower temperatures than in the other five households that explains the lowest mean temperature in the sample. In the free-running season, two factors seem to underpin the NV1 indoor environment: architectural design (e.g., high thermal mass, relatively small window area) and residents' practices of keeping both internal bedroom doors open, thus allowing cross-ventilation into other cool spaces. Further analysis of the factors other than window opening within the bedrooms spaces is beyond the scope of this paper. The contribution of door opening is not represented in the window opening time analysis; however, it is possible that open doors together with the unsealing of the windows, not registered by the reed switch, is sufficient to ensure high air exchange in the room. The second room located on the ground floor is the NV2 bedroom, where the average temperature in the analyzed periods is close to 22 °C. Here, however, the sealing of the room for the night (both internal doors and windows) prevents proper air exchange, which is confirmed by the other recorded parameters. Similar internal temperatures in both seasons were maintained in the HV1 facility (approx. 22 °C). Bedrooms MV1, MV2, MV3 and HV1 are located on the 1st floor, of which MV2 is the only room with a window facing south. This results in the highest average temperature of 23.7 °C for both periods. The most pronounced differences in the measurement results for individual seasons were observed for bedroom MV1—the average of the warm period was 23.1 °C, and for the cold period was 20.8 °C.

3.3. Relative Humidity

Figure 4 shows the relative humidity in bedrooms recorded in two measurement periods, between 11 p.m. and 7 a.m.

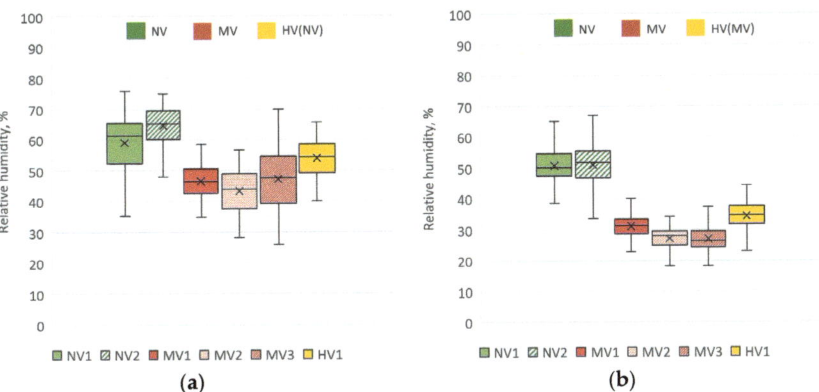

Figure 4. Bedroom relative humidity, hours 11 p.m.–7 a.m.: (**a**) summer–autumn; (**b**) winter.

Both in the warm and cold season, naturally ventilated bedrooms were characterized by higher values of relative humidity. This is most clear in the case of the NV1 and NV2 buildings, where the average relative humidity in the warm season was 59% and 65%, and 51% in the cold season for both buildings. The EN 16798-1 classifies rooms with normalization of relative humidity to the lowest, III category, if RH levels of 20–25% and 60–70% are observed. The lower and higher values went beyond the lowest class. NV1 and NV2 bedrooms would belong to the 3rd room category for 52% and 53% respectively in the warm season, while the time of moving beyond the lowest category was 5% and 24%. In winter, the relative humidity in these buildings was lower—exceeding 60% was observed for 2% and 3% of the time, and no measurements exceeding 70% was observed. There was also no relative humidity lower than 30% in any of the periods. In the HV1 building, which was also naturally ventilated at that time, the average relative humidity in summer was 54% and was higher than in buildings with the MV system. The period of exceeding the relative humidity of 60% in the warm season was 18% for HV1, and no measurements exceeding 70% were recorded.

In the MV bedrooms, the average relative humidity in the warm season ranged from 44% to 47%, and in the cold season from 27% to 31%. Category III in this period was observed only in the MV3 building for 13% of the warm season, and in the cold season in the MV2 and MV3 buildings for 2% of the measurement period. However, the time of failure to maintain the relative humidity of 30% was long—in the MV1, MV2 and MV3 buildings, it was, respectively, 34%, 80% and 80% of the time. In the HV1 building, mechanically ventilated at that time, the RH was higher and amounted to 35%. No relative humidity <20% was observed, and the values of 20–30% represented only 13% of the measuring points. The MV1, MV2 and MV3 buildings were equipped with air handling units with plate heat exchangers; only in the HV1 building was there an air handling unit with a rotary heat exchanger, which enables partial transfer of moisture between air streams.

3.4. Residents' Behavior

Figure 5 shows the share of registered window opening cycles in two analyzed measurement periods. The applied reed switches register whether a window is in the closed position for each 5 min cycle. They record "alarm" whenever a window is open wide, slightly tilted or only its air tightness is released. No indication of opening time is recorded. Presented results show how many of all cycles were those during which the window opening state was recorded. It is not a precise information about the duration of the opening, but it illustrates the trend in the behavior of householders, which affects the above-mentioned measurement results.

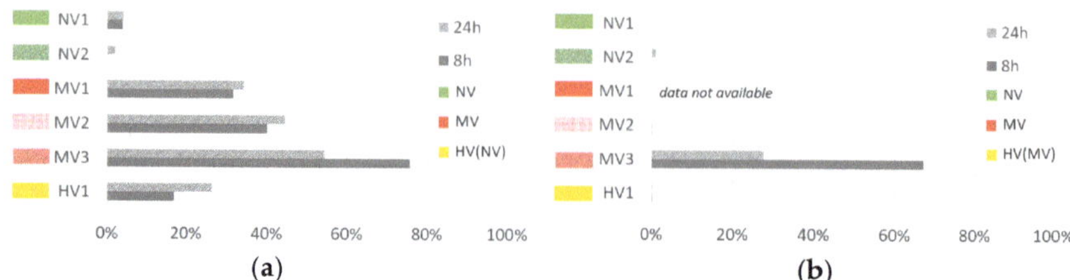

Figure 5. Share of cycles with an open window: (**a**) summer–autumn; (**b**) winter.

In the warm period, the lowest number of window opening cycles was recorded in bedrooms NV1 and NV2. The inhabitants of these buildings also did not open their windows in the winter season. These buildings are naturally ventilated and the windows are not equipped with air inlets. NV1 and NV2 bedrooms have the largest cubature of all analyzed rooms and are the only ones on the ground floor. In the warm season, only 2% of opening cycles were registered in the NV2 building throughout the day, almost none of which occurred at night (<0.1%). The behavior of the inhabitants is clearly reflected in the measurement results, especially in the concentration of CO_2 and relative humidity. In the NV1 building, window opening cycles in the warm season accounted for only 4%, but most of it took place at night. During the cold season, the inhabitants of the NV1 building occasionally opened their windows during the day (<0.1%), but no opening was recorded during the night. The results of CO_2 concentration measurements in this room did not differ so drastically from other buildings, but the level of relative humidity was high. In the HV1 facility, which was naturally ventilated in the warm season, the share of window opening cycles throughout the day was 26%, while 17% were at night. In addition, no drastic differences in the concentration of CO_2 were observed here, and the relative humidity level, compared to other measurements, had an average value. The largest number of window opening cycles was recorded in continuously mechanically ventilated buildings. In the warm season, in MV1 and MV2 buildings, window opening was recorded in 33% and 45% of measurement cycles during the day and in 32% and 40% of night cycles. Interesting results were observed in the MV3 building. The share of window opening time here was the highest of all buildings, and mostly concerned the night time. In the warm season, the opening cycles accounted for 55% of the entire day and 76% of the night time. In winter, it was 28% and 67%, respectively. The behavior of users is reflected in all previous results of internal parameters measurements—the lowest observed CO_2 concentrations and its slight fluctuations, the largest temperature fluctuations and its drops to a value deviating from what is commonly considered comfortable, and significant fluctuations in relative humidity in the warm season. In the cold season, smaller fluctuations in the internal relative humidity result from slight fluctuations in the moisture content in the external air. In summer, fluctuations in the moisture content in the outside air are greater, and these phenomena are typical for the Polish climate.

3.5. Daily Variability of Registered Parameters

Figures 6–9 show the detailed variability of the recorded parameters in selected periods lasting three consecutive days. The selection of the dates was based on the assessment of internal conditions. Periods selected:

- warm season, high value of the external temperature T_e and significant fluctuations between the time of day and night: $T_{e,max} = 29.2\ °C$, $T_{e,min} = 7.6\ °C$ (Figure 6),

- warm season, smaller fluctuations in external temperature T_e between day and night: $T_{e,max} = 24.5\ °C$, $T_{e,min} = 12.1\ °C$ (Figure 7),
- mild season, small fluctuations in external temperature between day and night: $T_{e,max} = 16.1\ °C$, $T_{e,min} = 2.6\ °C$ (Figure 8),
- cold season, the lowest recorded temperature values: $T_{e,max} = 3\ °C$, $T_{e,min} = -8.3\ °C$ (Figure 9).

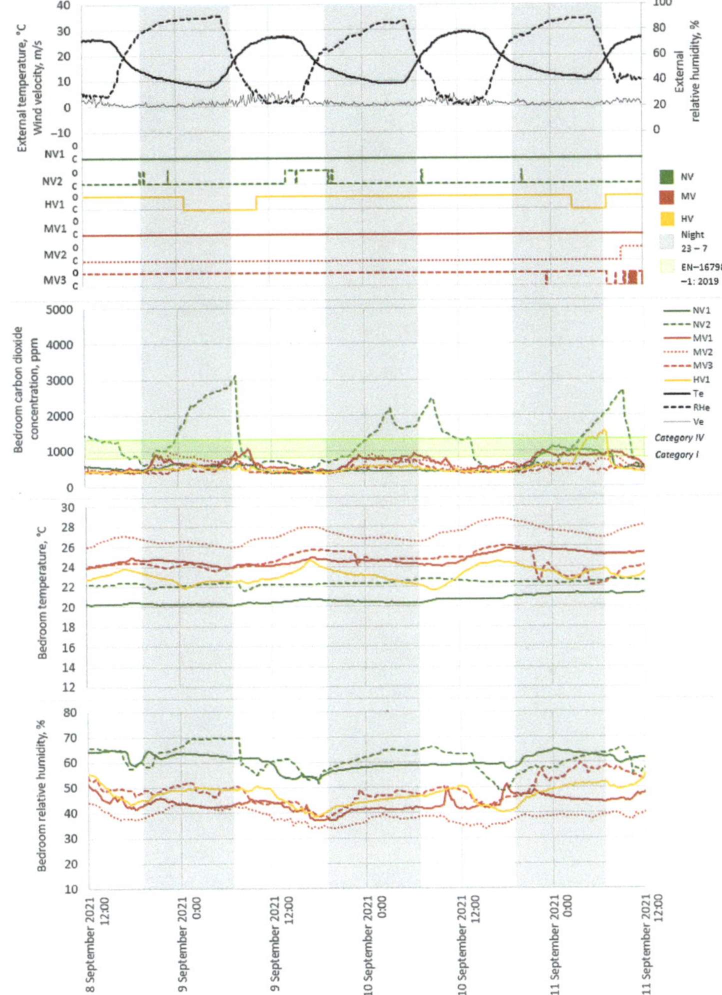

Figure 6. Detailed variability of the measured parameters over the period 8–11 September 2021.

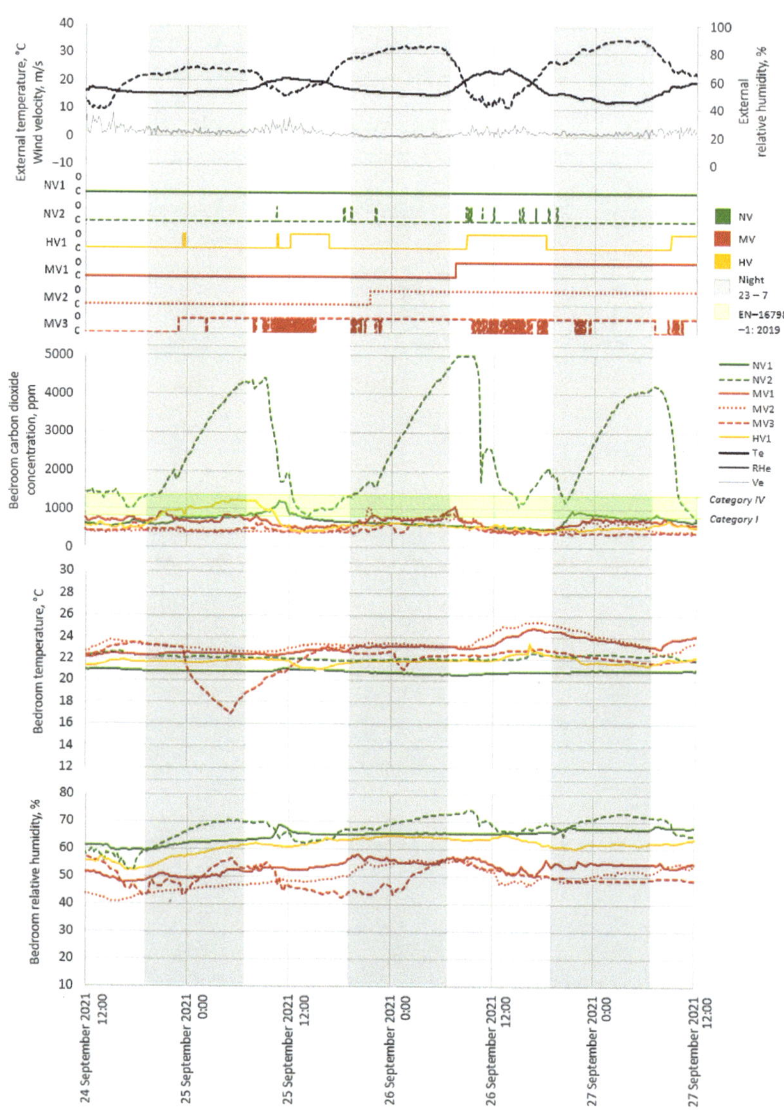

Figure 7. Detailed variability of the measured parameters over the period 24–27 September 2021.

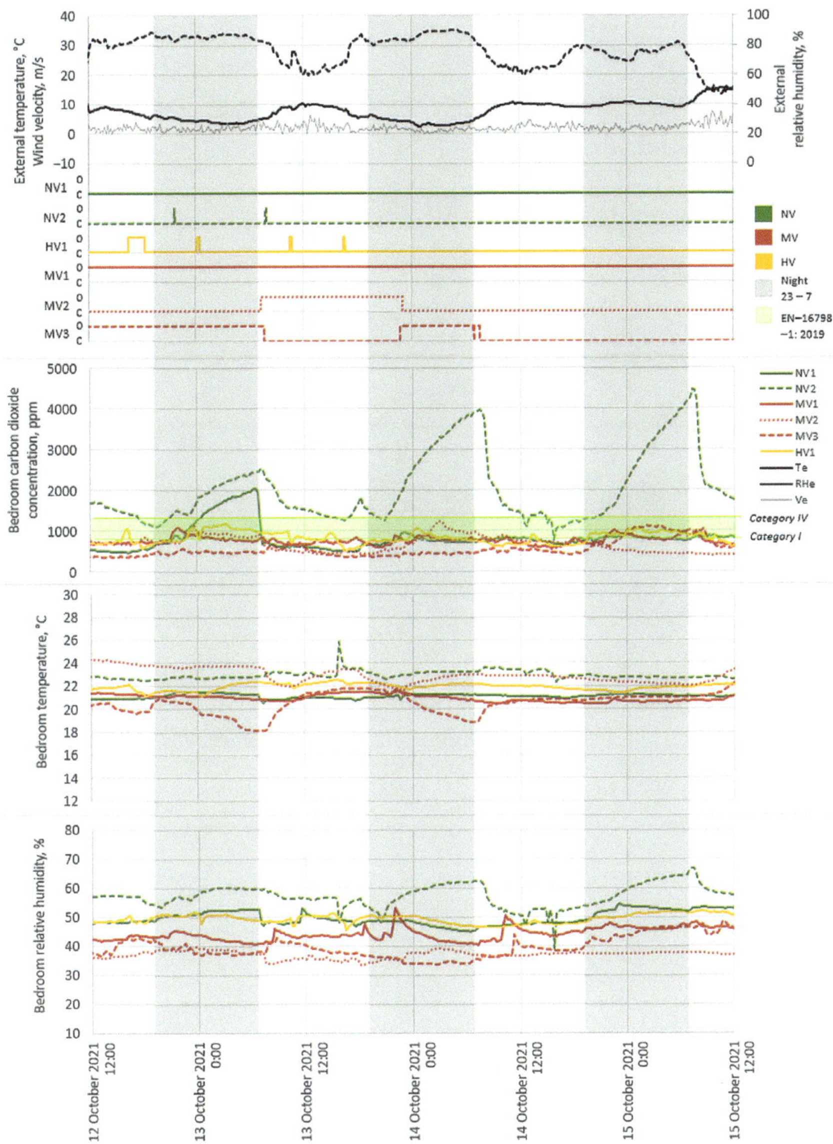

Figure 8. Detailed variability of the measured parameters over the period 12–15 October 2021.

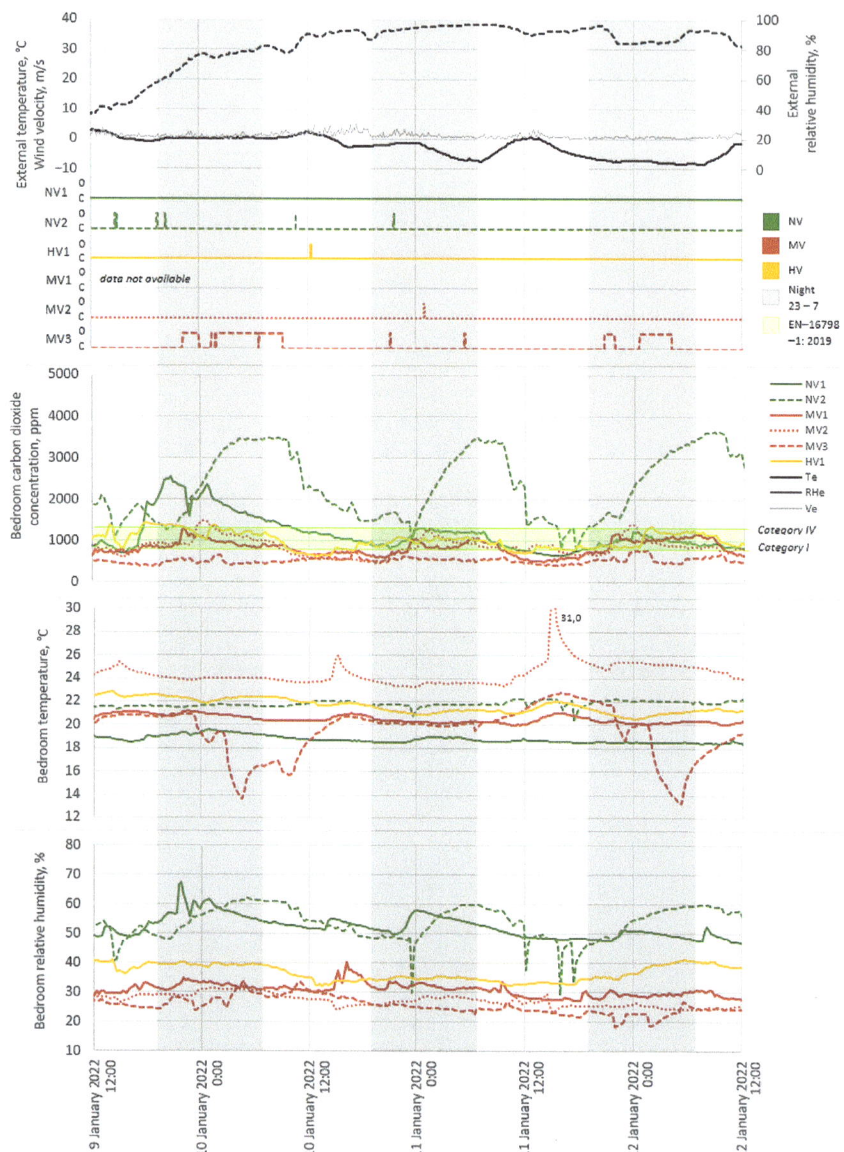

Figure 9. Detailed variability of the measured parameters over the period 9–12 January 2022.

For selected periods, the reaction of residents to changing external and internal conditions was analyzed. Symbols on the chart "O" and "C" next to the names of objects mean the position of the window "open" and "close".

During the period of the highest recorded values of the outside air temperature (Figure 6), the windows in the NV1, MV1 and MV2 buildings were closed. The residents' reaction to changing external conditions were not observed. The residents of MV2 did not use the window to cool the bedroom naturally, despite the high temperature inside. In the NV2 bedroom, a temporary window opening was observed during the night periods and one longer period of opening during the daytime. The inhabitants undertook actions that were likely to improve the internal conditions, but the procedure did not bring tangible results. Windows remained closed even at high CO_2 and relative humidity levels. In the HV1 building, the window was open for most of the period, but the reaction of the residents to the drop in temperature in the room was recorded—when the internal temperature dropped to approx. 22 °C at night, the window was closed. This resulted in a significant increase in CO_2 concentration on the morning of 11 November 2021. In the MV3 building, the window was mostly open. In the morning hours of 11 November 2021 there were problems with the reed switch. This period was excluded from the window position data analysis.

During the warm period with a lower amplitude of the external air temperature (Figure 7), the windows in the NV1, MV1 and MV2 buildings still remained in the closed position. The temperature in the rooms decreased compared to the previous period, so the energy accumulated in the building during warm periods was partially discharged. In the NV2 building, no changes in the behavior of users were observed—short periods of window opening occurred mainly during the day, with no reaction of residents to deteriorating internal conditions. In the HV1 building, the residents actively reacted to the changing external conditions. They chose not to open the window at night, but the CO_2 concentration was in the range of category IV. The MV3 reed switch indications were once again partially excluded from the analysis, but it can be noticed that a significant drop in internal temperature did not cause any reaction in the inhabitants.

During the mild period with a small amplitude of the outside temperature (Figure 8), most of the windows remained closed. Longer periods of opening were registered only in the buildings MV2 and MV3. Concentrations of CO_2 in closed spaces, apart from NV2, increased, but still allowed at least category IV to be maintained. The conditions in the NV2 bedroom did not change significantly, but a reduction in the number of cycles of airing the room was observed.

During the period of the lowest recorded values of outside air temperature (Figure 9), most windows remained closed. Due to the failure of the reed switch, no data on the opening of the MV1 bedroom window are available, but the lack of significant drops in the internal temperature suggests that it was also closed. During this period, the HV1 building operated in the mechanical ventilation mode, which had a noticeable effect on the reduction of window opening times. In the MV3 mechanically ventilated bedroom, the windows were still partially open despite periodic drops in the internal temperature to approx. 14 °C, but these periods were significantly shortened. In addition, no apparent changes in user behavior were recorded.

4. Discussion

Case studies were conducted on a small group of carefully selected sample buildings [24]. As such, they provided a chance to explore the rich context underpinning observed measurements, but they do not allow for drawing universal conclusions that can be fully translated into the entire typology of single-family buildings. However, the dependencies observed during the analyses confirm and build on the results recorded by scientists in other countries and other climates, concerning the significant impact of the way the building is used on the internal conditions it achieves.

Research on internal conditions in bedrooms ventilated with various systems, both mechanical and natural, was presented by Sekhar, Bivolarova et al. [25]. On the basis of measurements carried out during the heating season, they noticed that in a naturally ventilated bedroom, the concentration of CO_2 was usually 2.5–3 times higher than in a mechanically ventilated bedroom. Mechanical ventilation guaranteed good air mixing and dilution of CO_2 concentration to the level of approx. 1000 ppm. They also analyzed the influence of door opening and closing on the air exchange rate. In a naturally ventilated bedroom, despite the use of air inlets, this coefficient was very low and amounted to <0.15 h^{-1} with the door closed and 0.3 h^{-1} with the door open. For a mechanically ventilated bedroom it was 0.6 h^{-1}.

The results of tests in the cold period presented in the article also covered the heating period. The CO_2 concentrations observed in the two naturally ventilated bedrooms (NV1, NV2) exceeded the values observed in all buildings with MV. In the NV1 building, the average value was approx. 1.5 times the value for MV1, MV2 and HV1 buildings, and the period of maintaining at least category IV lasted over 70% of the registered time. In the NV2 building, the average value of CO_2 concentration was approx. 2.5 times the value for buildings with MV1, MV2 and HV1, and the maintenance period of at least category IV was only approx. 20% of the registered time. In mechanically ventilated bedrooms, the maintenance period of at least category IV was over 99% of the time, which proves the higher efficiency of air exchange by MV systems. The values observed in the MV3 building differ from other mechanically ventilated rooms, which is related to the frequent opening of the windows, as shown in Figure 5. Apart from the higher CO_2 concentration in the NV buildings, there are also diametrical differences between recorded values in NV1 and NV2. The observed discrepancies confirm the fundamental influence of the way the interior is used and the awareness of the inhabitants. The broad range of CO_2 concentrations in naturally ventilated interiors confirms the results also observed in other studies [26–29].

Information on the internal environment and the opening of windows in naturally ventilated bedrooms was also provided by Heide, Skyttern and Georges [30]. For 10 bedrooms located in six detached houses in Trondheim, they measured temperature, relative humidity, CO_2 concentration, particulate matter, formaldehyde and TVOC. The research was conducted in March and April. Most of the bedroom windows were open during the research. CO_2 concentrations exceeding the external concentration by over 950 ppm (category IV) were observed only in two bedrooms, and it lasted for 70% and 80% of the night period (the analyzed time range was 23:00–6:00). In the remaining bedrooms, the exceedance time was shorter and amounted to 10% or less. Six bedrooms had an average daily temperature of <18 °C, and the remaining four were >21 °C. Among the cooler bedrooms, the mean RH varied between 32% and 49%, while in three of the four warmer rooms it was <20%.

The research results for the warm period presented in the article also show a low level of CO_2 in the bedrooms, regardless of the ventilation system. The maximum observed time of exceeding the IV IAQ category was 3%. The exception is the NV2 facility, where the lowest category was not met for almost 80% of the time, and the problem was probably the lack of awareness of its inhabitants. The values of temperature and relative humidity do not meet the comparative conditions due to the discrepancy in the analyzed seasons. These parameters can be compared with the summary of the analysis by Sekhar, Akimoto et al. [6]. The authors found that in the analyzed bedrooms, the internal temperature range in the heating season was from 20 °C to 25 °C, and in the cooling season from 25 °C to 30 °C, with the greatest temperature variation occurring in facilities ventilated in a natural way. Relative humidity ranged from 40% to 80%, and the differences between the heating and cooling seasons were less pronounced. In the presented case study, the average indoor temperatures ranged from approx. 19 °C to approx. 24 °C in the heating season and from approx. 21 °C to approx. 24 °C in the warm season (but mostly outside the cooling season). Average values of relative humidity for the heating period were from 27% to approx. 50% in the heating season and from 43% to 65% in the warm season.

Satisfactory results of hybrid ventilation bedroom measurements provide the basis for further research on the validity of using such a solution, as global studies [31–36] show a significant impact of the use of natural ventilation in various climates on reducing energy demand and internal comfort.

5. Conclusions

This field study covered six energy efficient houses in and around Wrocław, Poland. Several key conclusions emerged:

- the fact that a bedroom is ventilated by natural means is not synonymous with the impossibility of maintaining a high-quality internal environment; other important factors are the heating season, bedroom size and door opening behavior,
- maintaining a high-quality internal environment using natural ventilation is, however, more reliable in warm and transitional periods,
- natural ventilation of rooms may increase the relative humidity to a level deviating from comfort standards, and carries the risk of the development of pathogenic organisms (e.g., fungi, mold),
- mechanical ventilation of rooms may cause the relative humidity to drop below the standard of comfort conditions,
- it is possible to effectively ventilate rooms with the use of hybrid ventilation systems, combining mechanical and natural systems in the "change-over" mode,
- effective natural ventilation of rooms requires knowledge, awareness and taking up activity by the residents,
- mechanical ventilation systems are more consistent in shaping the internal environment of a bedroom and are more resistant to the passivity of residents,
- the habits and preferences of the residents in many cases do not correspond to the activities expected for a given building standard.

The conducted analyses also allowed for the formulation of questions, and the answers to them will be sought during further exploration of the issue:

- Is it possible to determine which of the internal parameters is the main trigger of the residents' adaptive behaviors?
- What is the impact of active residents' responses to improve the quality of the indoor environment on the buildings' energy consumption?
- How does a hybrid ventilation system perform compared to mechanical or natural one in terms of household internal air quality and energy consumption?
- What is the influence of the location of the room in the building, and its location in relation to CO_2 emission sources, on the observed measurement results?
- What non-technical factors underpin residents' practices related to IAQ control in the bedrooms?
- Does the use of a rotary heat exchanger for heat recovery avoid unfavorable drops in internal relative humidity during the heating season?

Author Contributions: Conceptualization, M.B.-N. and M.K.; methodology, M.B.-N. and M.K.; software, M.K.; validation, M.K.; formal analysis, M.B.-N. and M.K.; investigation, M.B.-N. and M.K.; resources, M.B.-N. and M.K.; data curation, M.B.-N.; writing—original draft preparation, M.B.-N. and M.K.; writing—review and editing, M.B.-N. and M.K.; visualization, M.B.-N. and M.K. All authors have read and agreed to the published version of the manuscript.

Funding: The research was supported by the National Science Centre Poland as part of the OPUS15 competition project entitled "The impact of inhabitant involvement in micro-climate control on thermal comfort and energy use in owner-occupier low energy homes—interdisciplinary mixed methods research". (No. 2018/29/B/HS6/02958). The APC was funded by the Faculty of Architecture and the Department of Air Conditioning, Heating, Gas Engineering and Air Protection Faculty of Environmental Engineering, Wrocław University of Science and Technology.

Data Availability Statement: Data not provided due to privacy restrictions.

Acknowledgments: The authors gratefully acknowledge the voluntary participation and time devoted to this study by the residents of the analyzed houses.

Conflicts of Interest: The authors declare no conflict of interest. The funders had no role in the design of the study; in the collection, analyses, or interpretation of data; in the writing of the manuscript, or in the decision to publish the results.

References

1. STAT. Available online: https://stat.gov.pl/en/ (accessed on 10 November 2022).
2. OECD. Available online: http://stats.oecd.org/Index.aspx?datasetcode=TIME_USE (accessed on 10 November 2022).
3. Fan, X.; Liao, C.; Bivolarova, P.; Sekhar, C.; Laverge, J.; Lan, L.; Mainka, A.; Akimoto, M.; Wargocki, P. A field intervention study of the effects of window and door opening on bedroom IAQ, sleep quality, and next-day cognitive performance. *Build. Environ.* **2022**, *225*, 109630. [CrossRef]
4. Liao, C.; Akimoto, M.; Bivolarova, M.P.; Sekhar, C.; Laverge, J.; Fan, X.; Lan, L.; Wargocki, P. A Survey of Bedroom Ventilation Types and the Subjective Sleep Quality Associated with Them in Danish Housing. *Sci. Total Environ.* **2021**, *798*, 149209. [CrossRef] [PubMed]
5. Xiong, J.; Lan, L.; Lian, Z.; De Dear, R. Associations of Bedroom Temperature and Ventilation with Sleep Quality. *Sci. Technol. Built Environ.* **2020**, *26*, 1274–1284. [CrossRef]
6. Sekhar, C.; Akimoto, M.; Fan, X.; Bivolarova, M.; Liao, C.; Lan, L.; Wargocki, P. Bedroom Ventilation: Review of Existing Evidence and Current Standards. *Build. Environ.* **2020**, *184*, 107229. [CrossRef]
7. Sekhar, C.; Akimoto, M.; Fan, X.; Bivolarova, M.; Liao, C.; Lan, L.; Wargocki, P. Corrigendum to "Bedroom Ventilation: Review of Existing Evidence and Current Standards". *Build. Environ.* **2021**, *201*, 107983. [CrossRef]
8. Marini, M.; Frattolillo, A.; Baccoli, R.; Carlo Mastino, C. Incidence of the Ventilation Holes and the Mechanical Ventilation Systems of Façade on the Noise Insulation. *Energy Procedia* **2016**, *101*, 265–271. [CrossRef]
9. Lam, B.; Shi, D.; Gan, W.S.; Elliott, S.J.; Nishimura, M. Active Control of Broadband Sound through the Open Aperture of a Full-Sized Domestic Window. *Sci. Rep.* **2020**, *10*, 10021. [CrossRef]
10. Christina, E. Mediastika Barrier Design Strategies to Control Noise Ingress into Domestic Buildings. *Dimens. J. Tek. Arsit.* **2003**, *31*, 52–60. [CrossRef]
11. Kim, M.K.; Barber, C.; Srebric, J. Traffic Noise Level Predictions for Buildings with Windows Opened for Natural Ventilation in Urban Environments. *Sci. Technol. Built Environ.* **2017**, *23*, 726–735. [CrossRef]
12. Pereira, P.F.; Ramos, N.M.M.; Almeida, R.M.S.F.; Simões, M.L.; Barreira, E. Occupant Influence on Residential Ventilation Patterns in Mild Climate Conditions. *Energy Procedia* **2017**, *132*, 837–842. [CrossRef]
13. McGill, G.; Oyedele, L.O.; McAllister, K. Case Study Investigation of Indoor Air Quality in Mechanically Ventilated and Naturally Ventilated UK Social Housing. *Int. J. Sustain. Built Environ.* **2015**, *4*, 58–77. [CrossRef]
14. Park, J.S.; Jee, N.Y.; Jeong, J.W. Effects of Types of Ventilation System on Indoor Particle Concentrations in Residential Buildings. *Indoor Air* **2014**, *24*, 629–638. [CrossRef] [PubMed]
15. Canha, N.; Lage, J.; Candeias, S.; Alves, C.; Almeida, S.M. Indoor Air Quality during Sleep under Different Ventilation Patterns. *Atmos. Pollut. Res.* **2017**, *8*, 1132–1142. [CrossRef]
16. Baborska-Narozny, M.; Stevenson, F.; Chatterton, P. Temperature in Housing: Stratification and Contextual Factors. In *Proceedings of the Institution of Civil Engineers: Engineering Sustainability*; Thomas Telford Ltd.: London, UK, 2016; Volume 169. [CrossRef]
17. Baborska-Narożny, M.; Stevenson, F.; Grudzińska, M. Overheating in Retrofitted Flats: Occupant Practices, Learning and Interventions. *Build. Res. Inf.* **2017**, *45*, 40–59. [CrossRef]
18. Baborska-Narozny, M.; Stevenson, F. Continuous Mechanical Ventilation in Housing—Understanding the Gap between Intended and Actual Performance and Use. *Energy Procedia* **2015**, *83*, 167–176. [CrossRef]
19. Canha, N.; Teixeira, C.; Figueira, M.; Correia, C. How Is Indoor Air Quality during Sleep? A Review of Field Studies. *Atmosphere* **2021**, *12*, 110. [CrossRef]
20. Batog, P.; Badura, M. Dynamic of Changes in Carbon Dioxide Concentration in Bedrooms. *Procedia Eng.* **2013**, *57*, 175–182. [CrossRef]
21. Baborska-Narozny, M.; Szulgowska-Zgrzywa, M.; Mokrzecka, M.; Chmielewska, A.; Fidorow-Kaprawy, N.; Stefanowicz, E.; Piechurski, K.; Laska, M. Climate Justice: Air Quality and Transitions from Solid Fuel Heating. *Build. Cities* **2020**, *1*, 120–140. [CrossRef]
22. Uniwersytet Wrocławski; Obserwatorium Meteorologiczne; Zakład Klimatologii i Ochrony Atmosfery, Instytut Geografii i Rozwoju Regionalnego; Wydział Nauk o Ziemi i Kształtowania Środowiska. *Dane klimatyczne*; Uniwersytet Wrocławski: Wrocław, Poland, 2021; (revised 2022).
23. *PN-EN 16798-1:2019*; Energy Performance of Buildings—Ventilation for Buildings—Part 1: Indoor Environmental Input Parameters for Design and Assessment of Energy Performance of Buildings Addressing Indoor Air Quality, Thermal Environment, Lighting a. Polish Committee for Standardization: Warsaw, Poland, 2019.
24. Yin, R.K. *Case Study Research: Design and Methods*, 4th ed.; Sage: Thousand Oaks, CA, USA, 2009; Volume 14. [CrossRef]
25. Sekhar, C.; Bivolarova, M.; Akimoto, M.; Wargocki, P. Detailed Characterization of Bedroom Ventilation during Heating Season in a Naturally Ventilated Semi-Detached House and a Mechanically Ventilated Apartment. *Sci. Technol. Built Environ.* **2020**, *27*, 158–180. [CrossRef]
26. Baborska-Narożny, M.; Laska, M.; Fidorów-Kaprawy, N.; Mokrzecka, M.; Małyszko, M.; Smektała, M.; Stefanowicz, E.; Piechurski, K. Thermal Comfort and Transition from Solid Fuel Heating in Historical Multifamily Buildings—Real-World Study in Poland. *Energy Build.* **2021**, *248*, 111178. [CrossRef]
27. Lei, Z.; Liu, C.; Wang, L.; Li, N. Effect of Natural Ventilation on Indoor Air Quality and Thermal Comfort in Dormitory during Winter. *Build. Environ.* **2017**, *125*, 240–247. [CrossRef]

28. Bekö, G.; Lund, T.; Nors, F.; Toftum, J.; Clausen, G. Ventilation Rates in the Bedrooms of 500 Danish Children. *Build. Environ.* **2010**, *45*, 2289–2295. [CrossRef]
29. Fernández-Agüera, J.; Dominguez-Amarillo, S.; Fornaciari, M.; Orlandi, F. TVOCs and PM 2.5 in Naturally Ventilated Homes: Three Case Studies in a Mild Climate. *Sustainability* **2019**, *11*, 6225. [CrossRef]
30. Heide, V.; Skyttern, S.; Georges, L. Indoor Air Quality in Natural-Ventilated Bedrooms in Renovated Norwegian Houses. In *E3S Web of Conferences*; EDP Sciences: Les Ulis, France, 2021; Volume 246. [CrossRef]
31. Scheuring, L.; Weller, B. An Investigation of Ventilation Control Strategies for Louver Windows in Different Climate Zones. *Int. J. Vent.* **2021**, *20*, 226–235. [CrossRef]
32. Schulze, T.; Gürlich, D.; Eicker, U. Performance Assessment of Controlled Natural Ventilation for Air Quality Control and Passive Cooling in Existing and New Office Type Buildings. *Energy Build.* **2018**, *172*, 265–278. [CrossRef]
33. Park, K.Y.; Woo, D.O.; Leigh, S.B.; Junghans, L. Impact of Hybrid Ventilation Strategies in Energy Savings of Buildings: In Regard to Mixed-Humid Climate Regions. *Energies* **2022**, *15*, 1960. [CrossRef]
34. Khalil, S.; Ghali, K.; Ghaddar, N.; Itani, M. Hybrid Mixed Ventilation System Aided with Personalised Ventilation to Attain Comfort and Save Energy. *Int. J. Sustain. Energy* **2020**, *39*, 964–981. [CrossRef]
35. Schulze, T.; Eicker, U. Controlled Natural Ventilation for Energy Efficient Buildings. *Energy Build.* **2013**, *56*, 221–232. [CrossRef]
36. Raji, B.; Tenpierik, M.J.; Bokel, R.; van den Dobbelsteen, A. Natural Summer Ventilation Strategies for Energy-Saving in High-Rise Buildings: A Case Study in the Netherlands. *Int. J. Vent.* **2020**, *19*, 25–48. [CrossRef]

Article

A Small Modular House as a Response to the Energy Crisis

Miroslaw Zukowski

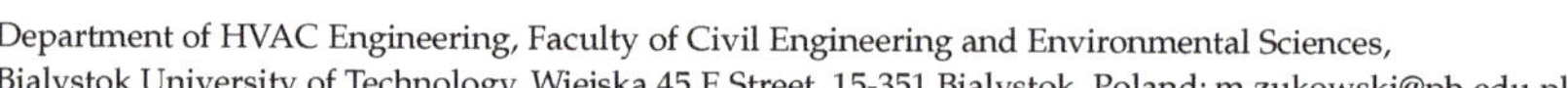

Department of HVAC Engineering, Faculty of Civil Engineering and Environmental Sciences, Bialystok University of Technology, Wiejska 45 E Street, 15-351 Bialystok, Poland; m.zukowski@pb.edu.pl

Abstract: Energy security is becoming one of the most important issues today. Continuous increases in the prices of fossil fuels, firewood and wood pellets have become commonplace in many countries. One positive effect of this situation is the greater focus on the development of renewable energy technologies and the search for solutions to reduce the heat demands of residential buildings. The purpose of this paper is to present a small modular building that can be a response to the energy crisis and Ukraine's wave of refugees in Poland. The results of the energy simulations performed in Design-Builder software showed that this type of house has a primary energy demand of 139.35 kWh/m^2. The calculations were performed for the climatic conditions of north-eastern Poland, assuming natural gas as the fuel. The use of a geothermal heat pump reduced this value to 90.14 kWh/m^2. In order to achieve a zero primary energy balance, 23.76 m^2 of PV panels and 4 m^2 of solar thermal collectors should be installed. In addition, the influence of the overhangs and the glazing area on the heat gain from the solar radiation was analyzed. A drop in temperature inside the house in the event of a continuous power failure was also investigated.

Keywords: renewable energy sources; ground source heat pump (GSHP); air-to-air heat pump (AAHP); zero-energy building (ZEB); energy simulations; small modular house

Citation: Zukowski, M. A Small Modular House as a Response to the Energy Crisis. *Energies* **2022**, *15*, 8058. https://doi.org/10.3390/en15218058

Academic Editors: Łukasz Amanowicz and Katarzyna Ratajczak

Received: 19 September 2022
Accepted: 26 October 2022
Published: 29 October 2022

1. Introduction

The events of the last two years, and in particular of recent months, have caused many people to change their life priorities and routine habits. The real threat associated with the lack of fossil fuels or their very high price has resulted in the need to reduce energy consumption in households. Building upgrades to improve the envelope energy characteristics will not result in a significant decrease in energy demands, especially when the house is located in a cold climate zone and has a large cubature. This is because the same amount of heat is still consumed to produce hot water and to heat the ventilation air. The use of mechanical ventilation with heat recovery in an existing building may cause collisions with the ceilings and walls. The aim of this article is to analyze one of the solutions that could, to some extent, increase energy security during the energy crisis.

In Poland, on 3 January 2022, a government program was introduced to mitigate the effects of the economic crisis. Among other things, a set of regulations was developed to enable the construction of a house without official authorization, a construction manager or a construction logbook. The investor can build a house with a built-up area on the external outline of up to 70 m^2. The building must be detached and dedicated only to one's own needs, so it cannot be rented out, for example. Depending on the local development plan or administrative requirements, it is possible to design a flat roof or a sloping roof with a small attic. Free architectural plans for this type of building should soon appear on the government websites. However, the investor will decide on the application of the heating, ventilation and air conditioning (HVAC) systems. This is a difficult decision due to the high competition in this industry and because a lot of the advertising materials can be misleading.

Recently, we have observed a rapid increase in the prices of building materials, and often a lack of or limited choices. In addition, the number of professional contractors continues to decline and the cost of their services is very high. Therefore, a good solution would be to buy a ready-made object.

1.1. Literature Review on Modular Building Technologies

Building a house based on modular technology is one of the most efficient off-site construction methods according to Kamala and Hewage [1]. Individual parts or the entire building, depending on the required floor space, are made together with HVAC systems in factories. Then they are transported to the construction site and placed on the previously made foundations.

Prefabricated houses can also be quickly used in the event of natural disasters. The level of thermal comfort in such facilities was analyzed by Wang et al. [2]. The simulation model was developed in the EnergyPlus software environment and validated by comparison with the experimental results. In one of the main conclusions, it was stated that the most effective measure to lower the air temperature inside this type of house was the use of external window blinds.

The use of a Sim (PLY) framing system to build a small house prototype for student communities was developed by Albright et al. [3]. This technology is based on the production of plywood prefabricates with the use of CNC machine tools. The building in the Sim (PLY) framing system can be assembled by the average person and then can be easily disassembled, and the construction elements can be reused.

Remote work has proven to be advantageous both in terms of energy savings and for many in terms of providing a new way of life. These new habits are now being promoted by many large companies because of the savings associated with them. Some remote work experiences have led some municipalities in Italy to support settlement in less popular locations. These conditions can lead to a kind of "climatic nomadism" that can be achieved through the use of mobile houses. The pandemic crisis in 2020 and 2021 inspired Roggeri et al. [4] to develop a portable modular building made of a timber frame system. The modular frame allows for its easy adjustment and provides various configurations that result from the individual needs of the user. Computer simulations performed using the BIM/BEM methodology showed that this type of building can be zero-energy in the climatic conditions of Italy.

Jiang et al. [5] formulated conclusions and recommendations related to the promotion of the use of prefabrication as an effective alternative to conventional construction. The researchers identified sixteen indicators in order to reliably compare and evaluate both methods of building houses. The real-coded accelerating genetic algorithm was used as a research tool, which demonstrated the sustainability of the prefabrication process.

Scientific research on the potential environmental benefits of using innovative modular design solutions was carried out by Gunawarden et al. [6]. In the first part of this work, the authors reviewed prefabricated modular technologies and compared them with conventional house-building methods. TRNSYS software was used for an energy analysis of structures made of steel, concrete and wood. The result of the analysis was that a precast concrete building is more than four times heavier than a precast steel building. However, the energy used for the production of steel and the construction of a house with this technology is about 50% higher compared to prefabricated concrete. On the other hand, the comparison of an object made of wooden prefabricated elements combined with a steel structure and a concrete building showed that the total embodied energy is about 10% higher in the first case. This analysis concerned only the building structure and did not include HVAC systems.

We are dealing with severe weather phenomena related to climate change more and more often. Hurricanes and tsunamis result in extensive damage to homes. One example is the effects of Hurricane Katrina, which left tens of thousands of families homeless. The answer to this type of crisis may be modular construction, offering houses with a small area but also at a low cost [7]. Homes such as the "Katrina cottage" and "coastal cabana" [8,9] can be used as alternatives to the trailers provided by the Federal Emergency Management Agency (FEMA trailers).

A similar project to develop compact and modular residential buildings is currently underway in Poland. The reason for their development is the influx of over 5.3 million refugees caused by the war in Ukraine (August 2022). The first containers for refugees from Ukraine were installed in Bialobrzegi (Poland) in April 2022. They consist of separate rooms, shared bathrooms and a kitchen. In April 2022, a container town for war refugees from various parts of Ukraine was established in Stryjski Park in Lviv (Ukraine). The housing estate consists of 88 accommodation units that are heated, furnished and connected to electricity. A similar container town was built in Borodzianka near Kiev (Ukraine) for 350 residents who lost their homes as a result of the Russian invasion.

A comparison of a small house of approximately 40 m^2 with a residence of approximately 225 m^2 (near the regional average) in California (USA) was made by Harkness [10]. This research covered energetic, technical and economic aspects. EnergyPlus software was used for the building's energy modeling. However, the simulation of the operation of HVAC systems was performed only in a simplified way. The results of the comparative analysis showed that tiny homes can have up to 85% lower energy requirements compared to standard buildings with an average regional floor space.

The Stockholm Tiny House Expo is planning to create an artificial island with several small houses in the vicinity of Stockholm (Sweden). This housing estate will be self-sufficient in terms of energy and the island should adapt to changes in sea level. Björnberg and Tarus [11] used the IDA ICE and HOMER Pro software programs to simulate the demand, production and distribution of energy on the island. As in the previously cited paper [10], this analysis also did not model the operation of HVAC systems in detail. Eight different energy supply and demand scenarios were simulated and assessed in terms of the technical, economic and environmental aspects. The results of this study showed that buildings can be energy self-sufficient using renewable sources of energy. However, this approach turned out not to be economically profitable.

Based on a very extensive review of the literature, it should be stated that a lot of research has been carried out in terms of the prefabrication methods and technologies. The research has focused primarily on the development of new technical solutions related to the construction of building partitions, the methods used for combining them into modules and ensuring the strength of the entire structure and their transport to the construction site. However, only a very small number of studies have covered the dimensioning of heating and ventilation systems in modular buildings and energy analyses of buildings together with HVAC systems under operating conditions.

1.2. Literature Review on Reductions in Heat Energy Consumption

Another broad and complex issue covered in this paper is the minimization of energy consumption in residential buildings. Climate change and energy crises have forced the governments of the most economically advanced countries to develop zero-energy strategies. Kibert and Fard [12] noted that both the United States and the European Union (EU) lack standard requirements for low-energy buildings and net-zero-carbon buildings, as well as definitions for net-zero-energy. In their article, they cited many publications that give different definitions of these terms and made many constructive comments related to them.

Eight single-family house modernization projects were analyzed by Galiotto et al. [13] in terms of minimizing energy consumption. Three houses in Denmark, three in Switzerland and two in Austria were selected for a multi-criteria overview of retrofitting possibilities. The authors distinguished passive and active measures that were taken to reduce the energy consumed. They included the first types of measures, such as the passive use of solar radiation, geothermal energy, thermal insulation for building envelopes, automatic control and heating system changes from high-temperature to low-temperature systems. The photovoltaic system and thermal solar collectors were classified as active measures.

As it turned out, the main motivations for the homeowners selected for this study were lowering their operating costs, improving their health and thermal comfort, wanting a sustainable home and wanting to expand their home. As it turned out, the cost side of the modernization projects was not the main decision-making factor.

Caruso et al. [14] analyzed the possibility of obtaining the net-zero-energy building (NZEB) standard for a seven-apartment residential house in Mediterranean climate conditions. It was a terraced house apartment building, which is typical for Italy, with an area of 435 m^2 and a cubature of 1670 m^3. The mean value of the heat loss coefficient for the external partitions was 0.55 $W/(m^2K)$. Dynamic simulations with EnergyPlus software were used as a research tool. In the first phase of this analysis, the year-round rate of primary energy consumption was determined, which was 79 kWh/m^2 for the whole building. In order to reduce the demand for electricity, the authors proposed the use of 87 m^2 of PV panels, 20 m^2 flat plate solar collectors and a demand-controlled ventilation system with heat recovery. Thanks to these projects, the apartment building achieved the NZEB standard theoretically. According to the authors, the case considered in their analysis was representative of the area of southern Italy, so the conclusions may be universal for building design in the whole region.

The test procedure for minimizing energy consumption in winter and summer without deteriorating the level of thermal comfort was performed by Ascione et al. [15]. The objects of the research were single-family, one-story buildings with an area of 140 m^2, consisting of 7 rooms and located in four cities: Madrid, Nice, Naples and Athens. Several variable parameters were assumed for the optimization process, such as the technology used for the external walls and roof, the optical and thermal characteristics of the glazing system, the window shading method and the window-to-wall area ratio. Building performance simulations were made in the EnergyPlus software environment. As a result of the calculations, numerous conclusions have been formulated, on the basis of which architects can design a nearly zero-energy building (nZEB). One of the criteria for reducing energy demands is the use of external walls with a thermal mass M_s greater than 250 kg/m^2, made of aerated concrete bricks or blocks with integrated thermal insulation. On the other hand, a brick concrete roof should have an even higher mass M_s equal to about 500 kg/m^2 and a heat transfer coefficient of 0.16 $W/(m^2K)$. Reductions of cooling demands of between 2% and 13% are possible thanks to the use of phase change materials as the finishing layers on the inner sides of the walls, with a melting temperature of 25 °C. This analysis did not take into account the influence of the type of heating and ventilation system on the energy performance of the building.

The aim of the work by Lobaccaro et al. [16] was to minimize greenhouse gas emissions and energy consumption in a zero-emission building in Oslo (Norway). An optimization analysis was performed with a graphical algorithm editor Grasshopper integrated with Rhino's 3-D modeling tools [17]. The results of the energy simulations made in the Design-Builder environment [18] were also used. The object of the research was a single-family, two-story building with dimensions of 8 m by 10 m, which was used to reach the level of ZEB-OM [19,20]. The ZEB-OM shortcut means that the production of energy from renewable sources can compensate for the emissions of greenhouse gases generated during the operation of the building and resulting from the production of building materials. The multi-objective optimization process involved changing building elements such as the shape, amount of building materials and dimensions of the structural elements. The environmental impact was determined for each variant by estimating the emissions and energy demands. The result of the analysis was the selection of the best ten house shapes at high latitudes that will enable the optimal use of solar radiation. The authors proposed a universal approach to designing energy-efficient buildings with minimal environmental impact that can be constructed in any climatic condition.

Soltaniehha et al. [21] analyzed the possibility of achieving the NZEB standard in US climatic conditions. The subject of the study was a two-story rectangular house without a basement or garage, which consisted of 3 bedrooms, 2 bathrooms and an office. The calculations were performed with a software package consisting of DesignBuilder, Autodesk Ecotect and Autodesk Vasari. The shape of the building, its orientation towards the direction of the world and the type and percentage of glazing were modified in order to achieve the minimum energy consumption. The roof surface was considered for the installation of systems for converting solar radiation into electricity and heat. The simulation results showed that the most economical solution is to use photovoltaic panels cooperating with a heat pump with a COP equal to or higher than 2. In some climatic conditions (for example in the immediate vicinity of Denver), southern glazing is effective in reducing the energy consumption, which was another conclusion of this analysis. On the other hand, in cold climates, the enlarged southern glazing causes greater heat loss than the heat gain from the solar radiation.

1.3. Main Goals, Hypothesis and Novelty Elements

The studies presented above concerned buildings with a relatively large area, as compared to the residential module analyzed in this article. The specificity of such facilities involves the different shares of energy consumption for heating, ventilation, domestic hot water (DHW) and lighting. The shares of heat gain from residents and solar radiation in the energy balance were also different. The main purpose of this article is to determine the characteristics of the small modular building developed by the author and the impact of various heating and ventilation technologies on energy consumption.

The energy simulations were performed for both the heat transfer through the building envelope and the operation of HVAC systems. In the vast majority of analyses of this type, this last aspect is ignored. The element of novelty involves the determination of the optimal slope angle of the solar collectors depending on the season in the considered location and the local climatic conditions. The influence of the supporting structure of the PV panels on the increase in energy consumption caused by the shading of the glazing on the southern facade of the building was also determined. The analysis of the thermal conditions in the modular house when the heating was turned off in winter can also be treated as a novelty. The recent events in Ukraine, related to the bombing of the infrastructure of the district heating system in mid-October 2022, prove that the content of this paper is very up-to-date.

The main objective of this study was to design a small modular house and a dedicated heating and ventilation system that would be characterized by minimal energy consumption. This article considers the research hypothesis that it is possible to obtain a zero balance of energy losses and gains in a small residential building in relatively severe winter climatic conditions.

2. Materials and Methods

Energy simulations of the building envelope and HVAC systems were used as a research tool to perform the analysis of a small modular house.

2.1. Energy Model of the Building and HVAC Systems

The modeling of transient heat transfer and the operation of HVAC systems in a modular building were achieved in the DesignBuilder Version 6.1.5.004 software environment. First, a three-dimensional model of the house envelope consisting of two modules of the same size was developed. The plane of division (south–north) running through the center of this structure is marked with an orange dashed line in Figure 1. The arrangement of the rooms in both parts can of course be optional.

The two modular building consist of six rooms, two of which, the vestibule and boiler room, are unheated and are not served by mechanical ventilation. Their characteristics are presented in Table 1. These types of modules can be combined in a number of pieces and also with different orientations. Of course, the layout of the windows in this situation must be changed.

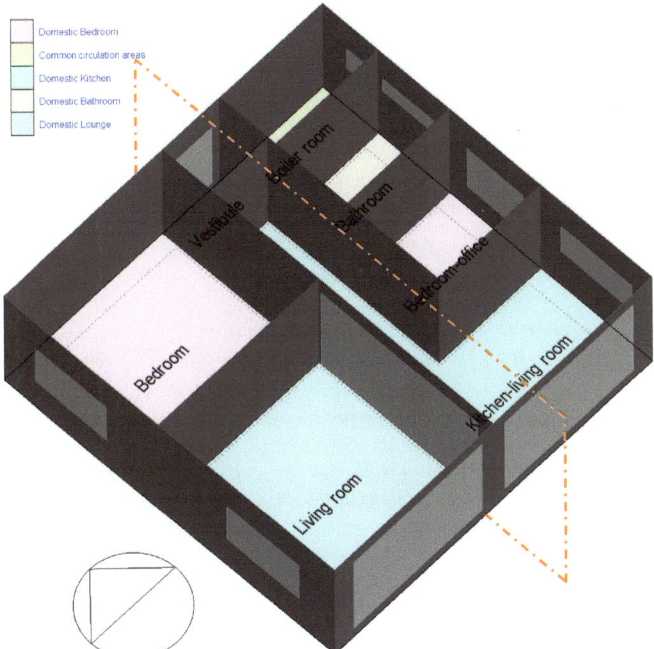

Figure 1. View of the modular building (horizontal cross-section) designed by the author.

Table 1. The area and volume of the rooms included in the modular house.

Room Name	Inside Area [m^2]	Volume [m^3]
Bedroom	12.97	42.81
Vestibule	1.04	3.44
Boiler room	2.15	7.11
Kitchen/Living room	14.78	48.76
Bathroom	3.79	12.50
Living room	14.85	49.01
Bedroom/Office	4.92	16.22
Conditioned Total	51.31	169.31
Unconditioned Total	3.20	10.54
Total	54.50	179.85

The heat transfer coefficients of this building partitions, calculated according to the standard EN ISO 6946:2017-10 [22], are the following: external walls—0.139 W/(m^{2}K); flat roof—0.107 W/(m^{2}K); ground floor—0.177 W/(m^{2}K); windows—0.78 W/(m^{2}K); total solar transmission—0.474; direct solar transmission—0.358. The structure is made of a wooden material insulated with extruded polystyrene. The building is located in Bialystok (north-eastern Poland), and its latitude is 53.1° and longitude is 23.17° (city center).

In order to achieve the lowest possible consumption of primary energy, one should not be limited only to the use of thick thermal insulation for the building envelope. As is well known, systems using renewable energy sources can also contribute to the achievement of this goal. Solar radiation, the energy accumulated in the ground and the outside air were taken into account when developing the energy supply concept for this single-family building. It was decided to consider the heating system and DHW system separately in order to make the optimal decision when choosing the variant requiring the lowest primary energy demand. This assumption complicated the analysis, but the decision was also caused by the different operational characteristics of both technologies. In the case of heating, the system supply temperature is a function of the outdoor air temperature, while DHW heating requires a constant supply temperature throughout the year. In the case of the heat pump operation, this parameter is very important and has a large impact on the seasonal coefficient of performance (SCOP). Thus, the following variants of energy supply systems were analyzed:

Variant I—Air-to-air heat pump (AAHP) for heating and ventilation.
Variant II—Ground source heat pump (GSHP) for heating and ventilation.

The following systems for preparing domestic hot water are proposed:

Variant III—Air-to-air heat pump.
Variant IV—Air-to-water heat pump (AWHP).
Variant V—Ground source heat pump.
Variant VI—Solar domestic hot water (SDHW) with 2 solar thermal collectors (STC).

In the technical solutions proposed above, the source of energy is electricity. Therefore, the amount of energy produced from the photovoltaic panels was analyzed:

Variant VII—Photovoltaic (PV) system—18 PV panels.

Selected diagrams of the systems using renewable energy sources are shown in Figures 2–4. It was assumed that the mechanical ventilation system would be equipped with a recirculation and recuperation system. The minimum outdoor flow rate was 0.015 m^3/s and the maximum outdoor flow rate was 0.02 m^3/s, while the minimum limit of the economizer dry-bulb temperature was 11 °C and the maximum limit was 25 °C. It was also assumed that the efficiency of the heat recovery system was 75%.

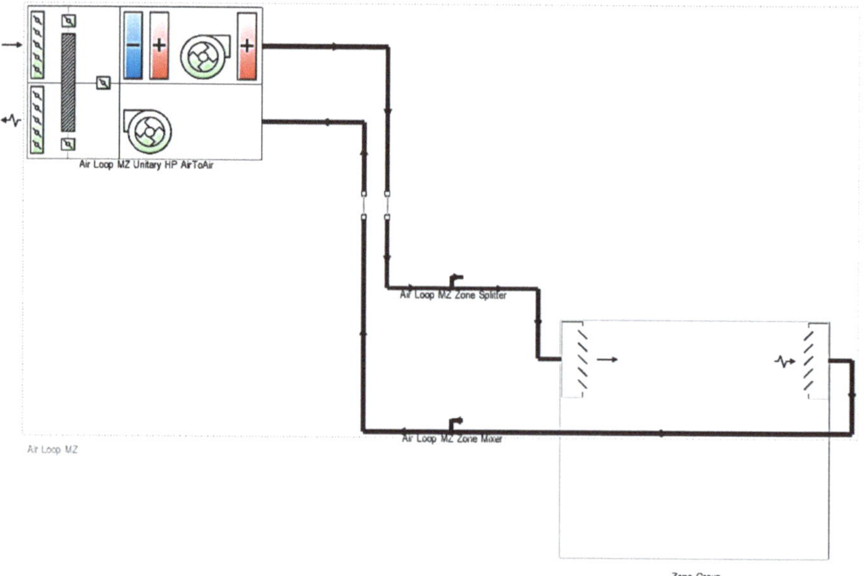

Figure 2. The unitary air-to-air heat pump with an electric supplementary heating coil and a heat recovery component.

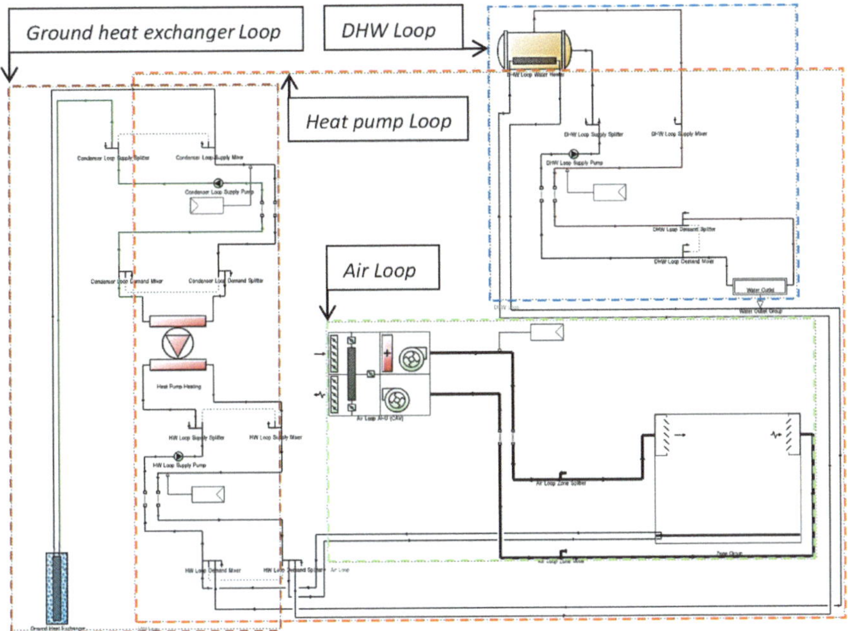

Figure 3. The ground source heat pump system used for heating applications cooperating with a mechanical ventilation system and a heat recovery component.

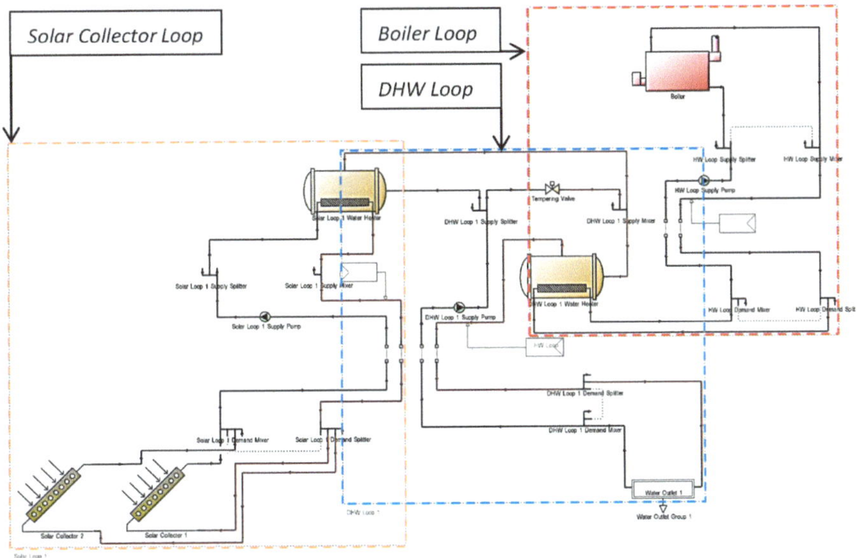

Figure 4. The solar domestic hot water system composed of solar thermal collectors, pumps, a water storage tank and a water heater.

The ground source heat pump was equipped with 4 U-tube boreholes, each 75 m long. The seasonal coefficient of performance was 4, the maximum loop temperature was 55 °C and the thermal conductivity of the ground was 0.693 W/(m K).

The solar domestic hot water system was equipped with two thermal collectors with an area of 2 m^2 each (in total 4 m^2). The maximum flow rate through the collectors was 0.00005 m^3/s, and the parameters for the efficiency equation were as follows: 0.75 (coefficient 1), -2.92 W/(m^2 K) (coefficient 2), -0.0178 W/(m^2 K^2) (coefficient 3). The capacity of the hot water storage tank was 100 liters.

The PV system consisted of 18 panels with an area of 1.32 m^2 each (in total 23.76 m^2). The characteristics of one module were as follows: cells in series—36; rated electric power output—250 W; transmittance absorbance product—0.9; heat loss coefficient—30 W/(m^2·K).

The method of arranging collectors for converting solar radiation into heat and electricity at an inclination angle of 30° is shown in Figure 5. In a simpler option, the PV panels can be arranged in two rows on a flat roof.

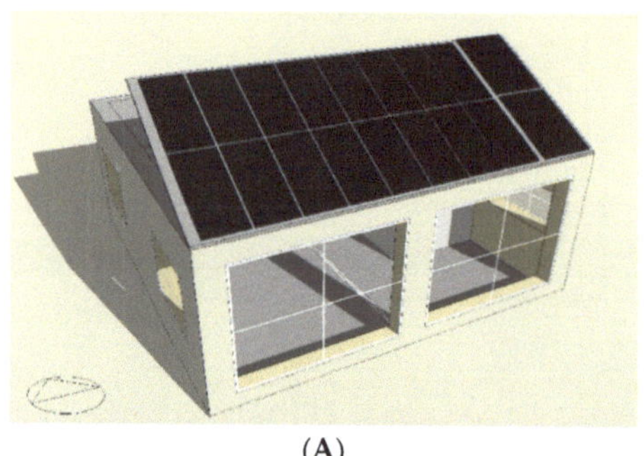
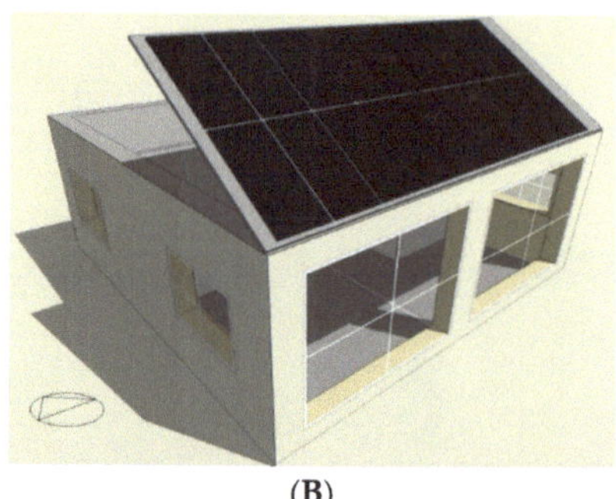

(A)   **(B)**

Figure 5. The arrangement of solar thermal collectors and photovoltaic panels: (**A**) the combination of both systems (16 PV panels and 2 STC collectors); (**B**) only the PV system (18 PV panels).

2.2. Main Assumptions for Modeling

The energy simulations for this house and its HVAC system were achieved with the most important assumptions listed below:

- The meteorological database POL_BIALYSTOK_IMGW (WMO station identifier 122950) was used for the calculations. The climate parameters are as follows: the annual average temperature of external air is 6.92 °C, the summer average temperature is 13.4 °C, the average winter temperature is 0.4 °C, the maximal difference monthly average external temperature is 22.17 °C and the solar radiation intensity on a horizontal surface is 882.3 kWh (702.33 kWh in summer, 179.98 kWh in winter). To describe the climatic conditions of this case study in more detail, the Köppen classification is used to quantify the climate variability. According to the Köppen weather types, north-eastern Poland is in a humid continental climate zone, denoted as Dfb. In the ASHRAE nomenclature, this is climate zone 6A;
- Two adults and a child live in the house and use 48.52 m^3 of DHW per year;
- The people's clothing parameters are 1.0 clo in winter and 0.5 clo in summer;
- The activity factor used to calculate the metabolic rate is 1 for men, 0.85 for women and 0.75 for children;
- The people's activity is declared according to schedules that depend on the type of room;
- The parameters of the indoor climate are calculated as a function of the residents' activity schedules, ventilation air flow rate and outdoor air parameters. It is only assumed that the heating system provides an indoor air temperature of 20 °C in the rooms and 24 °C in the bathroom. Other factors characterizing the indoor climate, such as the air humidity, CO_2 concentration and operating temperature, are calculated by the software from the energy and mass balance equations;
- The cooling of the building during the summer is considered only as free cooling.

3. Results and Discussion

The energy simulations of the building envelope and HVAC systems of a small modular house were performed using many variants. The results of these calculations and the discussion are presented below.

3.1. Determining the Energy Performance of a Building in the Base Variant

The primary energy E_P is defined as the energy harvested directly from the natural resources in the energy assessment of the buildings. The annual non-renewable primary energy demand for technical systems is determined by the following formula:

$$E_P = E_{P-H} + E_{P-DHW} + E_{P-E} \tag{1}$$

where E_{P-H} is the annual demand for non-renewable primary energy for the heating system [kWh], E_{P-DHW} is the annual demand for non-renewable primary energy for the domestic hot water heating [kWh] and E_{P-E} is the annual demand for non-renewable primary energy for electricity [kWh].

However, in the first calculations, it is necessary to estimate the demand for the final energy, i.e., the total energy consumed by end users. In the base variant (variant 0), it was assumed that the house is powered in the currently most popular way in Poland, i.e., from a dual-function gas boiler and from the power grid.

The final energy consumption in this case is E_{F-H} = 2692.38 kWh (heating and ventilation needs), E_{F-DHW} = 2389.93 kWh for the domestic hot water (DHW) and E_{F-E} = 623.81 kWh electricity for the lighting and pump drive. To convert the final energy demand E_F into the primary energy E_P, the following relationship should be used:

$$E_P = E_{F-H} \cdot PEF_G + E_{F-DHW} \cdot PEF_G + E_{F-E} \cdot PEF_E \tag{2}$$

where the conversion coefficient *PEF* is the primary energy factor. The list of conversion factors for primary energy in the current practice to estimate the energy performance of a building was presented in the article by Sartori et al. [23] for various countries. The current value of PEF_E for electricity generation and the grid supply is 2.5 [24] in the countries associated with the European Union. This value should probably decrease to 2.1 in the near future. The value of the conversion factor PEF_G for natural gas as a fossil fuel is assumed to be 1.1. Thus, the amount of primary energy consumed by this modular building in the base variant E_{P-W0} is calculated from Equation (2) and equals 7150.07 kWh. The conditioned area of the rooms is 51.31 m^2, i.e., the primary energy use per square meter of the floor area is 139.35 kWh/m^2.

The building energy rating (BER) allows a specific assessment of the energy performance of a building. In this analysis, the following levels of primary energy consumption in kWh/m^2 were used to estimate those six variants:

$E_P < 15$—passive building standard;
$E_P < 25$—class A1;
$E_P > 25$—class A2;
$E_P > 50$—class A3;
$E_P > 75$—class B1;
$E_P > 100$—class B2;
$E_P > 125$—class B3;
$E_P > 150$—class C1;
$E_P > 175$—class C2;
$E_P > 200$—class C3.

Therefore, the building under consideration can by classified as B3. This level of energy use is relatively low, despite the use of highly resistant thermal insulation for the external partitions. Changing the traditional technological solutions to those that use renewable energy sources will allow us to significantly reduce this rating level. It should be noted that in this case, the relatively high energy demand results from the low outside air temperature in winter, where the design temperature for sizing a heating system is $-22\,°\text{C}$.

3.2. Estimation of Energy Demands Using Renewable Energy Technologies

As mentioned earlier, six variants were proposed for which the final energy demand was first estimated and then converted into the primary energy value. A series of calculations were also made that allowed us to determine the dependence of the energy converted into electricity on the angle of inclination of the photovoltaic panels. This relationship is shown in Figure 6. All three functions are second-order polynomials with the convexity facing upwards, so it is not difficult to determine the maximum or optimal value.

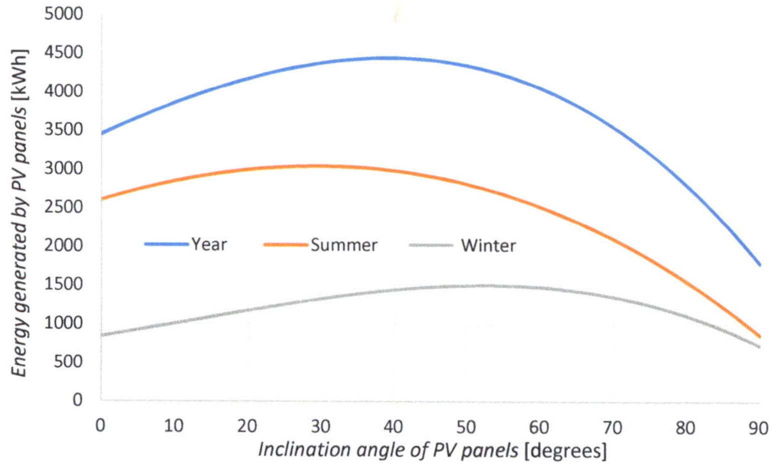

Figure 6. Influence of the slope angle of the collectors on the amount of produced energy depending on the season.

For variants V and VI, calculations were made for the three collector inclination angles of 39°, 30° and 53° for the optimal values for the whole year, summer and winter, respectively. The final results of the energy simulations are presented in Table 2.

The analysis of multi-variant simulations shows that the use of a ground source heat pump allows the energy consumption in the building to be minimized. In this case, the sum of the primary energy demands for heating and DHW purposes is 3065.33 kWh, and after taking into account the energy for lighting and the pump drive (1559.53 kWh), the total value is 4624.86 kWh. The energy demand rate is 90.14 kWh/m^2 (class B1). This represents a reduction of about 35% compared to the baseline.

As we can see, this is not a satisfactory result. In order to obtain a further reduction in the primary energy consumption, it will be necessary to use photovoltaic panels and solar thermal collectors. This assumption of using renewable energy technologies (16 PV panels and 2 STC collectors) in the most effective variant, i.e., the seasonal change of the slope angle, will allow us to save 4820.13 kWh. Thus, we can obtain a slight overproduction of electricity equal to about 195 kWh during the year. In this case, the building can be called zero-energy.

A drawback of the heat pump cooperating with the ground heat exchanger is the inability to move it together with the building. If we plan to move the house, the use of an air source heat pump is a better although less economical solution. Additionally, the AAHP can be placed on a flat roof to reduce noise and save space.

Table 2. The estimations of energy demands for variants I–V and the energy produced for variants VI and VII.

	E_F	E_P
	kWh/a	kWh/a
Energy consumption for heating and ventilation		
Variant I	935.41	2338.53
Variant II	455.75	1139.38
Energy consumption for DHW purposes		
Variant III	1968.42	4921.05
Variant IV	1941.74	4854.35
Variant V	770.38	1925.95
Amount of energy converted into heat		
Variant VI—30°	964.57	964.57
Variant VI—39°	982.42	982.42
Variant VI—53°	942.67	942.67
Variant VI—seasonal change of slope angle	1028.52	1028.52
Amount of energy converted into electricity		
Variant VII—30° (constant annual value)	4381.69	4381.69
Variant VII—39° (constant annual value)	4447.99	4447.99
Variant VII—53° (constant annual value)	4283.20	4283.20
Variant VII—seasonal change of a slope angle	4549.93	4549.93

3.3. Estimating the Impact of the Southern Shading Elevation on the Energy Demand

There is no cooling option in this budget building for economic reasons. Therefore, the use of all shading techniques is highly recommended. The supporting structure of the photovoltaic panels could also serve as a shading element for the southern facade of the building (Figure 7). The use of this type of solution would ensure a reduction in heat gains from solar radiation in the summer. Figure 8 shows the effect of the overhang projection on the heat requirement of a building and the amount of solar energy entering through the glazing. If the length of the shading element is less than 0.25 m, it practically does not affect the energy balance. On the other hand, a nearly linear increase in the heat demand can be observed, caused by the limitation of the heat gains from the sun when the overhang projection is greater than 0.25 m. We can achieve more than a 32% decrease in heat gains with a reach of 2 m. This will significantly reduce the temperature inside this small building. A further shift of the shading element was not analyzed, which was caused by design considerations.

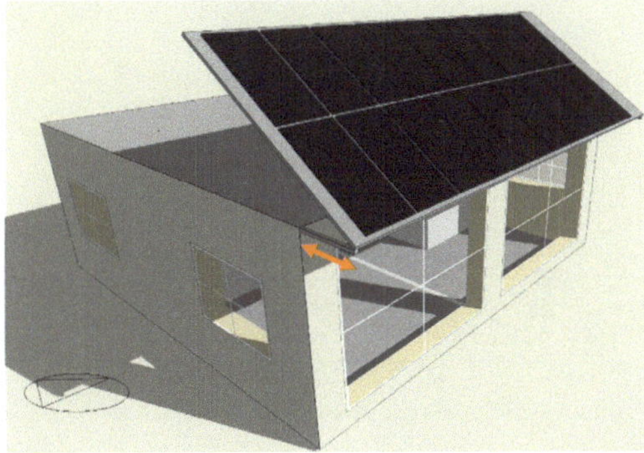

Figure 7. An example of the shading of the southern facade of the modular house with a sliding sun shade element in the form of PV panels on 15 March at 11:00.

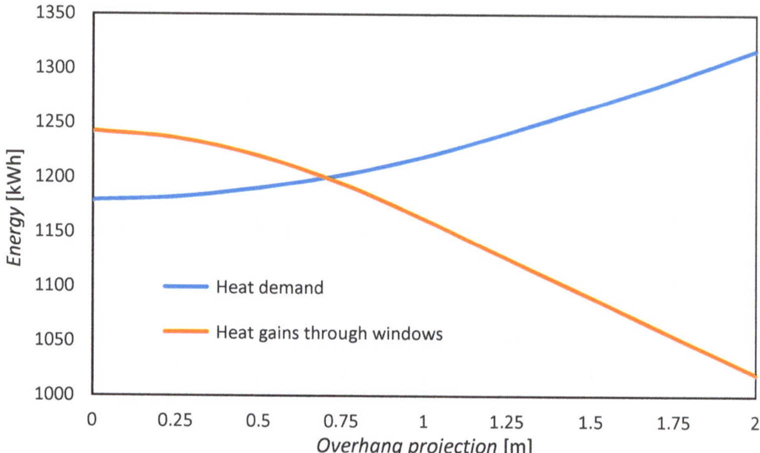

Figure 8. The dependence of heat gains and losses on the overhang length during the heating season in the analyzed modular house.

3.4. Determining the Influence of the Glazing Area on the Energy Demand

Climatic zones similar to the one discussed in this article are characterized by low outside air temperatures and low solar radiation during winter. Therefore, it should be checked whether the use of the largest possible glazing area on the south side will not increase the heat demand. For this purpose, the influence of this parameter was analyzed by reducing the height of the southern windows by 30 cm in the subsequent calculations. The area of the other windows was not changed because this ensured the correct level of lighting in the rooms.

Based on the diagrams shown in Figure 9, we can conclude that the southern side of the building should have as much glazing as possible. The heat gains from insolation are higher than the energy losses through windows in the winter. In the extreme case, i.e., by reducing the southern glazing by 90%, the demand for heating energy increases by 30%. For this reduction, there will be a decrease in solar gain by almost 70%.

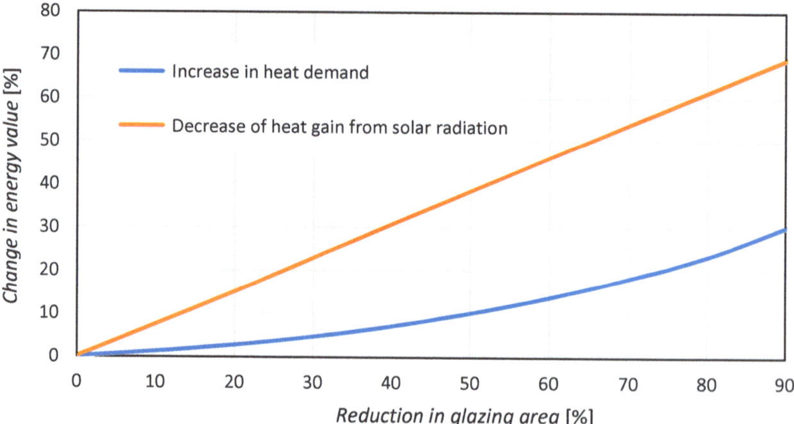

Figure 9. Influence of the glazing area of the southern windows on the heat gains and losses in the analyzed modular house.

3.5. Analysis of Temperature Changes inside Rooms without Heating

Another analysis was carried out in the extreme case of a deep energy crisis in which the building would not be heated. It was assumed that natural ventilation would be limited to the hygienic minimum, i.e., 0.2 air changes per hour (ACH) (case I). In case II, the fresh air stream provides comfortable room ventilation at 1.2 ACH. In both cases, the ACH is 1.5 over the summer term. In Figure 10, it can be seen that the operative indoor temperature in the extreme case in January (case II) can drop to its lowest value of about 3 °C on average in the whole building. Therefore, in such climatic conditions, even very good thermal insulation of walls, triple-glazed windows and solar heat gains will not provide at least the minimal thermal comfort in the winter. The temperature in the house would drop to around 10 °C with the ventilation reduced to a hygienic minimum (Case I). This value is still uncomfortable but acceptable in for short periods.

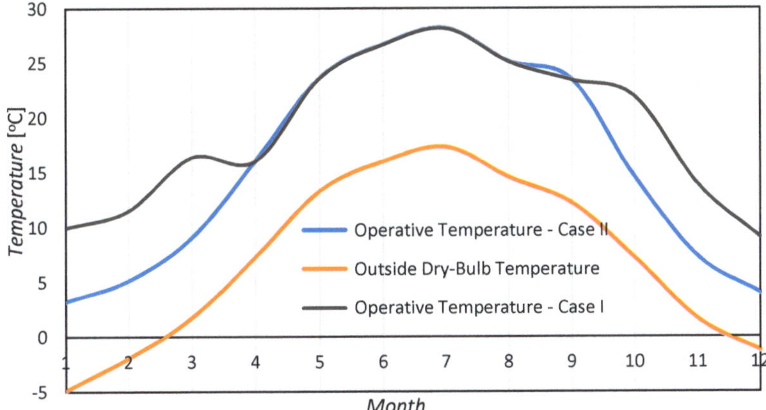

Figure 10. Changes in indoor and outdoor temperature in the absence of heating in the analyzed modular house.

3.6. Analysis of the Influence of the Building Rotation Angle on the Energy Consumption

It often happens that a building plot is not perfectly oriented to the cardinal directions of the world. As such, the house should be rotated in order that its walls are parallel to the sides of the plot. The *CRE* rotation coefficient was used to analyze the influence of the building's position in relation to the directions of the world on the final energy demand E_{F-H}:

$$CRE = \frac{E_{F-H,\alpha}}{E_{F-H,0}} \tag{3}$$

As one might expect, turning the building 180 degrees is the least favorable option in terms of the energy demand. In this case, the final energy consumption may increase by more than 11 percent, as shown in the graph in Figure 11.

3.7. Analysis of the Influence of the Building Height on the Energy Consumption

Buildings with a small living area may have a higher height in order to provide a higher cubature of the rooms. Another analysis examined how the changes in the height of the modular building would affect the final energy consumption. Another coefficient *CHE* was introduced, defined as:

$$CHE = \frac{E_{F-H,h}}{E_{F-H,h=3m}} \tag{4}$$

The reference value for these calculations was the energy consumption for the heating and ventilation $E_{F-H,\,h=3\,m}$ of the building analyzed in this study, which was 3 m high.

From the course of the chart shown in Figure 12, it can be concluded that an increase in height of the rooms by 10 cm increases the heat loss of the modular house by about 3.5%.

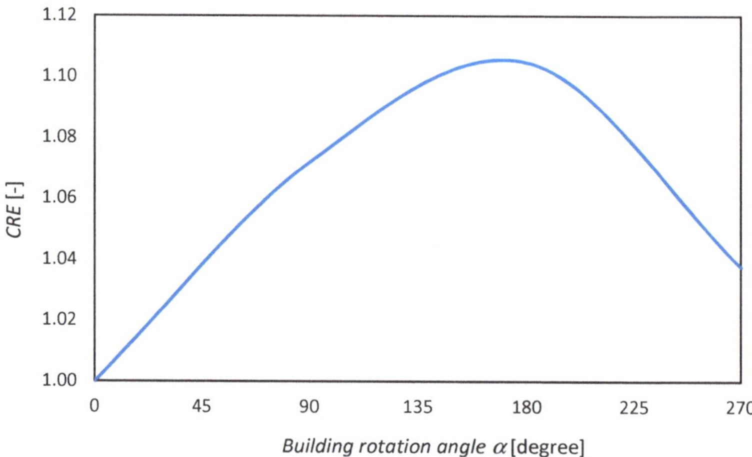

Figure 11. Dependence of the final energy demand on the building's orientation towards the directions of the world.

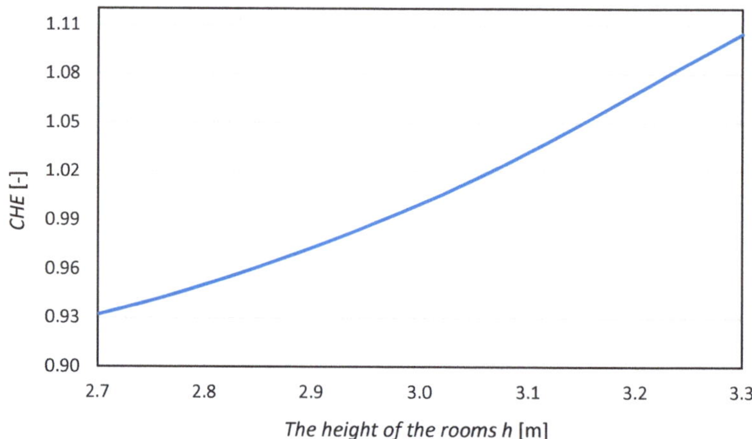

Figure 12. Influence of the height of a modular house on the final energy demand.

4. Summary and Conclusions

This article presents the results of the preliminary energy simulations of a modular single-family building. The aim of this project was to test the feasibility of designing a tiny home with a zero-energy balance over an annual period. The first reason for dealing with this topic was the constantly growing energy crisis, which is caused, among other factors, by the rising prices of fossil fuels. The second and the most important factor was the interruption of natural gas supplies to Poland as a result of the breach of contract agreements with Russia.

The energy simulations of the proposed research object were not limited only to the calculations of the heat exchange between the building envelope and the external environment. The calculations also take into account the operation of heating and ventilation systems supported by renewable energy sources. The most important conclusions resulting from the analysis are presented below:

1. The case building has an energy consumption rate of 139.35 kWh/m^2, so it can be classified as B3. Such a high consumption of primary energy per square meter of the floor area is due to the fact that a gas-fired boiler was used and the electricity was supplied from the power grid;
2. The ground source heat pump turned out to be the best technological solution for heating and DHW production. The primary energy demand was reduced by 35% compared to the baseline. The energy consumption rate decreased to 90.14 kWh/m^2 (class B1);
3. The use of renewable energy technologies (16 PV panels and 2 STC collectors) allowed about 4820 kWh of energy to be produced. Thanks to this solution, the house can be classified as a zero-energy building;
4. This type of building may overheat in summer. The use of an overhang on the south side of the facade may reduce the solar heat gains by up to 30%, which will increase the thermal comfort at this time of the year. It was assumed that a sliding support structure for the PV panels could be used as the shading option;
5. A large window area on the south facade is recommended in moderately cool climate conditions. The gains from solar radiation are higher than the heat losses through glazing. However, the windows must be characterized by a low value of the heat transfer coefficient (triple-glazed, low-emissivity coatings and insulated frames);
6. This article also analyzed a scenario in which the heating system would not work. In the first case, when the natural ventilation is limited to the minimum (0.2 ACH), the indoor air temperature may drop to 10 °C. However, if the conditions for comfortable ventilation are met (1.2 ACH), the internal temperature will be around 3 °C in January;
7. The optimal angle of inclination of the solar collectors located in north-eastern Poland was determined. The values are 39°, 30° and 53° for the year, summer and winter, respectively.

Performing an economic analysis based on the selection of the optimal heat source would be a valuable addition to this work. However, in this time of increasing energy crises, the price of HVAC systems continues to rise in Poland. For example, in the first quarter of 2022, there were increases in the price of heat pumps by over 30% and the cost of making ground heat exchangers by about 25%. A more rapid increase in prices occurs for fuels used for heating. The purchase price of the most popular type of coal in Poland increased almost three-fold during the year. A similar tendency applies to wood pellets; moreover, a large deficit appeared at the end of this winter. In addition, in the second quarter of 2022, there was also an unfavorable change in the government policy related to the development of electricity producers from PV panels. To sum up, performing a proper economic analysis under these conditions does not make much sense because its results will be unreliable.

The sample housing module presented in this article should be a very effective solution in times of energy crises, weather anomalies and humanitarian disasters. The practical aspect of this study is the presentation of a comprehensive technical solution, primarily for use by designers. Based on the information presented in this article, it will be possible to design similar buildings that integrate energy-efficient housing envelopes with HVAC systems using renewable energy sources.

The main limitations of this analysis are listed below:

- Limiting the energy simulations to only one climate zone;
- The assumption of a fixed number of inhabitants;
- The determination of the efficiency of a ground source heat pump for one type of soil;
- The assumption of a single value of the DHW design temperature;
- Not taking into account four-pane windows;
- No consideration of hybrid photovoltaic–thermal (PVT) collectors.

However, it should be noted that increasing the number of analyzed parameters would generate a very large number of variants. This could significantly worsen the readability of this article, and at the same time would not have a significant impact on the quality of the analysis.

This study is the first step in a larger project. Its purpose is to prepare initial assumptions related to the construction of a modular show house. The results of this analysis will be used to develop a grant application within the Horizon Europe Program together with the Hamburg University of Applied Sciences, involving a local producer of residential modules, which is one of Europe's largest producers.

Funding: This work was supported by the Bialystok University of Technology and financed by the Ministry of Science and Higher Education of the Republic of Poland (grant number WZ/WB-IIS/7/2022).

Data Availability Statement: The study did not report any data.

Conflicts of Interest: The author declares no conflict of interest.

Nomenclature

CHE	building height coefficient (-)
CRE	building rotation coefficient (-)
E_F	annual demand for final energy (kWh)
$E_{F\text{-}H}$	annual demand for final energy for the heating system (kWh)
$E_{F\text{-}DHW}$	annual demand for final energy for domestic hot water heating (kWh)
$E_{F\text{-}E}$	annual demand for non-renewable primary energy for electricity (kWh)
E_P	annual demand for non-renewable primary energy (kWh)
$E_{P\text{-}H}$	annual demand for non-renewable primary energy for the heating system (kWh)
$E_{P\text{-}DHW}$	annual demand for non-renewable primary energy for domestic hot water heating (kWh)
$E_{P\text{-}E}$	annual demand for non-renewable primary energy for electricity (kWh)
PEF	primary energy factor (-)
AAHP	air-to-air heat pump
ACH	changes per hour
AWHP	air-to-water heat pump
BER	building energy rating
DHW	domestic hot water
FEMA	Federal Emergency Management Agency
GSHP	ground source heat pump
HVAC	heating, ventilation and air conditioning
NZEB	net-zero-energy building
PV	photovoltaic
SCOP	seasonal coefficient of performance
SDHW	solar domestic hot water
STC	solar thermal collector
ZEB	zero-energy building

References

1. Kamali, M.; Hewage, K. Life cycle performance of modular buildings: A critical review. *Renew. Sustain. Energy Rev.* **2016**, *62*, 1171–1183. [CrossRef]
2. Wang, Y.; Long, E.; Deng, S. Applying passive cooling measures to a temporary disaster-relief prefabricated house to improve its indoor thermal environment in summer in the subtropics. *Energy Build.* **2017**, *139*, 456–464. [CrossRef]
3. Albright, D.; Blouin, V.; Harding, D.; Heine, U.; Huette, N.; Pastre, D. Sim[PLY]: Sustainable Construction with Prefabricated Plywood Componentry. *Procedia Environ. Sci.* **2017**, *38*, 760–764. [CrossRef]
4. Roggeri, S.; Olivari, P.; Tagliabue, L.C. Green and Transportable Modular Building: A prefabricated prototype of resilient and efficient house. *J. Phys. Conf. Ser.* **2021**, *2042*, 012163. [CrossRef]
5. Jiang, Y.; Zhao, D.; Wang, D.; Xing, Y. Sustainable Performance of Buildings through Modular Prefabrication in the Construction Phase: A Comparative Study. *Sustainability* **2019**, *11*, 5658. [CrossRef]
6. Gunawardena, T.; Mendis, P.; Ngo, T.; Aye, L.; Alfano, J. Sustainable Prefabricated Modular Buildings. In Proceedings of the 5th International Conference on Sustainable Built Environment, Kandy, Sri Lanka, 12–15 December 2014. [CrossRef]
7. Tiefenbacher, P.J. (Ed.) *Approaches to Disaster Management-Examining the Implications of Hazards, Emergencies and Disasters*; IntechOpen: London, UK, 2013. [CrossRef]
8. Talen, E. New Urbanism, Social Equity, and the Challenge of Post-Katrina Rebuilding in Mississippi. *J. Plan. Educ. Res.* **2008**, *27*, 277. [CrossRef]

9. Evans-Cowley, J.S.; Gough, M.Z. Evaluating New Urbanist Plans in Post-Katrina Mississippi. *J. Urban Des.* **2009**, *14*, 439–461. [CrossRef]
10. Harkness, J.A. Building Energy Modeling and Techno-Economic Feasibility Analysis of Zero Net Energy Tiny Homes in Coastal Humboldt County, California. Master's Thesis, Cal Poly Humboldt, Arcata, CA, USA, 2019. Available online: https://digitalcommons.humboldt.edu/etd/252 (accessed on 25 October 2022).
11. Björnberg, I.; Tarus, A. Off-Grid Tiny Housing: An Investigation of Local Sustainable Heat and Power Generation for an Artificial Island in Stockholm. Student Thesis, KTH Royal Institute of Technology, Stockholm, Sweden, 2021.
12. Kibert, C.J.; Fard, M.M. Differentiating among low-energy, low-carbon and net-zero-energy building strategies for policy formulation. *Build. Res. Inf.* **2012**, *40*, 625–637. [CrossRef]
13. Galiotto, N.; Heiselberg, P.; Knudstrup, M.A. Examples of Nearly Net Zero Energy Buildings through One-step and Stepwise Retrofits. In Proceedings of the Nordic Passive House Conference, Trondheim, Norway, 21–23 October 2012. [CrossRef]
14. Caruso, G.; Evola, G.; Margani, G.; Marletta, L. Design standards for residential N-ZEBs in mild Mediterranean climate. In Proceedings of the International Conference on Renewable Energies and Power Quality (ICREPQ'13), Bilbao, Spain, 20–22 March 2013.
15. Ascione, F.; De Masi, R.F.; De Rossi, F.; Ruggiero, S.; Vanoli, G.P. Optimization of building envelope design for nZEBs in Mediterranean climate: Performance analysis of residential case study. *Appl. Energy* **2016**, *183*, 938–957. [CrossRef]
16. Lobaccaro, G.; Wiberg, A.H.; Ceci, G.; Manni, M.; Lolli, N.; Berardi, U. Parametric design to minimize the embodied GHG emissions in a ZEB. *Energy Build.* **2018**, *167*, 106–123. [CrossRef]
17. Grasshopper Algorithmic Modeling for Rhino. Available online: https://www.grasshopper3d.com/ (accessed on 6 June 2022).
18. DesignBuilder Software. Available online: https://designbuilder.co.uk/ (accessed on 6 June 2022).
19. Dokka, T.H.; Wiberg, A.H.; Georges, L.; Mellegard, S.; Time, B.; Haase, M.; Maltha, M.; Lien, A.G. *A Zero Emission Concept Analysis of an Office Building*; ZEB Project Report No. 9; The Research Centre on Zero Emission Building: Trondheim, Norway, 2013.
20. Wiberg, A.H.; Georges, L.; Dokka, T.H.; Haase, M.; Time., B.; Lien, A.G.; Mellegård, S.E.; Maltha, M. A net zero emission concept analysis of a single-family house. *Energy Build.* **2014**, *74*, 101–110. [CrossRef]
21. Soltaniehha, M.; Gardzelewski, J.A.; Tan, G.; Denzer, A. Passivhaus and net zero energy residential designs in a cold climate: A simulation based design process for the next generation of green homes. In Proceedings of the SimBuild 2012, Fifth National Conference of IBPSA-USA, Madison, WI, USA, 1–3 August 2012.
22. *EN ISO 6946:2017-10*; Building Components and Building Elements—Thermal Resistance and Thermal Transmittance—Calculation Methods. ISO: Geneva, Switzerland, 2017.
23. Sartori, I.; Napolitano, A.; Voss, K. Net zero energy buildings: A consistent definition framework. *Energy Build.* **2012**, *48*, 220–232. [CrossRef]
24. Directive (EU) 2018/2002 of the European Parliament and of the Council of 11 December 2018 Amending Directive 2012/27/EU on Energy Efficiency. Available online: https://eur-lex.europa.eu/eli/dir/2018/2002/oj (accessed on 6 June 2022).

Article

Statistical Analysis of the Variability of Energy Efficiency Indicators for a Multi-Family Residential Building

Anna Życzyńska [1], Zbigniew Suchorab [2,*], Dariusz Majerek [3] and Violeta Motuzienė [4]

[1] Faculty of Civil Engineering and Architecture, Lublin University of Technology, 40 Nadbystrzycka Str., 20-618 Lublin, Poland; a.zyczynska@pollub.pl
[2] Faculty of Environmental Engineering, Lublin University of Technology, 40B Nadbystrzycka Str., 20-618 Lublin, Poland
[3] Faculty of Fundamentals of Technology, Lublin University of Technology, 38 Nadbystrzycka Str., 20-618 Lublin, Poland; d.majerek@pollub.pl
[4] Department of Building Energetics, Vilnius Gediminas Technical University, 11 Saulėtekio Str., 10223 Vilnius, Lithuania; violeta.motuziene@vilniustech.lt
* Correspondence: z.suchorab@pollub.pl; Tel.: +48-81-538-4756

Abstract: During the building design phase, a lot of attention is paid to the thermal properties of the external envelopes. New regulations are introduced to improve energy efficiency of a building and impose a reduction of the overall heat transfer coefficient; meanwhile, this efficiency is more influenced by the efficiency of the heating system and the type of fuels used. This article presents a complex analysis including the impact of: heat transfer coefficient of the envelope, efficiency of building service systems, the type of energy source, and the fuel. The analysis was based on the results of simulation tests obtained for an exemplary multi-family residential building located in Poland that is not equipped with a cooling system. The conducted calculations gave quantitative evaluation of the influence of particular parameters on building energy performance and showed that the decrease of heat transfer coefficient of building boundaries, in accordance to the Polish regulation for 2017 and 2021, gave only 11% of reduction on usable energy demand index. On the other hand, it was found that modification of the heating system and heat source can significantly influence the values of the final and primary energy consumption at the level of 70%. The application of heat pumps has a greater influence on the final and primary energy consumption for heating indices than other parameters, such as the building's envelopes.

Keywords: energy indicators; thermal retrofitting; primary energy; final energy

Citation: Życzyńska, A.; Suchorab, Z.; Majerek, D.; Motuzienė, V. Statistical Analysis of the Variability of Energy Efficiency Indicators for a Multi-Family Residential Building. *Energies* **2022**, *15*, 5042. https://doi.org/10.3390/en15145042

Academic Editor: Katarzyna Ratajczak

Received: 14 June 2022
Accepted: 6 July 2022
Published: 11 July 2022

Publisher's Note: MDPI stays neutral with regard to jurisdictional claims in published maps and institutional affiliations.

1. Introduction

The European Union (EU) is committed to developing a sustainable, competitive, secure, and decarbonized energy system by 2050. This is because the building stock is responsible for approximately 36% of all CO_2 emissions in the EU and its share is continuously increasing within the past decades [1–3]. Thermal performance of the building envelope is an important part of the overall heating energy efficiency. Hence, as emphasized in Directive (EU) 2018/844 [4], it is important to ensure that the measures to improve the energy performance of buildings do not focus only on the building envelope, but include all relevant elements and technical systems in a building [5,6]. Member States shall set additional requirements for the technical systems of existing and new buildings as well as their optimization.

According to EU Buildings Datamapper [7], residential buildings in different countries constitute 59–89% of the building stock. Poland is one of the largest EU countries where total building floor area is 1511 Mm^2 [8] and residential buildings account for about 67% of the entire building stock.

As the residential sector is one of the largest heat recipients on a country scale, the issues related to the energy consumption in housing sector are still relevant, as demonstrated by the conduct of a common energy policy of states, improvement of existing and introduction of new legal regulations at international and national level both for energy and waste policy [9,10]. This sector still faces many unsolved energy performance-related problems. Half of the residential buildings were built before 1980 and the majority of buildings are uninsulated or insulated at sub-optimal levels. The (deep) renovation rate is currently very low. Energy Performance Certificates (EPCs) are mandatory in Poland, yet the share of issued EPCs for the residential stock is just about 4%. Energy audits are rarely conducted in single-family houses due to their high cost. Heating energy consumption of residential buildings is still around 40% above the average of the EU and share of nearly zero-energy buildings (nZEB) [11,12] in new construction for residential buildings is one of the lowest (about 50% lower than the EU average).

In Poland, many multi-family residential buildings without mechanical ventilation with heat recovery are still being built. In this type of the buildings, the heating and ventilation needs account for the largest share of heat demand. On the other hand, different measures to solve these problems are applied in the case of newly designed buildings, including continuous improvement of requirements in the field of thermal insulation of building partitions and energy efficiency of the applied technical systems and devices, as well as the implementation of the modern technologies. Moreover, in the present geopolitical situation, when fossil fuels are becoming increasingly expensive and their combustion is not environmentally friendly, there is a growing interest in the use of renewable energy sources for heat production, while obtaining electricity, e.g., from photovoltaic cells or wind farms [13–16]. It should be emphasized here that in Poland, nearly all buildings are equipped with heat meters. However, it is required that energy consumption should be monitored in detail, technical systems should be properly managed and their performance parameters controlled, which should ultimately lead to rational energy management in the building [17,18].

The analysis of the methodology to prepare the building energy certificates and related standards indicate that many parameters affect the value of the building indicator for non-renewable annual primary energy consumption for heating (PE_H). The value of the PE_H index depends on the value of the building indicators of usable, final energy and primary energy factor [3,19–21]. The building usable energy demand, expressed through the usable energy demand, is influenced by: climatic conditions related to the location of the building, its orientation relative to world directions, internal heat gains and building envelope parameters, such as thermal insulation of building partitions, protection properties against wind, building glazing, shading of transparent partitions and their ability to transmit solar radiation [18,22–24].

The value of the building final energy consumption for heating (FE_H) is directly affected by the total efficiency of the heating system and the demand of auxiliary energy for pumps (parasitic energy) [3]. However, for a given value of the FE_H index, the value of the primary energy PE_H directly depends on the method of supplying energy to the building and the energy source [25]. This article attempts to analyze the magnitude of the impact of changing individual parameters on the values of energy indicators, based on simulation calculations.

Contrary to the considerations which take into account only the factors related to the shape of the building or only heat sources and the efficiency of heating systems, the goal of the article is a complex and quantitative assessment of the influence of the building parameters such as overall heat transfer coefficients (U-values) resulting from legal acts in a given period of erecting facilities, cardinal direction, glazing degree, shadow affect and solar heat gain coefficient (g-value) on the building energy indicators (EU_H, FE_H and PE_H). In addition, as part of a comprehensive analysis, the impact on the energy efficiency of heating systems with different heat sources, fuel types, heating medium temperatures, as well as the use of renewable heat sources, was checked.

2. Materials and Methods

The building considered within the presented investigation is presented in Figure 1. It has a cuboid shape, three heated floors above ground and the full unheated basement. Its heated area is A_f = 841.5 m², storey height—2.8 m³, and the A/V shape ratio—0.51 1/m. The analysis concerns only the energy demand required to cover the energy demand related to the heating of the object.

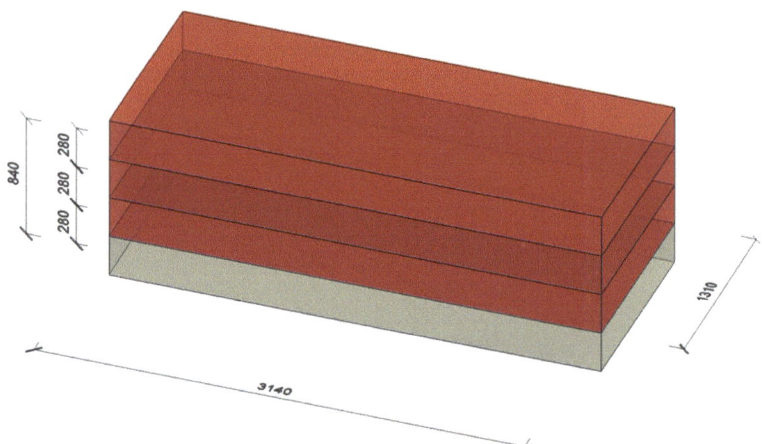

Figure 1. Schematic view of the investigated building (red color—heated area, grey color—unheated basement).

Calculations of the heating energy demand were carried out for the standard design weather data (average monthly temperatures from long period observations) and standard indoor conditions based on Polish technical and construction regulations (20 °C). Using these values, the number of degree-days (presented below) was defined. The calculations were based on the algorithm used to determine the energy performance of a building contained in the Polish legal acts [26,27] which implement the EPBD directive [28].

Knowing the value of the annual useable energy demand for heating and ventilation ($Q_{H,nd}$), calculated using various combinations of building orientation and parameters of partitions forming its external envelope, the following relationships were used in further calculations:

Annual, final energy consumption for heating and ventilation [26]:

$$Q_{k,H} = \frac{Q_{H,nd}}{\eta_{H,tot}} \; [\text{kWh/a}] \tag{1}$$

where: $Q_{H,nd}$—annual useable energy demand for heating and ventilation [kWh/a], $\eta_{H,tot}$—average seasonal total efficiency of the heating system [-].

Average seasonal total efficiency of the heating system calculated as a product of the seasonal efficiencies [-] of: heat generation ($\eta_{H,g}$), heat distribution ($\eta_{H,d}$), heat accumulation ($\eta_{H,s}$), heat adjustment and utilization ($\eta_{H,e}$).

Average seasonal total efficiency of the heating system calculated as a product of the seasonal efficiencies [-] of: heat generation ($\eta_{H,g}$), heat distribution ($\eta_{H,d}$), heat accumulation ($\eta_{H,s}$), heat adjustment and utilization ($\eta_{H,e}$) [26]:

$$\eta_{H,tot} = \eta_{H,g} \cdot \eta_{H,d} \cdot \eta_{H,s} \cdot \eta_{H,e} \; [-] \tag{2}$$

Annual demand for final auxiliary energy for the heating system [26]:

$$E_{el,aux,H} = \sum q_{el,H,i} \cdot t_{el,i} \cdot A_f \cdot 10^{-3} \; [\text{kWh/a}] \tag{3}$$

where: $q_{el,H,\,i}$—demand for unit power [W/m^2], $t_{el,i}$—number of operating hours during the year, A_f—surface of the rooms with controlled temperature.

Annual unit demand for final auxiliary energy for the heating system:

$$E_{el,aux,H,jed.} = \frac{E_{el,aux,H}}{A_f} \left[\text{kWh}/\left(\text{m}^2{\cdot}\text{a}\right)\right] \tag{4}$$

Annual demand for non-renewable primary energy for the heating system, including the auxiliary energy and the coefficients of the non-renewable primary energy input w_H and w_{el} [26]:

$$Q_{p,H} = Q_{k,H} \cdot w_H + E_{el,aux,H} \cdot w_{el} \; [\text{kWh}/\text{a}] \tag{5}$$

Annual usable energy demand index for heating and ventilation referred to the unit of the area with a design air temperature equal to 20 °C [26]:

$$UE_H = \frac{Q_{H,nd}}{A_f} \left[\text{kWh}/\left(\text{m}^2{\cdot}\text{a}\right)\right] \tag{6}$$

This index represents the building energy demand for heat conductivity through the external envelope and ventilation air heating. It does not cover the type and efficiencies of the building services, but is essential for the evaluation of the final and primary energy consumption indices.

Annual final energy consumption index for heating and ventilation referred to the unit of the area with a design air temperature [26]:

$$FE_H = \frac{Q_{k,\,H}}{A_f} \left[\text{kWh}/\left(\text{m}^2{\cdot}\text{a}\right)\right] \tag{7}$$

Annual primary energy index for heating and ventilation referred to the unit of the area with design air [26].

$$PE_H = \frac{Q_{p,H}}{A_f} \left[\text{kWh}/\left(\text{m}^2{\cdot}\text{a}\right)\right] \tag{8}$$

Assumptions for the calculation of the annual usable energy consumption index for heating (UE_H) are the following:

- Building location in the Eastern Poland, where outdoor design parameters are:
- Average temperature of the winter period months equal 2.8 °C.
- Number of days of winter period duration equal 222 days.
- Number of degree-days for the location equals to 3825.2 (day·K)/a.
- Solar radiation zone: 1200 kWh/m^2.
- Values of the overall heat transfer coefficients of the partitions (U) were considered in two groups—A and B. In the case of group, the U-values are boundary values according to the national legal acts, in force in Poland until 2021 (since 2017), in the case of B group—in force since 2021 (Table 1).

Table 1. Boundary values of the overall heat transfer coefficients U according to [29].

Type of the Partition	Value of the Overall Heat Transfer Coefficient U W/(m^2·K)	
	A Group	B Group
external walls	0.23	0.20
flat-roof	0.18	0.15
ceiling above the basement	0.25	0.25
windows	1.1	0.9
doors	1.5	1.3

- Values of the overall heat transfer coefficients of the partitions not covered by the requirements in the discussed case are the following, in the unheated zone of the building, covered by the calculation: wall in the ground—0.481 W/(m^2·K), floor on the ground—0.880 W/(m^2·K), external wall—0.484 W/(m^2·K)

- Unit internal heat gains 6.5 W/m^2.
- The proportion of the transparent surface in the entire window 0.7.
- Orientation relative to cardinal directions: N-S and E-W.
- Building glazing: minimum P1 and maximum P2, determined according to the requirements of Polish technical and building regulations [29]. Both values were calculated as the available values for the analyzed building model. P1 represents the building glazing level that provides minimal room lightning; on the other hand, P2—maximal allowed glazing level combined with the solar heat gains.
- No moving shading devices.
- Shading coefficient from external elements: 0.9 (e.g., building in the city center) and 1.0 (e.g., in the open air).
- Solar transmittance (g-value) for glazed surfaces: 0.75 (e.g., double glazing), 0.7 (e.g., triple glazing), 0.64 (e.g., glazed unit with one coating and argon space), 0.50 (e.g., special glass).

As a result of the combination of the individual parameters, 32 cases were obtained in each of the group A and B requirements in the field of the overall heat transfer coefficients of the building partitions. For each of them, simulation of the energy demand for usable energy was carried out and the UE$_H$ indicator of the building was determined.

For the calculation of FE$_H$ and PE$_H$ indicators of the building, seven variants (v1–v7) of the heating were considered. The description of heating systems is included in Table 2.

Table 2. Characteristics of the heating systems.

Variant Symbol	Variant Description
v1	individual, compact heating station with housing with nominal power up to 100 kW; heating with panel radiators with central and individual control, thermostatic valves with 1 K proportional band; heating from a local heat source, pipes well insulated in unheated space.
v2	condensing gas boiler with a nominal power of up to 50 kW, operating parameters 70/55 °C; heating with panel radiators with central and individual control, thermostatic valves with proportional integration with adaptive and optimization function; heating from a local heat source, pipes well insulated in unheated space.
v3	condensing gas boiler with a nominal power of up to 50 kW, operating parameters 55/45 °C; underfloor water heating with central and individual control with a proportional controller; individual gas boiler in each apartment (boiler located in the kitchen).
v4	compressor ground/water heat pump, electrically driven—80% coverage; peak condensing gas boiler—20% coverage; operating parameters 55/45 °C; underfloor water heating with central and local control with a proportional controller; heating from a local heat source, pipes well insulated in the heated space.
v5	biomass boiler, automatic, power of up to 100 kW; heating with panel radiators with central and individual control, thermostatic valves with proportional-integrating performance and adaptive and optimization function; heating from a local heat source (boiler room, common for all flats) in the basement of the building, pipes well insulated in unheated space.
v6	condensing gas boiler with a nominal power of up to 50 kW, operating parameters 55/45 °C; heating with panel radiators with central and individual control, thermostatic valves with proportional-integrating performance as well as adaptive and optimization function; heating from a local heat source (boiler room, common for all flats), pipes well-insulated in the heated space.

Table 2. *Cont.*

Variant Symbol	Variant Description
v7	absorption glycol/water heat pump, gas-powered—80% coverage; peak condensing gas boiler—20% coverage; operating parameters 55/45 °C; underfloor water heating with central and individual control with proportional-integrating controller; heating from a local heat source (boiler room, common for all flats), pipes well insulated in the heated space.

- values of average seasonal partial efficiency of the individual heating system elements and seasonal average total efficiency of the heating system in the individual variants are presented in Table 3.

Table 3. Average seasonal efficiencies of the heating system, coefficients of non-renewable primary energy input and auxiliary electric energy input.

Variant Symbol	Average Seasonal Efficiency of				Coefficient of the Non-Renewable Primary Energy Input	Coefficient of Auxiliary Electric Energy	
	Production	Regulation	Transfer	Total		Final	Primary
	$\eta_{H,g}$	$\eta_{H,e}$	$\eta_{H,d}$	$\eta_{H,tot}$	w_H	kWh/(m²·a)	
v1	0.98	0.89	0.90	0.785	0.8	1.13	3.39
v2	0.91	0.93	0.90	0.762	1.1	1.29	3.87
v3	0.94	0.89	1.00	0.837	1.1	4.06	12.18
v4	2.99	0.89	0.96	2.555	2.62	5.06	15.18
v5	0.70	0.93	0.90	0.586	0.2	1.29	3.87
v6	0.94	0.93	0.96	0.839	1.1	1.29	3.87
v7	1.31	0.89	0.96	1.119	1.1	4.66	13.98

- average seasonal accumulation efficiency in each variant $\eta_{H,s} = 1.0$
- values of coefficients of non-renewable primary energy input for individual heating systems were determined according to [26] and are given in Table 3
- value of the coefficient of non-renewable primary energy input for the purposes of determining the primary electric energy factor $w_{el} = 3.0$
- values of unit coefficients of the final and primary auxiliary electric energy demand for individual variants are presented in Table 3.

3. Results

In the first stage of investigation, the usable energy demand was evaluated for 32 cases (n). In order to determine the influence of U-value on usable energy demand, two parameters were considered ($Q_{H,nd}$ and U_{EH}). All measures of position, such as minimum, maximum, and mean are lower for the 2021 technical specification than for 2017 (see Table 4). Comparisons of means of both indicators between technical specifications show statistical significance of differences, based on t-Student test for independent samples. Dispersion of energy consumption are similar for two policies and distributions of them are normal (see Figure 2).

Table 4. Descriptive statistics of $Q_{H,nd}$ and U_{EH} for the 2017 and 2021 technical specification.

Technical Specifications	Variable	n	Min	Mean	Max	SD
2017	$Q_{H,nd}$	32	38,978	43,171	49,179	2488
2017	U_{EH}	32	46.32	51.30	58.44	2.96
2021	$Q_{H,nd}$	32	34,594	38,369	43,729	2220
2021	U_{EH}	32	41.11	45.60	51.97	2.64

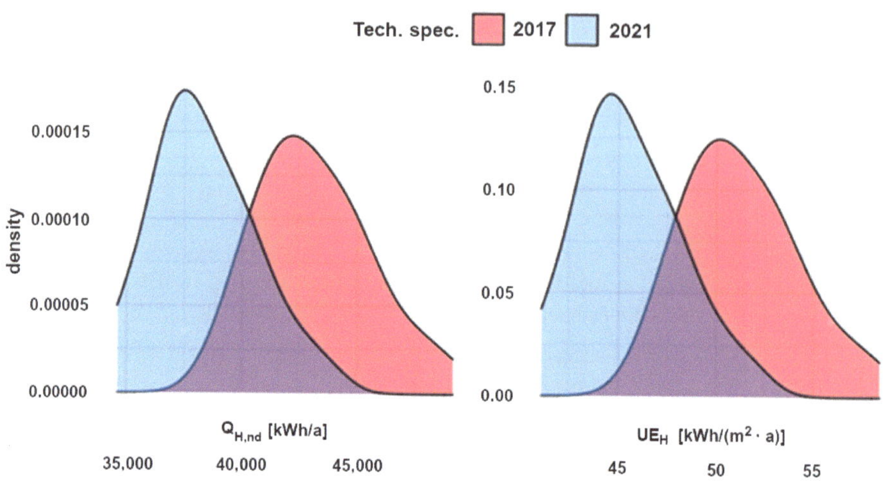

Figure 2. Distribution of $Q_{H,nd}$ and U_{EH} for 2017 and 2021 technical specification.

Regardless which value was examined, it can be noted that there is quite a wide range of energy demand in each technical specification. For the regulations of 2017, the relative difference calculated as $\Delta Qu = \frac{maxQu - minQu}{maxQu} \cdot 100\%$ is 20.74% of the annual usable energy demand. For the 2021 technical specification, the relative difference of both values is similar (20.89%).

Then, the question what indicators affect such a wide range of energy consumption arises. For further analysis, the linear models were applied to show the relationship of $Q_{H,nd}$ and UE_H with some indicators characterizing the building. Two statistical tests were used to verify whether the relationships of $Q_{H,nd}$ and UE_H to cardinal direction, glazing degree, shadow effect and solar gain coefficient are correct: the Ramsey RESET test [30] and the RAINBOW test [31].

Table 5, which summarizes all estimated models, shows that each indicator is statistically significant. Moreover, it can be said that cardinal direction N-S is better than E-W in terms of energy consumption, while higher glazing degree (P2) causes higher energy consumption. It can also be said that less shading for a residential building is better and higher g-value has a positive effect. The best combination of building parameters in terms of energy consumption is the building arrangement in the north-south system, with the smallest glazing surface, not shaded and with the highest energy transmission of solar radiation. The gain, consisting in lower energy demand (expressed in both $Q_{H,nd}$ and UE_H) caused by the selection of appropriate building parameters, is shown in Table 6.

One fact is also worth mentioning, namely that the relative difference in terms of energy demand between two policies reaches 11.12% (difference of means was compared to the mean level for 2017 policy—calculation based on Table 4).

A similar analysis was performed for annual final energy factor for heating and annual primary energy factor for heating, but this time heating variants (7) were attached to the list of differentiating factors. That is why the amount of variants increased to 224 (32 × 7).

Again, energy consumption (expressed by FE_H and PE_H) is higher for the 2017 policy than 2021 (see Table 7) and the difference is statistically significant in each case. This time, variances of both indicators are higher, which can be seen in Figure 3 and distributions deviate from normality. The relative differences (calculated in the same way as for $Q_{H,nd}$ and UE_H) for FE_H and PE_H reached 77.05% and 77.89%, respectively, for the 2017 technical specification. For the 2021 policy relative, differences are quite similar, i.e., 76.49% and 77.76%, respectively. Such a large increase in relative differences between FE_H, PE_H and $Q_{H,nd}$, UE_H is caused by inclusion of the heating systems into the analysis. Nevertheless, to ensure that the heating system is actually such an important factor, it is necessary to fit the models describing the relationship between energy consumptions (FE_H, PE_H) and parameters of building.

Table 5. The linear models describing the dependence of $Q_{H,nd}$ and UE_H on set of predictors.

	Dependent Variable			
	$Q_{H,nd}$	UE_H	$Q_{H,nd}$	UE_H
	2017		2021	
	(1)	(2)	(3)	(4)
cardinal direction (N-S)	−1930.75 ***	−2.29 ***	−1810.69 ***	−2.15 ***
	(166.04)	(0.20)	(157.64)	(0.19)
glazing degree (P2)	1986.63 ***	2.36 ***	1360.94 ***	1.62 ***
	(166.04)	(0.20)	(157.64)	(0.19)
shadow effect (1)	−1381.38 ***	−1.64 ***	−1202.69 ***	−1.43 ***
	(166.04)	(0.20)	(157.64)	(0.19)
solar heat gain coefficient, g-value (0.64)	−2868.13 ***	−3.41 ***	−2717.00 ***	−3.23 ***
	(234.82)	(0.28)	(222.94)	(0.26)
solar heat gain coefficient, g-value (0.70)	−4006.75 ***	−4.76 ***	−3585.50 ***	−4.26 ***
	(234.82)	(0.28)	(222.94)	(0.26)
solar heat gain coefficient, g-value (0.75)	−4919.63 ***	−5.85 ***	−4656.63 ***	−5.53 ***
	(234.82)	(0.28)	(222.94)	(0.26)
constant	46,782.50 ***	55.59 ***	41,935.09 ***	49.83 ***
	(219.65)	(0.26)	(208.54)	(0.25)
Observations	32	32	32	32
R^2	0.97	0.97	0.97	0.97
Adjusted R^2	0.96	0.96	0.96	0.96
Residual Std. Error (df = 25)	469.63	0.56	445.88	0.53
F Statistic (df = 6; 25)	140.78 ***	140.78 ***	123.90 ***	123.90 ***

Note: *** $p < 0.01$.

Table 6. Changes in energy demand expressed by $Q_{H,nd}$ and UE_H for the 2017 and 2021 technical specification.

Factor	Gain [1] for 2017 Policy	Gain [1] for 2021 Policy
cardinal direction%	4.37%	4.61%
glazing degree%	4.50%	3.49%
shadow effect%	3.15%	3.09%
g-value%	10.70%	11.30%

[1] The gain was calculated as proportion of change of energy demand to the maximum energy demand.

Table 7. Descriptive statistics of FE_H and PE_H for the 2017 and 2021 technical specification.

Technical_Specifications	Indicator [kWh/(m²·a)]	n	Min	Mean	Max	SD
2017	FE_H	224	23.19	61.05	101.02	18.42
2017	PE_H	224	19.68	62.56	88.99	18.74
2021	FE_H	224	21.2	53.80	89.97	16.03
2021	PE_H	224	17.90	56.50	80.47	16.86

For all models, the first heating variant (v1) was set as a reference, the rest of the indicators have the same references as in previous models.

According to Table 8, all indicators are statistically significant. Still, N-S cardinal direction, smaller glazing degree, and shadow effect with the highest energy transmission of solar radiation are the most beneficial setups for building parameters in each policy. Marginal effects of all mentioned indicators are smaller than the heating variant impact. The greatest difference in FE_H for the 2017 technical specification is between v4 and v5 variants, equals 63.69 kWh/(m^2·a). This value referred to maximal annual final energy factor for heating is equal to 71.70%, which confirms the heating system importance. Similar analysis for PE_H shows the gain exceeds 73.10% between the v3 and v5 variants. It is interesting that the v5 heating system is the worst for FE_H and the best for PE_H in both policies (see Table 9).

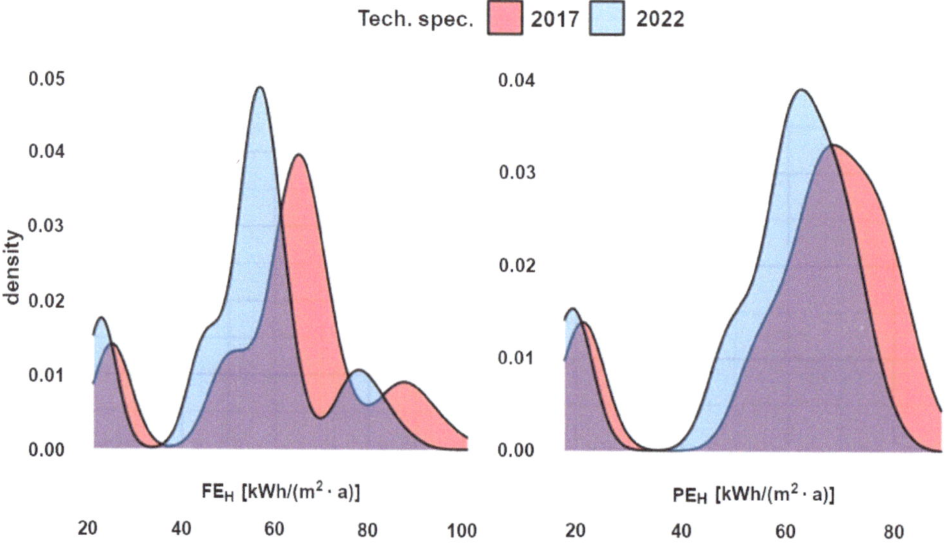

Figure 3. Distribution of FE_H and PE_H for the 2017 and 2021 technical specifications.

Table 8. The linear models describing the dependence of FE_H and PE_H on set of predictors.

	Energy Indicator (Dependent Variable)			
	FE$_H$	**PE$_H$**	**FE$_H$**	**PE$_H$**
	2017		2021	
cardinal direction (N-S)	−2.61 ***	−2.44 ***	−2.41 ***	−2.29 ***
	(0.17)	(0.15)	(0.15)	(0.14)
glazing degree (P2)	2.69 ***	2.51 ***	1.81 ***	1.72 ***
	(0.17)	(0.15)	(0.15)	(0.14)
shadow effect (1)	−1.87 ***	−1.74 ***	−1.60 ***	−1.52 ***
	(0.17)	(0.15)	(0.15)	(0.14)
g-value (0.64)	−3.88 ***	−3.62 ***	−3.62 ***	−3.43 ***
	(0.24)	(0.22)	(0.21)	(0.20)
g-value (0.70)	−5.42 ***	−5.06 ***	−4.78 ***	−4.53 ***
	(0.24)	(0.22)	(0.21)	(0.20)
g-value (0.75)	−6.65 ***	−6.21 ***	−6.20 ***	−5.88 ***
	(0.24)	(0.22)	(0.21)	(0.20)
heating system (v2)	2.13 ***	22.26 ***	−3.45 ***	19.83 ***
	(0.32)	(0.29)	(0.28)	(0.26)
heating system (v3)	−1.13 ***	23.93 ***	−0.68 **	22.25 ***
	(0.32)	(0.29)	(0.28)	(0.26)
heating system (v4)	−41.34 ***	12.11 ***	−36.31 ***	12.08 ***
	(0.32)	(0.29)	(0.28)	(0.26)
heating system (v5)	22.35 ***	−34.29 ***	19.88 ***	−30.43 ***
	(0.32)	(0.29)	(0.28)	(0.26)
heating system (v6)	−4.05 ***	15.46 ***	−3.58 ***	13.79 ***
	(0.32)	(0.29)	(0.28)	(0.26)

Table 8. *Cont.*

	Energy Indicator (Dependent Variable)			
	FE_H	PE_H	FE_H	PE_H
heating system (v7)	−15.98 ***	8.74 ***	−13.81 ***	8.94 ***
	(0.32)	(0.29)	(0.28)	(0.26)
Constant	71.37 ***	60.23 ***	63.96 ***	54.36 ***
	(0.31)	(0.28)	(0.27)	(0.25)
Observations	224	224	224	224
R^2	1.00	1.00	1.00	1.00
Adjusted R^2	1.00	1.00	1.00	1.00
Residual Std. Error (df = 211)	1.27	1.15	1.13	1.04
F Statistic (df = 12; 211)	3894.70 ***	4932.10 ***	3710.08 ***	4868.05 ***

Note: *** $p < 0.01$, ** $p < 0.05$.

Table 9. Changes in energy consumption expressed by FE_H and PE_H indices for the 2017 (A group) and 2021 (B group) technical specification.

Factor	Gain [1] of FE_H for 2017 Policy %	Gain [1] of FE_H for 2021 Policy %	Gain [1] of PE_H for 2017 Policy %	Gain [1] of PE_H for 2021 Policy %
cardinal direction	4.19	4.391	3.821	3.97
glazing degree	4.30	3.31	3.93	3.00
shadow effect	3.01	2.93	2.75	2.65
g-value	10.20	10.80	9.37	9.81
heating system	71.70	71.00	73.10	73.00

[1] The gain was calculated as proportion of change of energy consumption to the maximum energy consumption.

In order to conduct the sensitivity analysis of the applied models, the one-at-a-time (OAT) approach was used [32]. The main idea of this approach is to estimate the effect of particular explanatory variable keeping the rest variable constant, changing variable of interest within the range of it, and monitoring changes in output. Since the variables present in the model are of different scales, they were standardized before performing the sensitivity analysis. Then, comparison of model parameters is possible (Table 10).

Table 10. Comparison of model parameters.

Effects	qu_2017	qu_2021	ue_2017	ue_2021
cardinal_direction N-S	−0.39	−0.41	−0.39	−0.41
glazing_degree P2	0.41	0.31	0.41	0.31
shadow_effect1	−0.28	−0.28	−0.28	−0.28
g0.64	−0.51	−0.54	−0.51	−0.54
g0.7	−0.71	−0.71	−0.71	−0.71
g0.75	−0.87	−0.92	−0.87	−0.92

Since $UE_H = \frac{Q_{H,nd}}{A_f}$ all standardized coefficients are the same as $Q_{H,nd}$. In both cases (technical specifications 2017 and 2021) the most important prediction is solar heat gain coefficient. Shadow effect is the least significant.

4. Discussion

In the examined model of the building, shape indicators and heat transfer coefficients of partitions, as well as the parameters affecting the radiation gains characteristic of many multi-family residential buildings were assumed. Simulations were conducted for the temperature zone covering about one-third of the Polish area. Therefore, the obtained results of simulation calculations can be referred to similar buildings located in the places with a similar number of degree days, not just in Poland.

The analysis of usable energy calculations indicates that the change of the U coefficients of the partitions from the values given in group A (2017 requirements) to the values given in group B (2021 requirements) leads to a decrease in the average value of the UE index by 11.12%. However, in a given group, the variability of the UE index due to other parameters is as follows: change of orientation from N-S to E-W means an increase in the UE index from 4.37% to 4.61%, change in glazing from the minimum (P1) to maximum (P2) in the group A caused an increase of 4.50% and in group B 3.49%, an increase in shading causes an increase in UE by about 3%, a change in the solar radiation transmission coefficient of the glass from 0.75 to 0.50 causes an increase in UE in group A by 10.7 % in group B by 11.30%. The above-mentioned solution reduces heat gains in the summer period, which improves the comfort of using the rooms, but it is irrelevant in the calculations of the energy consumption index in buildings not equipped with air conditioning.

With different combinations of parameters in a given group of U coefficients, the following extreme values of the EU index were obtained: in group A the minimum value is 46.32 kWh/(m²·a), and the maximum is 58.44 kWh/(m²·a), while in group B, the minimum value is 41.11 kWh/(m²·a), maximum 51.97 kWh/(m²·a), which is about 20% in both groups.

The charts presented in Figure 2 indicate that the difference in the UE value due to the improvement of thermal insulation of building partitions with some combinations of other parameters can be maintained on the same level.

The considered variants of heating systems are the solutions that can be used in multi-family residential buildings. The assumed partial efficiencies are typical for individual variants described in Table 2. In most cases, the total efficiency of the heating system is less than 1.0, which causes the index of the annual final energy demand for heating and ventilation (FE_H) to be higher than the UE indicator. Only in the case of variants v4 and v7, in which heat pumps were used, the value of the $\eta_{H,tot}$ coefficient, is greater than 1.0, which means that the FE_H index is smaller than the UE indicator. The analyzed range of the $\eta_{H,tot}$ coefficient, *tot* from a minimum of 0.586 in the v5 variant to a maximum of 2.555 in the v4 variant causes very high variability of the FE_H index (including auxiliary electricity). In group A, this ratio varies from a minimum value of 23.19 kWh/(m²·a) to a maximum of 101.02 kWh/(m²·a), while in group B from a minimum value of 21.15 kWh/(m²·a) to a maximum of 89.97 kWh/(m²·a). The type of heating system can generate a variation of around 71%. The calculation results indicate that the use of technical systems with higher efficiency can level the differences resulting from the building envelope parameters. This means that with a less favorable combination of parameters affecting the UE indicator, the use of a system with correspondingly higher efficiency may cause that the FE_H index will be lower than in the case of a more favorable combination of parameters. Such high variability of the FE_H indicates a very significant impact of the energy efficiency of a given technical solution of a heating system on one of the key indicators of the energy performance of a building. It also means that in a building with good thermal insulation of building partitions, but equipped with a low-efficiency heating system, low final energy demand indicators (FE_H) cannot be obtained.

In the national guidelines, the energy standard of a building is defined by the non-renewable primary energy index PE. In a residential building without a cooling system, the most significant for the PE index value is the energy component necessary to meet the demand for heating and ventilation, i.e., PE_H. For a given FE_H index value, the PE_H value depends on the coefficient of non-renewable primary energy input w_H. The variability of this factor is very large and results from the method of energy production, the fuel used, and the energy supply system to the building. Designers usually have no influence on the value of this factor. First of all, their values are given in national legal acts. Second, in many cases the designer cannot choose any heat source or centralized system, much less the type of fuel used due to formal and technical barriers. The analysis shows how much influence the w_H factor has on the PE_H index, with the same standard of building thermal insulation (in a given group A or B) in the considered variants

With the same standard of thermal insulation of the building (in a given group A or B), the variability of the considered variants was approx. 73%. In many cases, low FE_H

value resulted in high PE_H and vice versa. The lowest values of the PE_H index are found in the v5 variant, in which biomass is used as fuel (then the value of w_H is 0.2). Similar observations were made in the previous research that the authors of this article presented in monograph [33]. In group A of U coefficients, when analyzing the seven adopted variants, value of this indicator varies from the minimum value of 19.68 kWh/(m²·a) to the maximum value of 88.99 kWh/(m²·a), while in group B from the minimum value of 19.90 kWh/(m²·a) to a maximum of 80.47 kWh/(m²·a). It ought to be noted that the lowest PE_H values are not associated with the lowest values in the FE_H value set. This means that a building properly designed in terms of thermal insulation and equipped with a high-efficiency heating system does not have to be characterized with a low PE_H index value. Nevertheless, a building is assessed in the national regulations by the value of the annual primary energy index and it is the key parameter of energy performance. In summary, the energy quality of designed or existing buildings should be assessed through the final energy index (FE_H), while considering the building in terms of its impact on the environment through the index of non-renewable primary energy.

5. Conclusions

A thorough analysis of the obtained results from the heat consumption measurement and theoretical calculations enables to draw the following conclusions:

- The change of policy for the value of the overall heat transfer coefficients considering the requirements for 2017 and 2021 reduces the demand for usable energy demand by only 11.12% This enables to conclude that the energy saving potential associated to the building envelope is practically exploited.
- The impact of the parameters influencing solar heat gains, such as building glazing, shading of transparent partitions, was only at the level of about 3–5%. The only exception where the changes in solar heat gain (ability to transmit solar radiation) coefficient that influenced UE_H index more significantly (11.30%).
- The heating system has the greatest impact, which in this case differentiated these primary and final energy indices at the level of approx. 70% for variant 5 (biomass boiler).
- The application of the heat pumps has a greater influence on FE_H and PE_H indices than other parameters, mainly the buildings envelopes.
- Final energy for heating (FE_H) index should be used to assess the planned or existing building in terms of energy quality, i.e., their solids together with equipment in a technical heating system. Obtaining a low value of this factor is primarily associated with the use of good architectural solutions and technical systems at the same time as the building equipment.

From the conducted investigation, it can be finally concluded that improvement of the thermal parameters of the building external partitions becomes ineffective after reaching the lower but close to the boundary level of heat transfer coefficients presented in legal acts. Taking into the consideration the environmental impact of the building performance, it seems more reasonable to improve the quality of the building services, mainly heating system and heat source.

In addition, it should be noted that an important aspect related to energy consumption is investment and operating costs, which should be analyzed in order to make the right decisions regarding the scope of technical solutions. The issue of profitability is influenced by many additional factors, such as fuel prices, costs of equipment, and technical solutions. In the case of specific solutions, energy analyses should be carried out in conjunction with analyses of economic profitability.

The obtained results of simulation calculations can be referred to similar buildings located not just in Poland, but also in the places with a similar number of degree-days.

Author Contributions: Conceptualization, A.Ż. and Z.S.; methodology, A.Ż.; software, D.M. and Z.S.; validation, D.M. and A.Ż.; formal analysis, A.Ż., V.M. and D.M.; investigation, A.Ż.; resources, A.Ż.; data curation, A.Ż. and D.M.; writing—original draft preparation, Z.S., A.Ż. and V.M.; writing—review and editing, D.M.; visualization, D.M. and Z.S.; supervision, V.M.; project administration, A.Ż.; funding acquisition, A.Ż. All authors have read and agreed to the published version of the manuscript.

Funding: This research was financially supported by Ministry of Science and Higher Education in Poland within the statutory research of scientific units under subvention for science programs.

Institutional Review Board Statement: Not applicable.

Informed Consent Statement: Not applicable.

Data Availability Statement: Data are contained within the article.

Conflicts of Interest: The authors declare no conflict of interest.

References

1. Pérez-Lombard, L.; Ortiz, J.; González, R.; Maestre, I.R. A review of benchmarking, rating and labelling concepts within the framework of building energy certification schemes. *Energy Build.* **2009**, *41*, 272–278. [CrossRef]
2. Reis, J.A.; Escórcio, P.C.C. Energy certification in St. António (Funchal)—Statistical analysis. *Energy Build.* **2012**, *49*, 126–131. [CrossRef]
3. Cao, X.; Dai, X.; Liu, J. Building energy-consumption status worldwide and the state-of-the-art technologies for zero-energy buildings during the past decade. *Energy Build.* **2016**, *128*, 198–213. [CrossRef]
4. European Union. Directive (EU) 2018/844 of the European Parliament and of the Council of 30 May 2018 amending Directive 2010/31/EU on the energy performance of buildings and Directive 2012/27/EU on energy efficiency. *Off. J. Eur. Union* **2018**, *156*, 75–91.
5. Ahmad, M.W.; Mourshed, M.; Mundow, D.; Sisinni, M.; Rezgui, Y. Building energy metering and environmental monitoring—A state-of-the-art review and directions for future research. *Energy Build.* **2016**, *120*, 85–102. [CrossRef]
6. Dall'O', G.; Galante, A.; Torri, M.D.C. A methodology for the energy performance classification of residential building stock on an urban scale. *Energy Build.* **2012**, *48*, 211–219. [CrossRef]
7. EU Buildings Datamapper. Share of Non-Residential in Total Building Floor Area; 2013. Available online: https://ec.europa.eu/energy/eu-buildings-datamapper_en (accessed on 13 June 2022).
8. iBRoad, Individual Building Renovation Roadmaps. Factsheet: Poland, Current Use of EPCs and Potential Links to iBRoad. 2020. Available online: http://ibroad-project.eu/wp-content/uploads/2018/01/iBROAD_CountryFactsheet_POLAND.pdf (accessed on 13 June 2022).
9. Piotrowska, E.; Borchert, A. Energy consumption of buildings depends on the daylight. *E3S Web Conf.* **2017**, *14*, 1029. [CrossRef]
10. Saleem, M.; Blaisi, N.I.; Alshamrani, O.S.D.; Al-Barjis, A. Fundamental investigation of solid waste generation and disposal behaviour in higher education institute in the Kingdom of Saudi Arabia. *Indoor Built Environ.* **2018**, *28*, 927–937. [CrossRef]
11. Zhang, S.-C.; Yang, X.-Y.; Xu, W.; Fu, Y.-J. Contribution of nearly-zero energy buildings standards enforcement to achieve carbon neutral in urban area by 2060. *Adv. Clim. Chang. Res.* **2021**, *12*, 734. [CrossRef]
12. Li, H.; Li, Y.; Wang, Z.; Shao, S.; Deng, G.; Xue, H.; Xu, Z.; Yang, Y. Integrated building envelope performance evaluation method towards nearly zero energy buildings based on operation data. *Energy Build.* **2022**, *268*, 112219. [CrossRef]
13. Cuevas-Figueroa, G.; Stansby, P.K.; Stallard, T. Accuracy of WRF for prediction of operational wind farm data and assessment of influence of upwind farms on power production. *Energy* **2022**, *254*, 124362. [CrossRef]
14. Zhang, B.; Xu, G.; Zhang, Z. A holistic robust method for optimizing multi-timescale operations of a wind farm with energy storages. *J. Clean. Prod.* **2022**, *356*, 131793. [CrossRef]
15. Helseth, L. Harvesting energy from light and water droplets by covering photovoltaic cells with transparent polymers. *Appl. Energy* **2021**, *300*, 117394. [CrossRef]
16. Zhang, F.; Han, C.; Wu, M.; Hou, X.; Wang, X.; Li, B. Global sensitivity analysis of photovoltaic cell parameters based on credibility variance. *Energy Rep.* **2022**, *8*, 7582–7588. [CrossRef]
17. Csoknyai, T.; Hrabovszky-Horváth, S.; Georgiev, Z.; Jovanovic-Popovic, M.; Stankovic, B.; Villatoro, O.; Szendrő, G. Building stock characteristics and energy performance of residential buildings in Eastern-European countries. *Energy Build.* **2016**, *132*, 39–52. [CrossRef]
18. Hajian, H.; Ahmed, K.; Kurnitski, J. Dynamic heating control measured and simulated effects on power reduction, energy and indoor air temperature in an old apartment building with district heating. *Energy Build.* **2022**, *268*, 112174. [CrossRef]
19. Holzmann, A.; Adensam, H.; Kratena, K.; Schmid, E. Decomposing final energy use for heating in the residential sector in Austria. *Energy Policy* **2013**, *62*, 607–616. [CrossRef]
20. Le Truong, N.; Dodoo, A.; Gustavsson, L. Final and primary energy use for heating new residential area with varied exploitation levels, building energy performance and district heat temperatures. *Energy Procedia* **2019**, *158*, 6544–6550. [CrossRef]
21. Życzyńska, A.; Suchorab, Z.; Majerek, D. Influence of thermal retrofitting on annual energy demand for heating in multi-family buildings. *Energies* **2020**, *13*, 4625. [CrossRef]

22. Wei, S.; Jones, R.; de Wilde, P. Driving factors for occupant-controlled space heating in residential buildings. *Energy Build.* **2014**, *70*, 36–44. [CrossRef]
23. Laskari, M.; de Masi, R.-F.; Karatasou, S.; Santamouris, M.; Assimakopoulos, M.-N. On the impact of user behaviour on heating energy consumption and indoor temperature in residential buildings. *Energy Build.* **2021**, *255*, 111657. [CrossRef]
24. Haq, M.A.U.; Rao, G.S.; Albassam, M.; Aslam, M. Marshall–Olkin Power Lomax distribution for modeling of wind speed data. *Energy Rep.* **2020**, *6*, 1118–1123. [CrossRef]
25. Ballarini, I.; Corrado, V. Application of energy rating methods to the existing building stock: Analysis of some residential buildings in Turin. *Energy Build.* **2009**, *41*, 790–800. [CrossRef]
26. Regulation of The Polish Minister of Infrastructure of 27 February 2015 Concerning the Methodology for Calculating the Energy Performance of the Building or Part of a Building and the Preparation of Certificates of Energy Performance. Available online: http://prawo.sejm.gov.pl/isap.nsf/DocDetails.xsp?id=WDU20150000376 (accessed on 13 June 2022).
27. Act of August 29, 2014 on the Energy Performance of Buildings. Available online: https://isap.sejm.gov.pl/isap.nsf/download.xsp/WDU20140001200/U/D20141200Lj.pdf (accessed on 13 June 2022).
28. European Union. Directive 2010/31/EU of the European Parliament and of the Council of 19 May 2010 on the energy performance of buildings. *Off. J. Eur. Union* **2010**, *153*, 13–35.
29. Regulation of the Polish Ministry of Transport, Construction and Maritime Economy of 13 August 2013 Amending the Regulation on the Technical Conditions to Be Met by Buildings and Their Location. Available online: http://isap.sejm.gov.pl/isap.nsf/download.xsp/WDU20130000926/O/D20130926.pdf (accessed on 1 February 2021).
30. Ramsey, J.B. Tests for specification errors in classical linear least-squares regression analysis. *J. R. Stat. Soc. Ser. B Methodol.* **1969**, *31*, 350–371. [CrossRef]
31. Utts, J.M. The rainbow test for lack of fit in regression. *Commun. Stat. Theory Methods* **1982**, *11*, 2801–2815. [CrossRef]
32. Murphy, J.M.; Sexton, D.M.H.; Barnett, D.N.; Jones, G.S.; Webb, M.J.; Collins, M.; Stainforth, D. Quantification of modelling uncertainties in a large ensemble of climate change simulations. *Nature* **2004**, *430*, 768–772. [CrossRef]
33. Zyczynska, A. *Wydawnictwo Analiza Zmiennosci Charakterystyki Energetycznej na Przykladzie Budynku Wielorodzinnego*; Wydawnictwo Politechniki Lubelskiej: Lublin, Poland, 2019; ISBN 978-83-7947-390-8.

Article

Thermal Comfort—Case Study in a Lightweight Passive House

Krzysztof Wąs *, Jan Radoń and Agnieszka Sadłowska-Sałęga

Department of Rural Building, Faculty of Environmental Engineering, University of Agriculture in Krakow, 30-059 Krakow, Poland; jan.radon@urk.edu.pl (J.R.); agnieszka.sadlowska@urk.edu.pl (A.S.-S.)
* Correspondence: krzysztof.was@urk.edu.pl

Abstract: Saving energy while maintaining a high-quality internal environment is an increasingly important scientific and technological challenge in the building sector. This paper presents the results from a long-term study on thermal comfort in a passive house situated in the south of Poland. The building was constructed in 2010 with the use of prefabricated, lightweight technology. The main energy source is a ground source heat pump which powers the floor heating and DHW. The building is also equipped with a mechanical ventilation system with heat recovery and a ground source heat exchanger. A lightweight building structure which has active systems with limited capabilities (especially for cooling) is a combination which increases the difficulty of maintaining a proper inner environmental condition. Extensive experimental investigations on hygrothermal performance and energy use have been carried out in the building for several years. The measurement results, such as inner air temperature and humidity, as well as the inner surface temperature of partitions, could be directly used to determine basic thermal comfort indicators, including *PMV* and *PPD*. Any missing data that has not been directly measured, such as the surface temperature of the windows, floors, and some of the other elements of the building envelope, have been calculated using WUFI®PLUS software and validated with the available measurements. These results are not final; the full measurement of thermal comfort as an applied methodology did not consider human adaptation and assumed constant clothing insulation. Nevertheless, in general, the results show good thermal comfort conditions inside the building under research conditions. This was also confirmed via a survey of the inhabitants: 2 adults and 3 children.

Keywords: thermal comfort; lightweight passive house; hygrothermal building simulations

Citation: Wąs, K.; Radoń, J.; Sadłowska-Sałęga, A. Thermal Comfort—Case Study in a Lightweight Passive House. *Energies* **2022**, *15*, 4687. https://doi.org/10.3390/en15134687

Academic Editors: Katarzyna Ratajczak and Łukasz Amanowicz

Received: 26 May 2022
Accepted: 24 June 2022
Published: 26 June 2022

1. Introduction

Obtaining high energy efficiency while providing adequate thermal comfort in all spaces and all seasons is becoming a challenge in passive buildings. Many energy-efficient buildings function at lower heating temperatures which might not usually provide a high quality of thermal comfort [1]. In addition to the radical minimization of heat loss, buildings with a very low energy demand obtain maximum solar gains thanks to the southern orientation of their windows, which can lead to significant overheating in the summer [2,3]. A review of the available literature around this matter of thermal comfort in energy-efficient buildings shows that the available papers are usually based on numerical analyses [4]. Most experimental research was carried out under laboratory conditions [5] or mainly concerns office-type spaces [2,6]. There are only a few examples of real residential buildings for which the monitoring results in terms of indoor living environment quality have been presented. Truong and Garvie [7] presented the results of monitoring the indoor climate and comfort outcomes of a three-bedroom single-storey detached passive house in Australia. Berr et al. [8] compared interviews, which were conducted 'face-to-face' in the resident's household regarding thermal comfort and energy use with yearly measurements of microclimate parameters. The literature also lacks information on the perception of comfort on a daily scale, let alone on an hourly scale. Under real conditions and concerning long-term experimental studies, determining the thermal sensations on the basis of a resident's survey answers is practically impossible (difficulties, e.g., in the hourly determination of the thermal insulation of clothing or the activity of inhabitants). The absence of empirical evidence documenting a resident's perceptions of their low-energy home shows that little is understood about whether residents enjoy living in them [8].

The building under research conditions fulfills PH standard requirements. Low energy use is the result of a high thermally-insulated envelope and active systems based on a ground source heat pump as well as mechanical ventilation with heat recovery. Lightweight structures have low heat buffering capacities. This factor has an impact on heating control in winter and increases the overheating risk in summer. No active cooling system is available. The heat recovery unit has a bypass, but night cooling is ineffective because of the lightweight structure. A basic way to avoid overheating is air cooling via a passive ground-coupled heat exchanger. Therefore, our main research aim was to examine the thermal comfort within this particular lightweight building equipped with active systems with limited capabilities, especially for cooling.

There are many different versions of the definition of comfort. Difficulties in defining this concept and determining the scope of its parameters result primarily from the subjective perceptions of users and the interrelationships between the parameters that define it. The basic condition for experiencing thermal comfort is a balance between body heat and the environment. This means that the excess energy produced by the body during metabolic processes can be freely released into the environment.

In order to define the comfort standard, a closer look at the relevant parameters is needed. They can be divided into values related to the thermal environment, such as temperature, humidity, air velocity, and radiation temperature, and parameters characterizing humans such as activity, age, and clothing. Air and envelope surface temperatures have always been regarded as the main comfort indicators within equivalent temperature [9,10], effective temperature [11], operative temperature [12], and standard effective temperature (SET*) [13]. The method of assessing comfort, as presented by European standards [14,15], includes the statistical indicators for assessing thermal comfort from the user's point of view:

- Predicted mean vote (*PMV*)—expressing, on a seven-point scale, the average thermal feeling rating of a large group of people;
- Predicted Percentage of Dissatisfied (*PPD*)—describing the percentage of people dissatisfied with the thermal conditions.

PMV is an index of thermal comfort, which is most widely used for assessing moderate indoor thermal environments. It predicts the expected comfort vote on the ASHRAE scale of subjective warmth (cold (-3), cool (-2), slightly cool (-1), neutral (0), slightly warm (1), warm (2), and hot (3)). It can be calculated for any combinations of human metabolic rate (M), clothing thermal insulation (I_{cl}), air temperature (t_a) and velocity (v_{ar}), mean radiant temperature (t_r), and partial pressure of water vapor (p_a) [14]. As the *PMV* index was developed on the basis of test results, which differed only slightly from the neutral state (*PMV* = 0), the standard [14] precisely specifies the scope of *PMV* application: $M = 46\text{–}232\ \text{W·m}^{-2}$ (0.8–4 met), $I_{cl} = 0\text{–}0.310\ \text{m}^2\text{·K·W}^{-1}$ (0–2 clo), $t_a = 10\text{–}30\ ^\circ\text{C}$, $t_r = 10\text{–}40\ ^\circ\text{C}$, $v_{ar} = 0\text{–}1\ \text{m·s}^{-1}$, and $p_a = 0\text{–}2700\ \text{Pa}$.

In recent years, opinions have been expressed about the inadequacy of comfort assessment using *PMV*. Humphreys and Nicol [16] showed that *PMV* was less closely correlated with comfort votes than the air temperature or the globe temperature, and that the effects of errors in the measurement of *PMV* were not negligible. An analysis of the ASHRAE database showed that *PMV* can be significantly misleading when used to predict the mean comfort votes of people in everyday conditions in buildings, particularly in warm environments [17]. Studies from other research centers also proved that the calculated value of *PMV* does not match the answers (Thermal Sensation Votes (*TSV*)) obtained in field studies [18–20].

Behavioral adaptation includes all of the conscious and unconscious behavior in daily life. These can be personal (e.g., changing clothing), technical (e.g., turning on an air conditioner or a fan), and cultural (e.g., an afternoon rest or nap taken during the hottest working hours of the day in a hot climate). Behavioral adaptation can also be influenced by social norms (injunctive and descriptive), such as reducing energy consumption. These behavioral indices were included in a more adequate comfort calculation model.

Concerning comfort during sleep, in addition to the thermal insulation of nightwear, the total thermal insulation of the bedding system should be considered. This depends on the thermal insulation of the bed itself, mattresses, pillows, body coverage of the quilt (considering thicknesses, fibre filling weights, and weight), and the air layer between the human body and the system [21–23]. The studies described in [24] indicate that sleeping posture also affects the total thermal insulation of a bedding system.

Some standards of so-called low energy-intensive construction differ from each other in terms of reducing the impact on the environment. They are often confused with each other [25]. Discussion in the literature [26,27] has focused on the positive and negative features of low-energy buildings, net-zero energy buildings (NZEB) [28,29], nearly zero-energy buildings (nZEB) [30–32], green buildings, solar houses, sustainable buildings, energy-plus buildings, and passive houses (PH) [33,34]. The energy-saving measures applied within passive houses ensure savings on heating- and cooling-related energy, reaching 90% compared to traditional buildings and over 75% in comparison to average new buildings [35]. The energy demand of a PH fulfils the requirements of the EU EPBD [36], which states that energy use should be as low as is practically achievable. To a large extent, this is due to its efficiency design, which is exemplified by a high level of thermal insulation, windows with low heat transfer, airtightness of the building envelope, and mechanical ventilation systems with heat recovery. On top of these, there is also significant attention paid to the elimination of thermal bridges. PHs require no more than $15\ kWh \cdot m^{-2} \cdot year^{-1}$ for heating or cooling, and the heating or cooling peak load does not exceed $10\ W \cdot m^{-2}$ [33,35]. For a building to be considered as being a PH, its conventional primary energy use cannot go beyond $120\ kWh \cdot m^{-2} \cdot year^{-1}$. This standard does not allow for excessive temperatures that exceed 10% of the cooling period in warmer climates [33]. With such a limited amount of energy supply, it is easier to meet the subsequent demand by means of renewable energy sources. Currently, a significant majority of studies on PHs focus on factors such as thermal performance under various climatic conditions [37–40], life-cycle assessment (LCA) and costing (LCC) [41–45], comparative assessment with zero-energy buildings [26,46,47], integration of renewable energy technologies (RET) [48,49], upgrading historic buildings to the standard [50,51], investigations of building material performance [52–55], and the indoor environment [3,56–58].

2. Materials and Methods

2.1. Case Study

Our research was conducted in a single-family building located in the south of Poland. The house was built with a technology of a prefabricated wooden frame structure, which rests on a reinforced concrete foundation slab isolated from the ground with a 50 cm layer of extruded polystyrene. The individual partitions (walls and ceilings) were made in the factory. Then they were transported and assembled at the construction site. The building has almost 120 m^2 of usable area. On the ground floor, there is a living room with kitchen, an office room, a toilet, and a technical room. The first floor includes three bedrooms and a bathroom (Figure 1). During the research, the building was inhabited by a family of five (parents and three children).

(a)   (b)

Figure 1. South-west facade (**a**) and south-east facade (**b**).

The building meets the requirements of PH standards. The values of the relevant parameters are presented in Table 1. The average value of the wall thermal conductivity coefficient is 0.08 W·m^{-2}·K^{-1}. Such a low U-value was achieved by using the skeleton structure. The major parts of the cross-section of the walls are filled with an insulating material. The interior of the skeleton is filled with 16 cm-thick mineral wool. Moreover, a 25 cm layer of insulation (polystyrene, wood wool, or mineral wool) was applied to the outer surface. Figure 2 presents the cross-section of the particular partitions. In total, eight variants of assemblies, differing mainly in their use of thermal insulation materials, various types of stiffening plates, and vapor barrier, were used to build the house. The intention behind such a design was to test various configurations with regards to their hygrothermal parameters [55]. The foundation interface, characterized by extreme thermal insulation, eliminating the influence of thermal bridges in floor area, was another specific solution. To obtain lower U-values and to ensure better tightness, non-opening windows were installed, with the exception of one opening terrace window on the ground floor.

Table 1. Relevant parameters of the building.

Parameter	Value
Average heat transfer coefficient of opaque, outer building walls	0.08 W·m^{-2}·K^{-1}
Heat transfer coefficient of windows (3 glass panes)	0.74 W·m^{-2}·K^{-1}
Solar heat gain coeficient (average)	0.6
Efficient heat recovery ventilation unit	93%
Airtightness, ACH	0.5 h^{-1}
Heating energy demand	7.5 kWh·m^{-2}·year^{-1}
Primary energy demand	104.4 kWh·m^{-2}·year^{-1}

The active systems included floor and an air heating (reheating the ventilation air) and a system for preparing domestic hot water. The heat is supplied by a ground heat source pump. The building is equipped with a mechanical ventilation system with heat recovery and a ground heat exchanger. The heat exchanger mitigates fluctuations in the temperature of the outside air let into the building in the winter and summer season. The systems are shown in Figures 3 and 4.

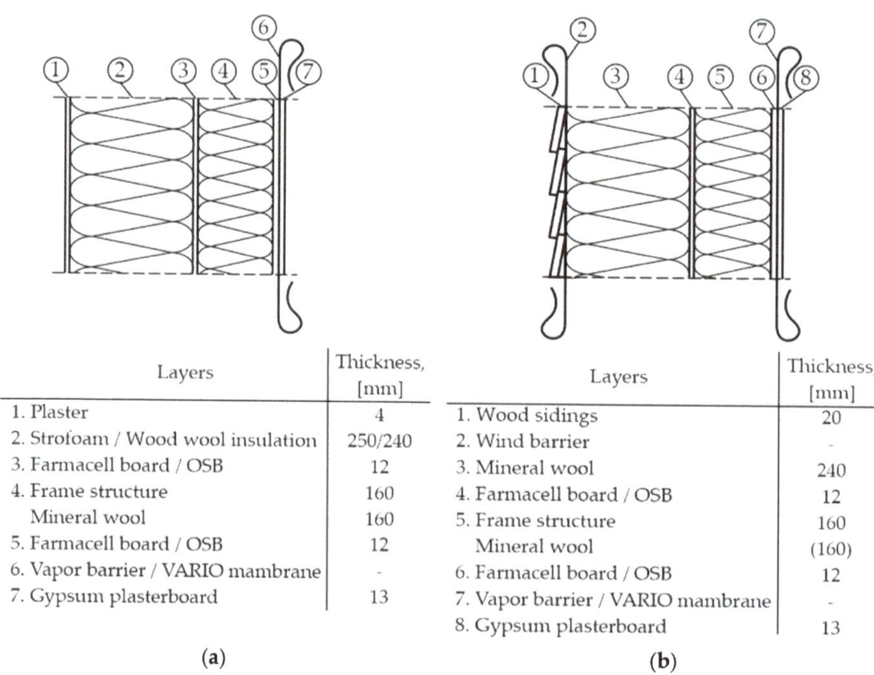

Layers	Thickness, [mm]	Layers	Thickness, [mm]
1. Plaster	4	1. Wood sidings	20
2. Strofoam / Wood wool insulation	250/240	2. Wind barrier	-
3. Farmacell board / OSB	12	3. Mineral wool	240
4. Frame structure	160	4. Farmacell board / OSB	12
Mineral wool	160	5. Frame structure	160
5. Farmacell board / OSB	12	Mineral wool	(160)
6. Vapor barrier / VARIO mambrane	-	6. Farmacell board / OSB	12
7. Gypsum plasterboard	13	7. Vapor barrier / VARIO mambrane	-
		8. Gypsum plasterboard	13

(a)   (b)

Figure 2. Assembly variants of the outer wall: external plaster finishing (**a**) and siding on the external surface (**b**).

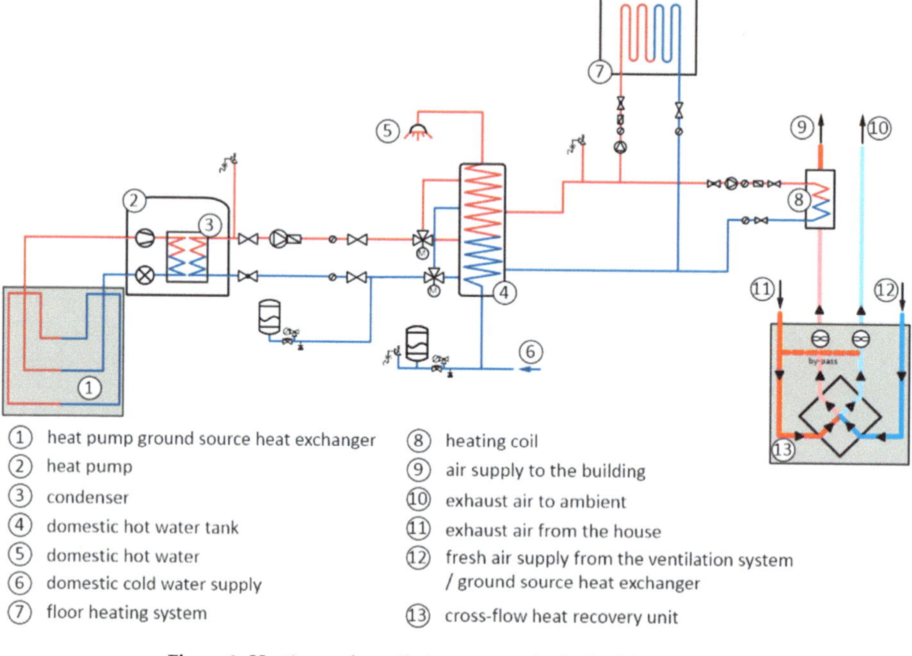

①	heat pump ground source heat exchanger	⑧	heating coil
②	heat pump	⑨	air supply to the building
③	condenser	⑩	exhaust air to ambient
④	domestic hot water tank	⑪	exhaust air from the house
⑤	domestic hot water	⑫	fresh air supply from the ventilation system / ground source heat exchanger
⑥	domestic cold water supply		
⑦	floor heating system	⑬	cross-flow heat recovery unit

Figure 3. Heating and ventilation systems in the building.

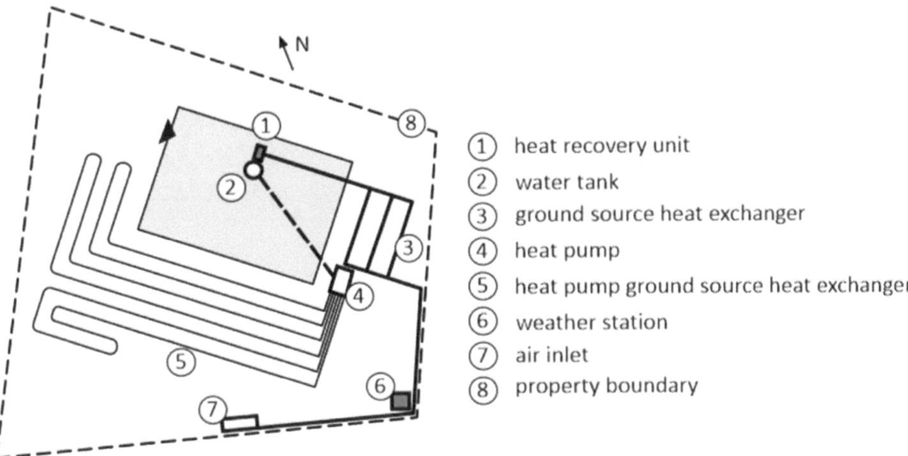

Figure 4. Heat pump, ground heat exchanger, and weather station arrangement.

The building was subjected to extensive experimental research for many years, starting in 2011. A total of 158 sensors were installed in the building structure and in the technical systems. A local meteorological station was installed next to the building. Measurements were recorded at 1 or 6 min intervals. The results, including the energy flows in particular elements of the active systems (heat pump, circulation pumps, and fans) and electricity use in the entire building, were analyzed. The temperature and humidity from within the assemblies were also measured. This allowed us to determine the hygrothermal performance in eight variants of the walls and two types of roof structures under real operating conditions [54,55]. Also, analysis and computational simulations of thermal conditions around the ground heat exchanger [59] were carried out.

2.2. Measurement of Microclimate Parameters in the Building

Inside the building, the equipment for monitoring of the microclimate parameters was installed. The air temperature and relative humidity were measured using the integrated LB-710HS thermo-hygrometer. The relevant parameters were as follows:

1. Temperature measurement:
   - Accuracy: 0.1 °C;
   - Measurement range: −40–85 °C;
   - Resolution: 0.1 °C.

2. Relative humidity measurement:
   - Accuracy: 2%;
   - Range: 0–100%;
   - Resolution: 0.1%.

Thermo-hygrometers were installed in the living room on the ground floor, in the bedroom and bathroom on the first floor, and in the non-functional attic (Figure 5). Radiant temperature was measured by black globe thermometers:

- Accuracy: 0.3 °C;
- Range: −50–200 °C;
- Resolution: 0.1 °C;
- Diameter: 150 mm, ball 150 mm;
- Material: matte, blackened, copper, diameter.

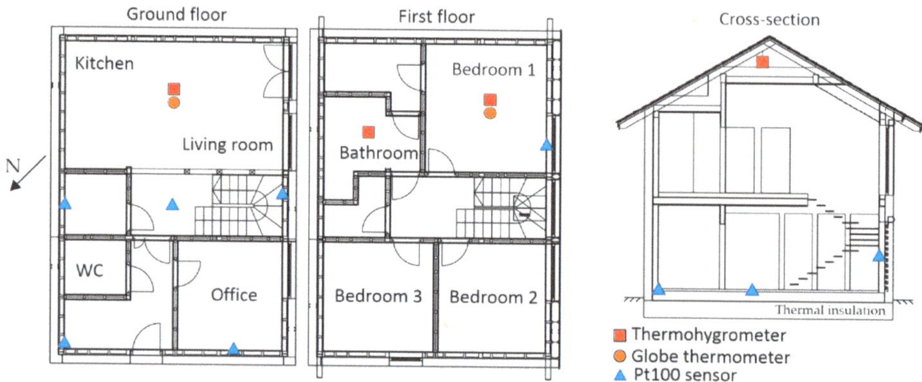

Figure 5. Location of measurement sensors.

Globe thermometers were located in the living room on the ground floor and in the bedroom on the first floor. The temperature of the inner surface of the partitions was also measured. The sensors (of type PT100 with an accuracy of 0.1 °C) installed close to the surface were used for this purpose (Figure 5).

2.3. Measurement of Outdoor Climate Parameters

The parameters of the external climate were measured in situ by the meteorological station located in the southern part of the plot (Figure 4). The following parameters were recorded: temperature, relative humidity, direct and diffuse radiation, wind speed, and direction.

2.4. Complementary Calculations

As not all the relevant parameters could be directly measured, e.g., inner surface temperature of all the partitions and floor, WUFI®PLUS software was used for supplementary calculation. Based on the blueprints, a 3D model of the building was created. Measured hourly values of temperature and relative air humidity inside the building were assumed as boundary conditions for the calculation of hygrothermal performance of partitions. The outdoor climate, based on measured parameters in the weather conditions near the building, was assumed as the external boundary condition. Calculation results were then validated with the available measured results, such as measured surface temperature (Figure 5).

Floor surface temperature is an essential component of thermal comfort. It is influenced by the transient building–soil thermal interaction [60]. The surface temperature was not measured directly. Temperature sensors were located in 3 positions (Figure 5) between the screed and reinforced slabs. Detailed 3D transient heat flow calculations were carried out to analyze the thermal conditions of the slab-on-grade with floor heating [59]. The measured maximal temperature difference between central and corner points was 3 K. This is because of the thick thermal insulation under the floor and the perimeter insulation. Based on the validated model mean floor surface, the floor temperature pattern was calculated and assumed for the comfort analysis.

2.5. Assumptions for the Comfort Analysis

The assessment of the microclimate in terms of residents' comfort was based on the analysis of indoor air parameters, i.e., temperature and moisture content, the inner surface temperature of partitions and their juxtaposition, such as operative temperature. Measured and complementary calculated parameters were compiled according to the comfort zones of Leusden and Freymark (based on air temperature and relative humidity) [61], Frank (based on air temperature and mean surface temperature) [62], and ASHRAE summer and winter comfort zones (based on operative temperature and humidity) [63].

The assessments of comfort *PMV* and the *PPD* index during the day—for the living room and at night—for the bedroom were made as follows:

1. During the day, two clothing insulation values were assumed: 0,5 clo (e.g., underwear, short-sleeved shirt, light pants, thin socks, and shoes [14]) and 1.0 clo (e.g., briefs, shirt, pants, jackets, socks, and shoes [14]), which correspond to the thermal insulation proposed for winter and summer as standard [15]. An activity of 1.0 met was assumed (seating, writing, and reading [63]).

2. In the case of sleeping comfort, two values of thermal bedding system insulation were adopted: 3.7 clo and 2.0 clo (based on [24]). Activity for sleeping was assumed to be 0.7 met (based on [63]).

The results were compared to the comfort category defined in the standard [14]:

*   I (A) category—*PMV* < ±0.2 and *PPD* < 6%;
*   II (B) category—*PMV* < ±0.5 and *PPD* < 10%;
*   III (C) category—*PMV* < ±0.7 and *PPD* < 15%.

The essential elements of the research and their interrelationship are presented in flowchart (Figure 6).

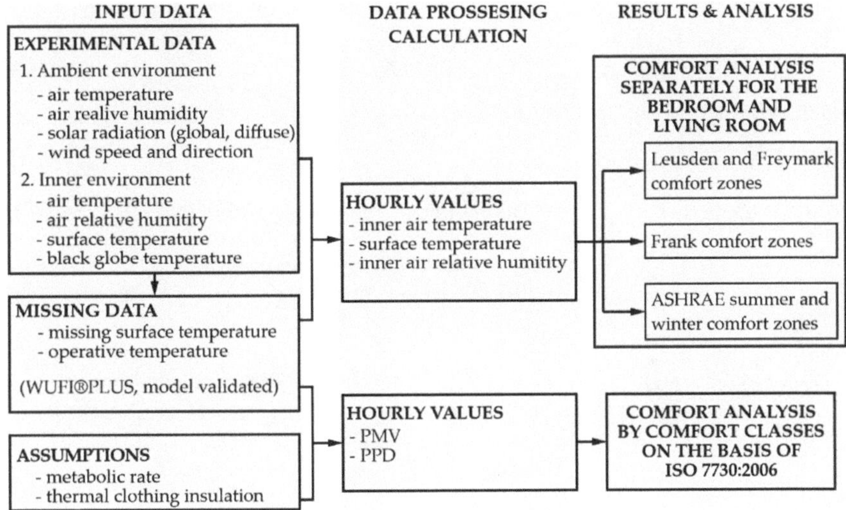

Figure 6. Flowchart of the research.

3. Results and Discussion

3.1. Outdoor Climate

In the year under study, the outside air temperature fluctuated within the range of -28.2–36.2 °C. The minimal temperature was recorded in January and the maximum in August. The mean annual temperature was 8.4 °C (Figure 7). Relative humidity varied from 16.5% to 96.9% (Figure 7), with an average of 79.8% and a median of 87.7%. South and south-easterly winds with speeds up to 5 km·h^{-1} prevailed (Figure 7). The intensity of solar radiation is shown in Figure 7. A comparison with the typical meteorological year (TMY) (climate POL_SL_Katowice, Intl.AP.125600_TMYx.epw [64]) shows that the climate in the year under review was characterized by a greater amplitude (in winter, it was much cooler than the statistical climate, and in summer, it was slightly warmer; the median in both cases differed by only 0.3 °C). The relative humidity was slightly higher than that resulting from the statistical climate.

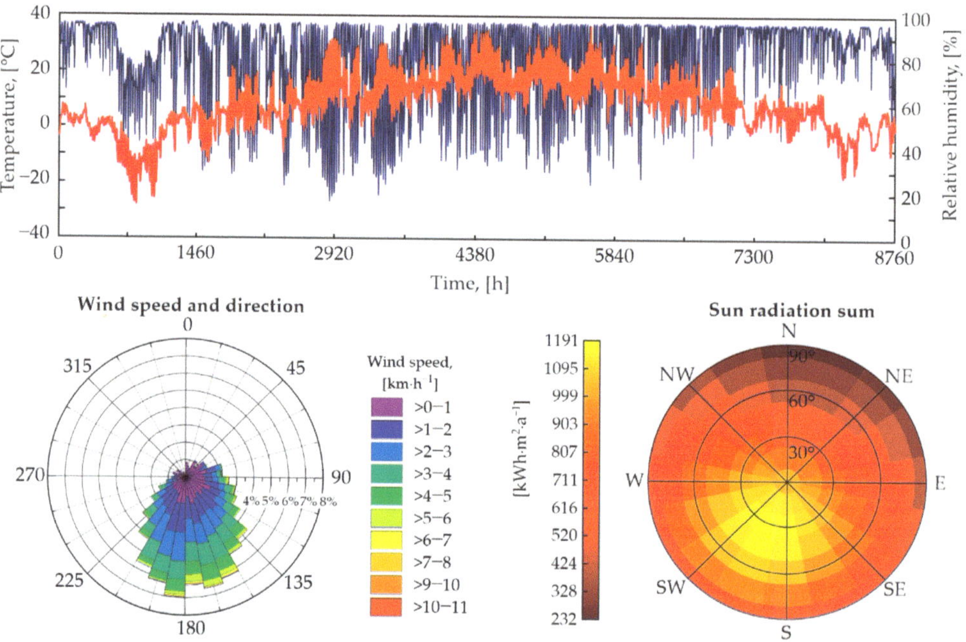

Figure 7. Variability of the measured outer climate.

3.2. Indoor Air Temperature and Relative Humidity

The temperature in the living room fluctuated within the range of 18.5–26.9 °C. For the bedroom, the range was 17.8–27.4 °C, whilst the bathroom ranged from 18.0–29.2 °C. The median was 22.4 °C in the living room, 21.9 °C in the bedroom, and 22.2 °C in the bathroom (Figure 8a). An example of the course of the temperatures during January from within the analyzed rooms is presented in (Figure 9). The temperature in the living room was characterised by a rather low variability (standard deviation 1.2 °C). For the bedroom and bathroom, the variation was similar (standard deviation 1.8 °C and 1.7 °C, respectively).

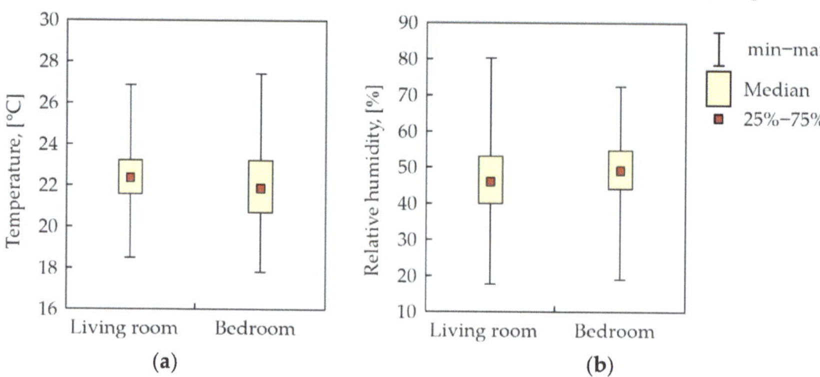

Figure 8. Extremes, mean value, and 25/75 percentiles for the temperature (**a**) and relative humidity (**b**) of the air in the living room and bedroom.

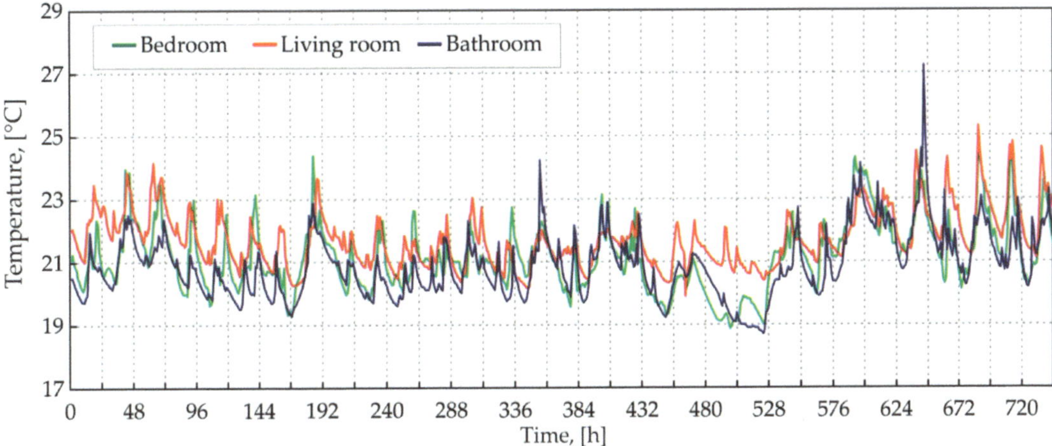

Figure 9. Air temperature during January in the analyzed rooms.

Relative humidity inside the building was measured in the same locations as the temperature. The ranges were: 17.5–80.2% for the living room, 18.8–72.4% for the bedroom, and 21.1–86.3% for the bathroom on the first floor (Figure 5). The measurements showed that the relative humidity in the bedroom was characterized by the lowest variability, with a standard deviation of 8.3%. The highest variability was observed in the living room, where the standard deviation was 10.5%. For the bathroom, the standard deviation was 8.9%. The measured ranges of relative humidity are presented in (Figure 8b).

Differences in microclimate parameters between the living room, bedroom, and bathroom were statistically compiled. Based on the measured patterns of temperature and relative humidity, an R–Spearman correlation coefficient between the rooms was calculated. The high correlation between the results was statistically significant. Detailed values are presented in Table 2.

Table 2. R-Spearman correlation coefficients.

	Temperature			Relative Humidity		
	Bedroom	**Livingroom**	**Bathroom**	**Bedroom**	**Livingroom**	**Bathroom**
Bedroom	1.000	0.680	0.878	1.000	0.891	0.872
Livingroom	0.680	1.000	0.686	0.891	1.000	0.858
Bathroom	0.878	0.686	1.000	0.872	0.858	1.000

The comfort assessment proposed by Leusden and Freymark [61] was carried out for the living room and bedroom. These are spaces with the longest time inhabited by the residents. As shown in Figure 10, hourly values in both rooms indicate that for most of the time, the conditions were "comfortable" or "almost comfortable". Only a small fraction of the measurements fell outside of the comfort zone.

The inner climate quality, in terms of operative temperature, humidity ratio, and standard clothing insulation (0.5 and 1.0 clo), was also analyzed. Figure 11 shows the results on an annual basis. While wearing a garment with an insulation value of 0.5 clo, the conditions were outside the comfort zone most of the time. When analyzing the monthly periods, it was observed that from January to May and from September to December, the tested values were practically entirely outside the comfort zone. On the other hand, with a garment insulation performance of 1.0 clo, most measurements were in the comfort zone. This is also confirmed by monthly analyses which show that exceeding the comfort zone occurs mainly in the summer months, i.e., June, July, and August.

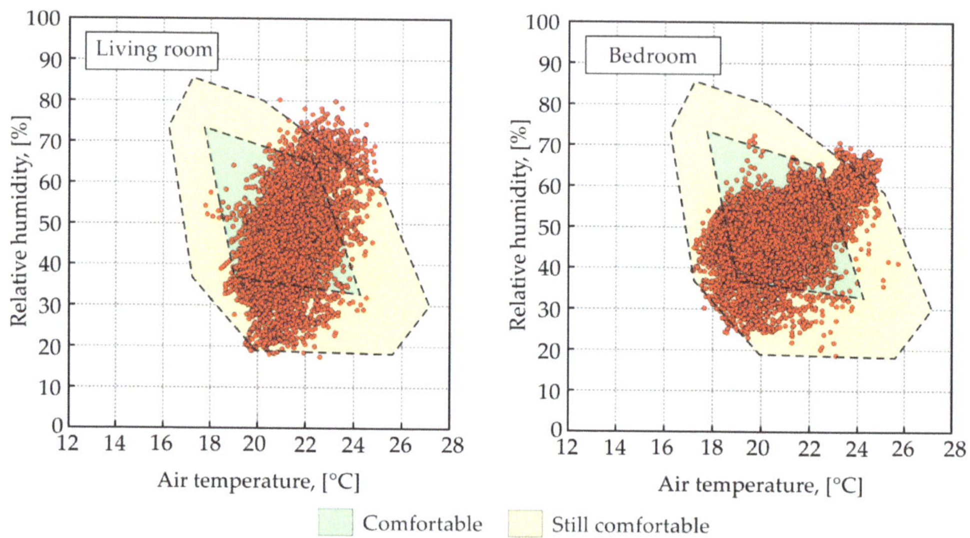

Figure 10. Internal air temperature and relative humidity against the comfort zones according to Leusden and Freymark [61].

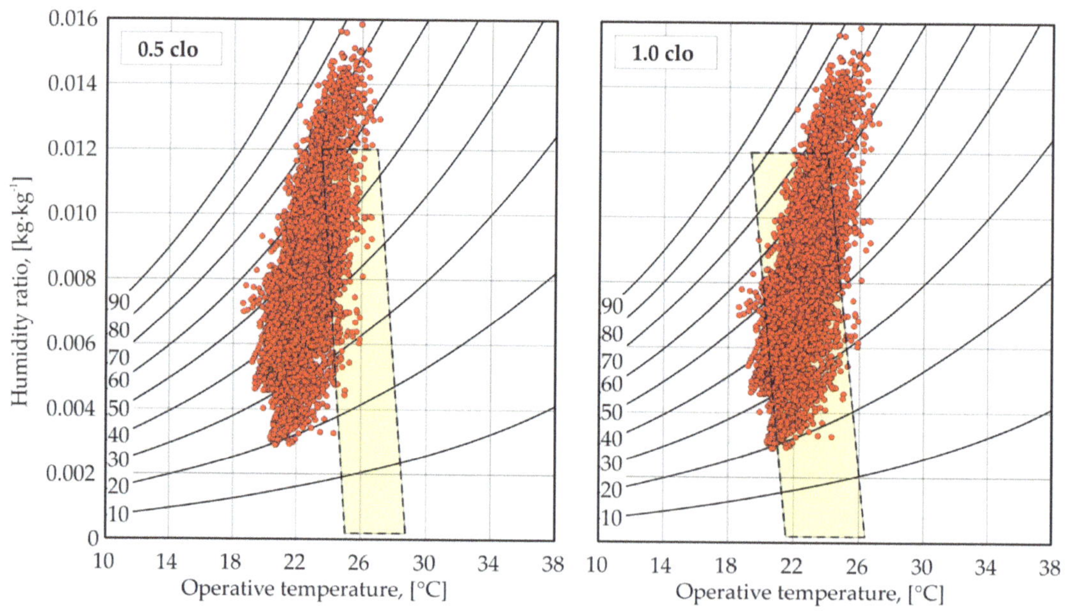

Figure 11. Operative temperature and humidity ratio against the comfort zones in winter and summer [62].

3.3. Surface Temperature

The inner surface temperature of two walls was directly measured. The average temperature was at a similar level, i.e., 20.9 °C in the living room, 21.1 °C in the bedroom and 21.2 °C in the bathroom. In the bedroom, the biggest fluctuation in temperature range (16.4–26.4 °C) was recorded on an annual basis, Figure 12a.

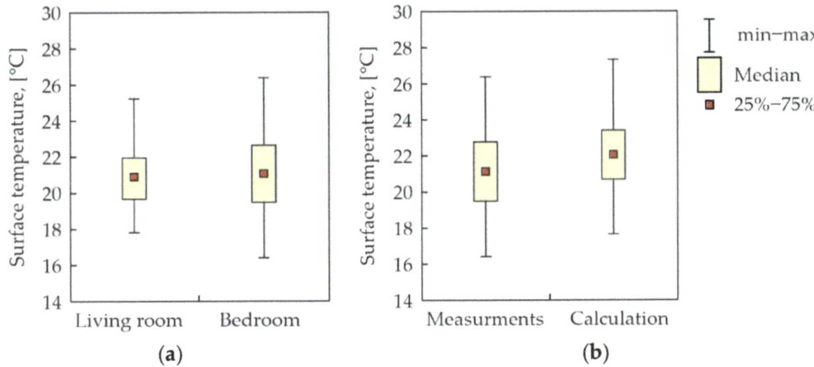

Figure 12. Measured surface temperature in rooms (**a**) and the measured and calculated surface temperature in the bedroom (**b**).

The surface temperature of the remaining building assemblies was determined by calculation. A high correlation coefficient (0.97) by absolute difference for the living room below 0.5 °C and bedroom at 1.0 °C was observed, when compared with the available measurement results (Figure 12b).

Based on the validated calculation results of surface temperature and measured air temperature, the comfort level, according to Frank [62], was determined. The nomograms (Figure 13) show the results for the living room and the bedroom on an annual basis. In both rooms, most of the surface-air temperature value pairs are in the "comfortable" or "almost comfortable" zone. Monthly analysis specified an exceedance that occurred mainly in the summer months. They are available from June to September for the living room and from June to August for the bedroom.

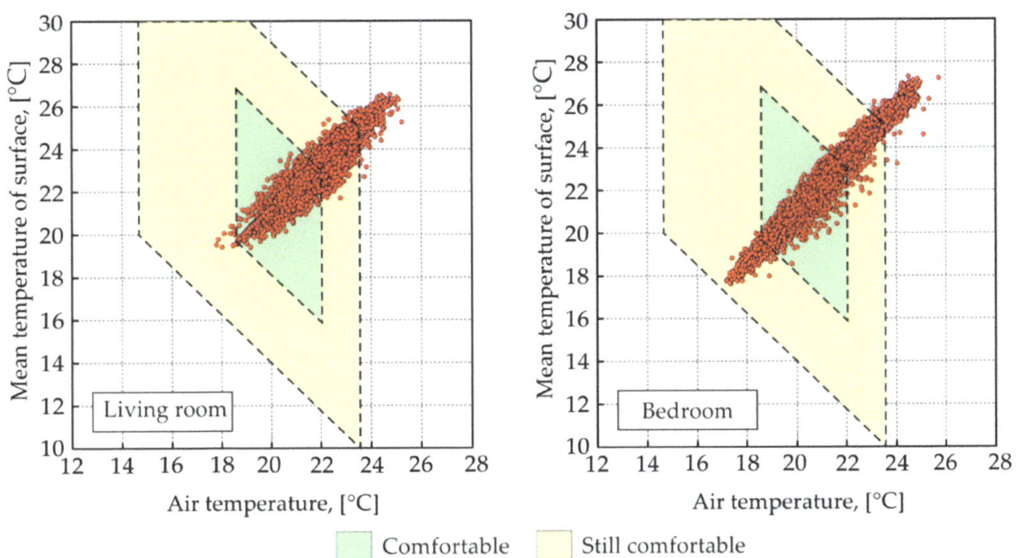

Figure 13. Comfort according to Frank [62] in living room and bedroom.

3.4. Radiant and Operative Temperature

The combined air and radiant temperatures were measured by a globe thermometer in the living room on the ground floor and the bedroom on the first floor, Figure 5. The average value for the living room was 22.4 °C, and the range was 18.6–27 °C. The mean value for the bedroom was 21.4 °C, ranging from 17.8–28.7 °C (Figure 14). The mean radiant temperature within the rooms was also calculated as a weighted value of room enclosure surface temperature. The surface temperature was determined by validated computer simulations (see Section 3.3).

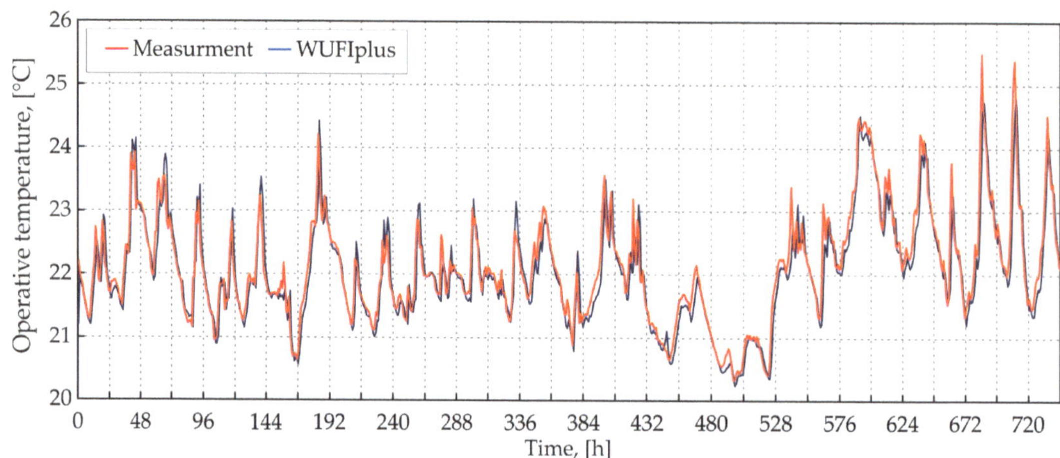

Figure 14. Measurement and calculation based on operative temperature in living room during January.

Based on air and radiant temperature, an operative temperature was determined. The hourly pattern for the living room during January is shown in Figure 14. The mean value for the measurement based on operative temperature was 22.1 °C and 22.0 °C for the calculated data. The series both differ mainly in their maximum values. The value range for the measurements was 17.8–28.7 °C, and for the calculations, it was 17.7–27.4 °C, as shown in Figure 15. For the measured and calculated mean operative temperature series, a correlation coefficient of 0.98 was obtained.

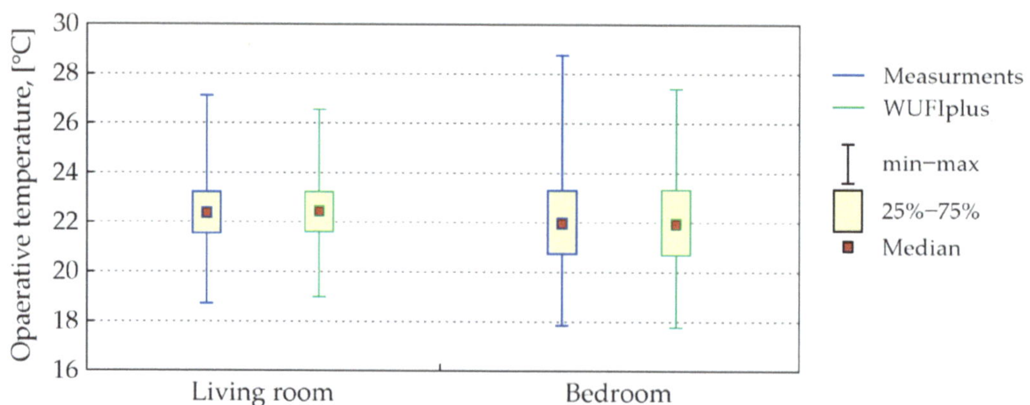

Figure 15. Operative temperature for the living room and bedroom, obtained from the measurements and computer simulations.

3.5. PMV and PPD Indicators

The analysis of *PMV* and *PPD* indicators was split into day and night, as described in the methodology. The adopted scenario assumed a human presence in the living room from 7 AM to 10 PM (16 h a day in total), with three variants of thermal insulation of clothing for activities for activities of 1.0 met. Two of them are standard values, i.e., 1.0 clo and 0.5 clo. Additionally, an intermediate value of 0.75 clo was considered. For bedroom presence time, the time from 11 PM to 6 AM was assumed. Two variants of thermal insulation, including bedclothes, were set: 3.7 clo and 2.0 clo, with an activity of 0.7 met.

The results of the calculations for daily comfort, including comfort categories for the living room, are presented in Figure 16. *PMV* fluctuates annually in the range −1.1–1.0, when an insulation value of 1.0 clo is assumed. The conditions of I category lasted a total of 2027 h, which is 34.7% of the period considered. The II category covers 2268 h (38.8%), and the III category an amount of 951 h (16.3%). It was found that 602 h occurred outside the limit of applicability toward the methodology, which is 10.3%. If 0.5 clo is assumed, the results indicate that the conditions are generally too cold. Most of time (5141 h, which is 88%) this is even beyond the area of the quantifiable parameters. In terms of comfort, time coverage was I—149 h (2.6%), II—255 h (4.4%), and III—300 h (5.1%). On an annual basis, the analyzed *PMV* values range from −2.6 to 0.3. In turn, the third case of thermal insulation of clothing of 0.75 was an intermediate variant. *PMV* values ranged from −1.7 to 0.7. The I, II, and III categories covered 10.3%, 23.0%, and 19.7% of the analyzed time, respectively. The percentage of time left which occurred outside quantifiability was 47.1%. Detailed values are presented in Table 3. Figure 17 shows the percentage of monthly-based *PMV* in terms of the comfort category for the considered thermal insulation properties of clothing. It confirms that the most optimal clothing value for the living room, for all analyzed cases, is 1.0 clo.

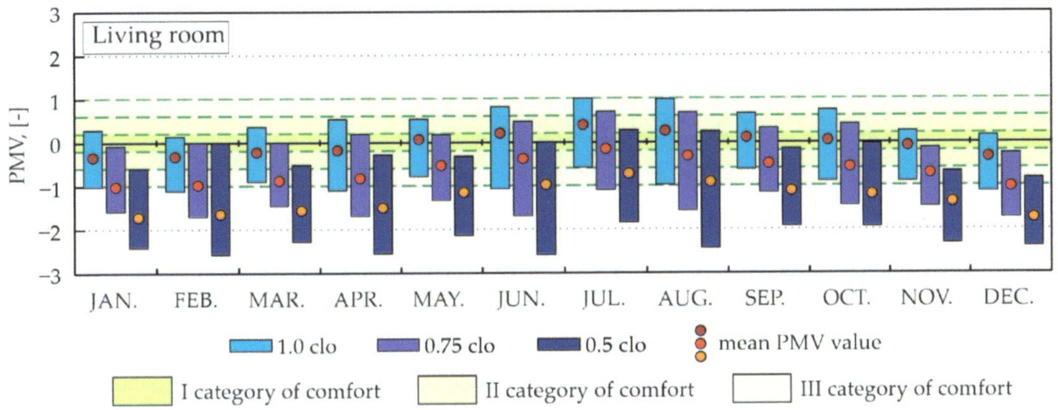

Figure 16. The range of *PMV* values depending on the clothing insulation for the living room.

Table 3. Daily comfort categories in living room.

Parameter	1 clo		0.75 clo		0.5 clo	
	hours	%	hours	%	hours	%
I category	2027	34.7	599	10.3	149	2.6
II category	2268	38.8	1343	23.0	255	4.4
III category	951	16.3	1153	19.7	300	5.1
Beyond applicability—cool	503	8.6	2750	47.1	5141	88.0
Beyond applicability—warm	99	1.7	0	0.0	0	0.0

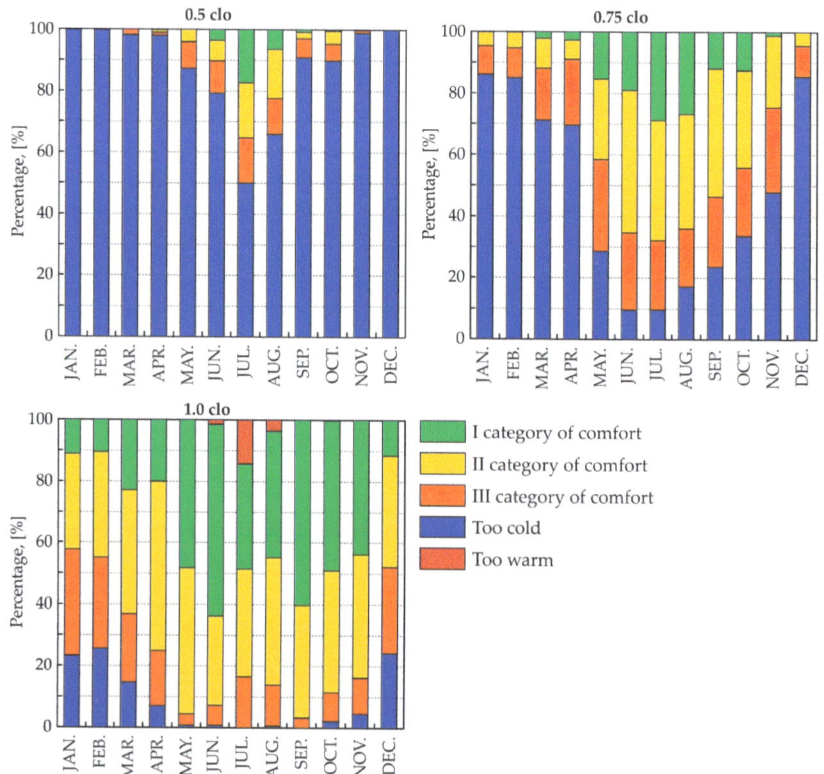

Figure 17. Monthly comfort category quota for the living room by clothing insulations of 1.0, 0.75, and 0.5 clo.

The results for night-time comfort in the bedroom are presented in Figure 18. When assuming a total insulation for clothing and bedding of 2.0 clo, the *PMV* values fluctuate annually in the range of −1.8–0.8. In the I category of comfort, 403 h were recorded, which constitutes 13.8% of the period considered. For 1524 h (52.2%), conditions outside the qualifiable range occurred, mainly reported as the feeling of being "too cold" during the winter half of the year. When assuming an insulation of 3.7 clo (bed set plus clothing), most of the *PMV* values fall within the comfort categories (71.5%). An overview of comfort for the bedroom is summarized in Figure 18 and Table 4. The percentage share for particular months is shown in Figure 19.

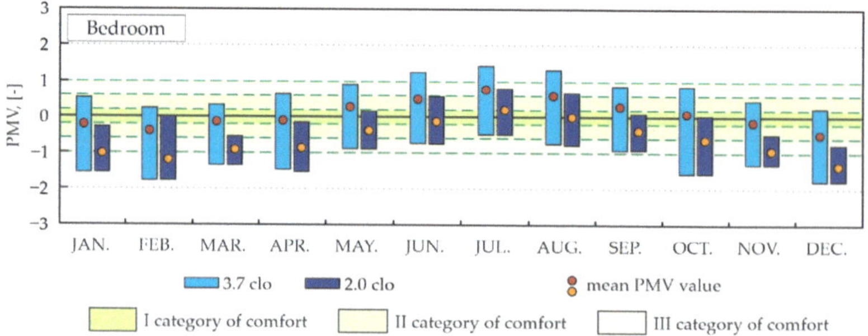

Figure 18. The range of *PMV* values depending on the clothing insulation for the bedroom.

Table 4. Night comfort categories in bedroom.

Parameter	2.0 clo		3.7 clo	
	hours	%	hours	%
I category	403	13.8	840	28.8
II category	590	20.2	857	29.3
III category	403	13.8	390	13.4
Beyond applicability—cool	1497	51.3	383	13.1
Beyond applicability—warm	27	0.9	450	15.4

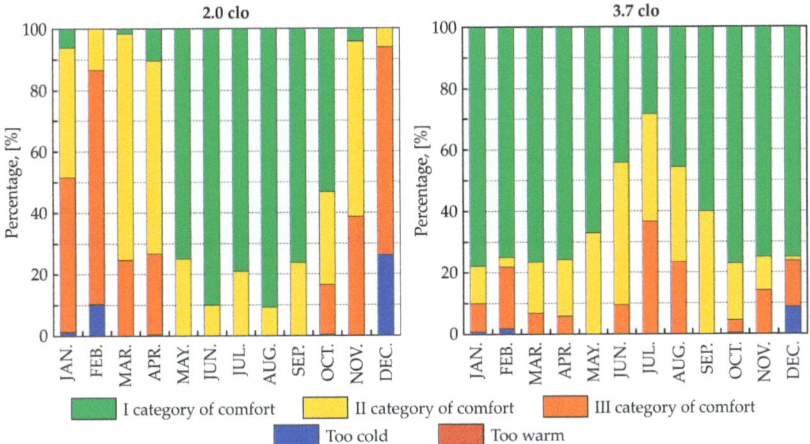

Figure 19. Monthly comfort category quota for the bedroom with clothing insulations of 2.0 and 3.7 clo.

3.6. Discussion

Achieving very low energy consumption is the highest priority when designing a low-energy building. It is assumed that active systems with dimensions based on the demand for standard heat outputs and passive cooling are sufficient to provide adequate thermal comfort. This assumption seems to be particularly relevant in the case of passive buildings, where the internal environment is better isolated from the influence of the external climate. There are very few publications verifying this thesis. Almost all of them indicate some comfort issues and a need for further research in residential passive buildings. A study of the literature, and the specific features of the building under study, prompted the undertaking of this research. In order to assess comfort as widely as possible, various criteria, according to current standards, were used.

Due to the long period needed for the measurements and the fact that the building was inhabited, not all of the parameters were measured. Missing data, like the surface temperature of most partitions, including the floor in the living room and windows, were determined via calculations using the WUFI®PLUS software. The results were validated by comparative calculations, with the use of experimental data both for the boundary conditions and the temperature on the surfaces and the internals of the assemblies. Very good compliance between the measurement and the calculation result allowed us to regard the input parameters as sufficiently accurate.

Obviously, the environmental conditions inside of buildings are strongly dependent on the outdoor climate. The measurement results for the chosen one-year research period showed a colder winter and a warmer summer than the statistical data. Therefore, it might be expected that the comfort parameters during the future use of the building will be in general within the obtained ranges. Measurements from the following years confirmed this thesis. However, the occurrence of different conditions due to climate change may significantly limit the applicability of anticipated comfort.

This research, even though very extensive, was carried out for only one building. Thus, extrapolation of the results to other buildings is limited to similar cases, i.e., lightweight-structure, non-opening windows, mechanical ventilation combined with ground heat exchanger, and underfloor heating powered with a ground source heat pump. Similar usage and external climate are also essential factors. No directly comparable case has been found in the literature. The most differences pertain to building structure and climate zone.

Nevertheless, similar conclusions can be found in some publications. Most papers pay attention to the overheating issue, e.g., the studies by Foster et al. [65] showed overheating above 30 °C in passive houses in Scotland. A study based on interviews with the inhabitants of 25 households (Berr et al. [8]) confirmed good comfort conditions; however, significant issues were identified in the reliability and usability of the energy technologies. Research conducted in Australia (Truonga [7]) and Berr [8] showed very good comfort conditions during the transitional periods (spring and autumn) yet worse but acceptable comfort conditions in winter and summer. Good comfort conditions in the building under study, similar to the results presented in these publications, occurred for the majority of the time. The problem of periodic overheating in summer was also observed. Assumed passive cooling based on the ground source heat exchanger was not sufficient for hot periods.

Despite a correctly selected heat pump, the feeling of cold occurred for relatively short periods of time. This happened when the outer air dropped below −20 °C. Since the heat pump power was calculated for −22 °C conditions, the relatively low heat buffering of the building and poor regeneration of a lower ground heat source could be the reasons for this. Underheating in passive houses is less documented in the literature.

The results presented are not the final measure of thermal comfort in the building as the methodology omits the human adaptation and assumes constant clothing insulation. The results show the time in which individual insulations give a specific category of comfort. Depending on current conditions, people dress according to individual needs. Human adaptation, omitted in the applied methodology, could be another factor improving individual sensing and thermal comfort assessment.

4. Summary and Conclusions

Air temperature and humidity, as well as radiative temperature, are basic input parameters for the calculation of thermal comfort indicators. As the building under research was investigated, mainly for hygrothermal performance and energy use, not all experimental data were available. The missing parameters were obtained by calculations, while measurement results were used for validation. High agreement allowed for reliable analysis based both on experimental and calculated data.

The combined air and radiant temperatures were measured by a globe thermometer in two rooms. This allowed us to determine the operative temperature from air and surface temperature and from directly measured results. Between the two series, a correlation coefficient of 0.98 was obtained.

The compilation between the differences in microclimate parameters for the living room, bedroom, and bathroom was statistically significant. This means that homogeneous conditions do not exist, even in lightweight buildings. Surprisingly higher correlations were obtained for the relative humidity patterns compared to those for the temperature. Mechanical ventilation dominates the changes in humidity, whereas different solar and inner gains in particular rooms cause higher temperature differences.

Based on the hourly and yearly patterns of experimental and complementary data obtained by computer simulations, basic comfort indicators in particular rooms were determined and statistically summarized. Even though the analysis was carried out under a certain calendar year and for a certain building type, and considering that the winter was slightly cooler and the summer warmer than the statistical climate, some general regularities in terms of thermal comfort, typical for lightweight passive buildings, could be observed.

Inner climate quality assessment, according to Leusden and Freymark [61], showed that in the living room, the majority (more than 50%) of hourly temperature and relative humidity value pairs fell within the "comfort zone", and more than 45% fell within the "still comfort zone". Less than 5% was estimated to be outside of the comfort range. Estimations, according to Frank [62], had similarities to this. Lightweight building structure and high thermal insulation cause little air and partition surface temperature differences. Thus, the windows have a greater impact when it comes to comfort criteria, including the inner thermal envelope surface temperature.

Thermal comfort depends strongly on clothing insulation. Analysis, according to ASHRAE methodology, showed that operative temperature and humidity ratio in combination with a standard insulation of 0.5 clo mostly fall outside of the comfort zone, whereas for 1 clo, they fall mostly inside of the comfort zone on an annual basis. Based on *PMV* and *PPD* indices, the best comfort in the living room was obtained assuming 1 clo for the whole year, which gives almost 90% of the time within I, II, and III categories. In the bedroom (night), more than 70% of the *PMV* values fall within the comfort categories when assuming 3.7 clo (bed set plus clothing).

The occurrence of periods in which comfort parameters fall outside of I category, or even beyond the applicability of the *PMV* methodology, is a measure of the price paid for energy-saving solutions on the side of the building structure and the active systems. The heat pump power was correctly quantified according to the heat load of the building in the appropriate climate zone. Nevertheless, during very cold times (sometimes a temperature below $-20\,°\text{C}$), the heat pump was not able to overcome these conditions. Similarly, the ground-coupled heat exchanger and bypass in the heat recovery unit were not sufficient to avoid overheating. Obviously, optimal conditions could be established with additional heating or active cooling. This, however, would have meant the loss of PH status and so was not used. Instead, the inhabitants dressed according to their individual needs. The results showed that clothing insulation can improve the comfort conditions up to I category for the most of time. As the survey confirmed, the inhabitants were generally satisfied with the microclimate conditions.

Author Contributions: Conceptualization, K.W. and J.R.; methodology, A.S.-S. and J.R.; software, J.R.; validation, K.W.; formal analysis, K.W.; investigation, A.S.-S.; resources, A.S.-S.; data curation, K.W.; writing—original draft preparation, K.W.; writing—review and editing, A.S.-S. and J.R.; visualization, A.S.-S. and K.W.; supervision, J.R. All authors have read and agreed to the published version of the manuscript.

Funding: This research was supported by European Union/European Regional Development Fund: "Passive house for everyone", Operational Program Innovative Economy, Action 1.4 Support for goal-oriented projects of priority axis 1: Research and development of modern technologies, and action 4.1 Support for the implementation of R&D results of priority axis 4: Investments in innovative projects (2012).

Institutional Review Board Statement: Not applicable.

Informed Consent Statement: Not applicable.

Data Availability Statement: Not applicable.

Acknowledgments: We give a special gratitude to the house inhabitants, the Pawliczek family, for their friendly attitude and great endurance shown during the many years of investigation.

Conflicts of Interest: The authors declare no conflict of interest. The funders had no role in the design of the study; in the collection, analyses, or interpretation of data; in the writing of the manuscript, or in the decision to publish the results.

References

1. Piasecki, M.; Fedorczak-Cisak, M.; Furtak, M.; Biskupski, J. Experimental Confirmation of the Reliability of Fanger's Thermal Comfort Model—Case Study of a Near-Zero Energy Building (NZEB) Office Building. *Sustainability* **2019**, *11*, 2461. [CrossRef]
2. Kisilewicz, T. Overheating of Low-Energy Buildings. Available online: http://beta.nis.com.pl/userfiles/editor/nauka/122013_n/Kisilewicz_12-2013.pdf (accessed on 9 June 2022).
3. Mlakar, J.; Štrancar, J. Overheating in residential passive house: Solution strategies revealed and confirmed through data analysis and simulations. *Energy Build.* **2011**, *43*, 1443–1451. [CrossRef]
4. O'Donovan, A.; Murphy, M.D.; O'Sullivan, P.D. Passive control strategies for cooling a non-residential nearly zero energy office: Simulated comfort resilience now and in the future. *Energy Build.* **2021**, *231*, 110607. [CrossRef]
5. Fedorczak-Cisak, M.; Furtak, M.; Gintowt, J.; Kowalska-Koczwara, A.; Pachla, F.; Stypuła, K.; Tatara, T. Thermal and Vibration Comfort Analysis of a Nearly Zero-Energy Building in Poland. *Sustainability* **2018**, *10*, 3774. [CrossRef]
6. Ballarini, I.; De Luca, G.; Paragamyan, A.; Pellegrino, A.; Corrado, V. Transformation of an Office Building into a Nearly Zero Energy Building (nZEB): Implications for Thermal and Visual Comfort and Energy Performance. *Energies* **2019**, *12*, 895. [CrossRef]
7. Truonga, H.; Garvie, A.M. Chifley Passive House: A Case Study in Energy Efficiency and Comfort. *Energy Procedia* **2017**, *121*, 214–221. [CrossRef]
8. Berr, S.; Whaley, D.; Davidson, K.; Saman, W. Near zero energy homes—What do users think? *Energy Policy* **2014**, *73*, 127–137. [CrossRef]
9. Matzarakis, A.; Mayer, H.; Iziomon, M. Applications of a universal thermal index: Physiological equivalent temperature. *Int. J. Biometeorol.* **1999**, *43*, 76–84. [CrossRef]
10. Höppe, P. The physiological equivalent temperature—A universal index for the biometeorological assessment of the thermal environment. *Int. J. Biometeorol.* **1999**, *43*, 71–75. [CrossRef]
11. Fabbri, K. A Brief History of Thermal Comfort: From Effective Temperature to Adaptive Thermal Comfort. *Indoor Therm. Comf. Percept.* **2015**, 7–23. [CrossRef]
12. Myhren, J.; Holmberg, S. Comfort temperatures and operative temperatures in an office with different heating methods. In Proceedings of the 8th International Conference and Exhibition of the Healthy Building, Lisboa, Portugal, 4–8 June 2006.
13. Zhang, S.; Lin, Z. Standard effective temperature based adaptive-rational thermal comfort model. *Appl. Energy* **2020**, *264*, 114723. [CrossRef]
14. *ISO 7730:2005*; Ergonomics of the Thermal Environment—Analytical Determination and Interpretation of Thermal Comfort Using Calculation of the PMV and PPD Indices and Local Thermal Comfort Criteria. International Organization for Standardization: Geneva, Switzerland, 2005.
15. *EN 16798-1:2019*; Energy Performance of Buildings. Ventilation for Buildings. Indoor Environmental Input Parameters for Design and Assessment of Energy Performance of Buildings Addressing Indoor Air Quality, Thermal Environment, Lighting and Acoustics. European Committee for Standardization: Brussels, Belgium, 2019.
16. Humphreys, M.; Nicol, F. The validity of ISO-PMV for predicting comfort votes in every-day thermal environments. *Energy Build.* **2002**, *34*, 667–684. [CrossRef]
17. Humphreys, M.; Nicol, J. Effects of measurement and formulation error on thermal comfort indices in the ASHRAE database of field studies. *ASHRAE Trans.* **2000**, *106*, 493–502.
18. Havenith, G.; Holmér, I.; Parsons, K. Personal factors in thermal comfort assessment: Clothing properties and metabolic heat production. *Energy Build.* **2002**, *34*, 581–591. [CrossRef]
19. Almeida, R.; Ramos, N.; de Freitas, V. Thermal comfort models and pupils' perception in free-running school buildings of a mild climate country. *Energy Build.* **2016**, *111*, 64–75. [CrossRef]
20. Gilani, S.; Khan, M.; Ali, M. Revisiting's Fanger's thermal comfort model using mean blood pressure as a bio-marker: An experimental investigation. *Appl. Therm. Eng.* **2016**, *109*, 35–43. [CrossRef]
21. Pan, D.; Lin, Z.; Deng, S. A mathematical model for predicting the total insulation value of a bedding system. *Build. Environ.* **2010**, *45*, 1866–1872. [CrossRef]
22. Lin, Z.; Deng, S. A study on the thermal comfort in sleeping environments in the subtropics—Measuring the total insulation values for the bedding systems commonly used in the subtropics. *Build. Environ.* **2008**, *43*, 905–916. [CrossRef]
23. Lan, L.; Pan, L.; Lian, Z.; Huang, H.; Lin, Y. Experimental study on thermal comfort of sleeping people at different air temperatures. *Build. Environ.* **2014**, *73*, 24–31. [CrossRef]
24. Lu, Y.; Nu, M.; Song, W.; Liu, Y.; Wang, M. Investigation on the total and local thermal insulation of the bedding system: Effects of filling materials, weights and body postures. *Build. Environ.* **2021**, *204*, 108–161. [CrossRef]
25. Berardi, U. Clarifying the new interpretations of the concept of sustainable building. *Sustain. Cities Soc.* **2013**, *8*, 72–78. [CrossRef]
26. Mahdavi, A.; Doppelbauer, E. A performance comparison of passive and low-energy buildings. *Energy Build.* **2010**, *42*, 1314–1319. [CrossRef]
27. Marszal, A.; Heiselberg, P.; Bourrelle, J.; Musall, E.; Voss, K.; Sartori, K.; Napolitano, A. Zero Energy Building—A review of definitions and calculation methodologies. *Energy Build.* **2011**, *43*, 971–979. [CrossRef]
28. Voss, K.; Musall, E.; Lichtmeß, M. From Low-Energy to Net Zero-Energy Buildings: Status and Perspectives. *J. Green Build.* **2011**, *6*, 46–57. [CrossRef]

29. Sartori, I.; Napolitano, A.; Vossc, K. Net zero energy buildings: A consistent definition framework. *Energy Build.* **2012**, *48*, 220–232. [CrossRef]
30. Altobello, A.; Misceo, M.; Russo, P.; Stefanizzi, P.; Vignola, I. Comparison of numerical and experimental performances of nZEB residential building in Putignano (Apulia Region). In *Materials Science and Engineering, Proceedings of the IOP Conference Series: Materials Science and Engine, Palembang, Indonesia, 5–16 October 2019*; IOP Publishing: Bristol, UK, 2019.
31. Kampelis, N.; Sifakis, N.; Kolokotsa, D.; Gobakis, K.; Kalaitzakis, K.; Isidori, D.; Cristalli, C. HVAC Optimization Genetic Algorithm for Industrial Near-Zero-Energy Building Demand Response. *Energies* **2019**, *12*, 2177. [CrossRef]
32. Xu, X.; Feng, G.; Chi, D.; Liu, M.; Dou, B. Optimization of Performance Parameter Design and Energy Use Prediction for Nearly Zero Energy Buildings. *Energies* **2018**, *11*, 3252. [CrossRef]
33. Passive House Institute (PHI). Available online: https://passivehouse.com (accessed on 24 May 2022).
34. Sadineni, S.; Madala, S.; Boehm, R. Passive building energy savings: A review of building envelope components. *Renew. Sustain. Energy Rev.* **2011**, *15*, 3617–3631. [CrossRef]
35. PASSIPEDIA. What Is a Passive House? Available online: https://passipedia.org/basics/what_is_a_passive_house (accessed on 24 May 2022).
36. The European Parliament and the Council of the European Union. *Directive (EU) 2018/844 of the European Parliament and of the Council of 30 May 2018 Amending Directive 2010/31/EU on the Energy Performance of Buildings and Directive 2012/27/EU on Energy Efficiency*; The European Parliament and the Council of the European Union: Brussels, Belgium, 2018.
37. Ridley, I.; Bere, J.; Clarke, A.; Schwartz, Y.; Far, A. The side by side in use monitored performance of two passive and low carbon Welsh houses. *Energy Build.* **2014**, *82*, 13–26. [CrossRef]
38. Ridley, I.; Clarke, A.; Bere, J.; Altamirano, H.; Lewis, S.; Durdev, M.; Farr, A. The monitored performance of the first new London dwelling certified to the Passive House standard. *Energy Build.* **2013**, *63*, 67–78. [CrossRef]
39. Rohdin, P.; Molin, A.; Moshfegh, B. Experiences from nine passive houses in Sweden–indoor thermal environment and energy use. *Build. Environ.* **2014**, *71*, 176–185. [CrossRef]
40. Schnieders, J.; Feist, W.; Rongen, L. Passive Houses for different climate zones. *Energy Build.* **2015**, *105*, 71–87. [CrossRef]
41. Dahlstrøm, O.; Sørnes, K.; Eriksen, S.; Hertwich, E. Life cycle assessment of a single-family residence built to either conventional-or passive house standard. *Energy Build.* **2012**, *54*, 470–479. [CrossRef]
42. Proietti, S.; Sdringola, P.; Desideri, U.; Zepparelli, F.; Masciarelli, F.; Castellani, F. Life Cycle Assessment of a passive house in a seismic temperate zone. *Energy Build.* **2013**, *64*, 463–472. [CrossRef]
43. Badea, A.; Baracu, T.; Dinca, C.; Tutica, D.; Grigore, R.; Anastasiu, M. A life-cycle cost analysis of the passive house 'POLITEHNICA' from Bucharest. *Energy Build.* **2014**, *80*, 542–555. [CrossRef]
44. Galvin, R. Are passive houses economically viable? A reality-based, subjectivist approach to cost-benefit analyses. *Energy Build.* **2014**, *80*, 149–157. [CrossRef]
45. Stephan, A.; Crawford, R.; Myttenaere, K. A comprehensive assessment of the life cycle energy demand of passive houses. *Appl. Energy* **2013**, *112*, 23–34. [CrossRef]
46. Chel, A.; Janssens, A.; De Paepe, M. Thermal performance of a nearly zero energy passive house integrated with the air–air heat exchanger and the earth?water heat exchanger. *Energy Build.* **2015**, *96*, 53–63. [CrossRef]
47. Kylili, A.; Fokaides, P. European smart cities: The role of zero energy buildings. *Sustain. Cities Soc.* **2015**, *15*, 86–95. [CrossRef]
48. Georges, L.; Skreiberg, Ø.; Novakovic, V. On the proper integration of wood stoves in passive houses: Investigation using detailed dynamic simulations. *Energy Build.* **2013**, *59*, 203–213. [CrossRef]
49. Rekstad, J.; Meir, M.; Murtnes, E.; Dursun, A. A comparison of the energy consumption in two passive houses, one with a solar heating system and one with an air-water heat pump. *Energy Build.* **2015**, *96*, 149–161. [CrossRef]
50. Moran, F.; Blight, T.; Natarajan, S.; Shea, A. The use of Passive House Planning Package to reduce energy use and CO2 emissions in historic dwellings. *Energy Build.* **2014**, *75*, 216–227. [CrossRef]
51. Leardini, P.; Manfredini, M.; Callau, M. Energy upgrade to Passive House standard for historic public housing in New Zealand. *Energy Build.* **2015**, *95*, 211–218. [CrossRef]
52. Sage-Lauck, J.; Sailor, D. Evaluation of phase change materials for improving thermal comfort in a super-insulated residential building. *Energy Build.* **2014**, *79*, 32–40. [CrossRef]
53. Němeček, M.; Kalousek, M. Influence of thermal storage mass on summer thermal stability in a passive wooden house in the Czech Republic. *Energy Build.* **2015**, *107*, 68–75. [CrossRef]
54. Radoń, J.; Wąs, K.; Flaga-Maryanczyk, A.; Schnotale, J. Experimental and theoretical study on hygrothermal long-term performance of outer assemblies in lightweight passive house. *J. Build. Phys.* **2018**, *41*, 299–320. [CrossRef]
55. Wąs, K. Wpływ Konstrukcji Ściany Lekkiej Szkieletowej na Zjawiska Cieplno-Wilgotnościowe w Budynku Pasywnym (Impact of Light-Frame Wall Structure on Hygrothermal Performance in a Passive House). Ph.D. Thesis, Cracow University of Technology, Kraków, Poland, 2016.
56. Kaklauskas, A.; Rute, J.; Zavadskas, E.; Daniuna, A.; Pruskus, V.; Bivainis, J.; Plakys, V. Passive House model for quantitative and qualitative analyses and its intelligent system. *Energy Build.* **2012**, *50*, 7–18. [CrossRef]
57. Langer, S.; Bekö, G.; Bloom, E.; Widheden, A.; Ekbe, L. Indoor air quality in passive and conventional new houses in Sweden. *Build. Environ.* **2015**, *93*, 92–100. [CrossRef]

58. Fokaides, P.; Christoforou, E.; Ilic, M. Performance of a Passive House under subtropical climatic conditions. *Energy Build.* **2016**, *133*, 14–31. [CrossRef]
59. Radoń, J.; Wąs, K.; Flaga-Maryańczyk, A.; Antretter, F. Thermal Performance of Slab on Grade with Floor Heating in a Passive House. *Tech. Trans. Civ. Eng.* **2014**, *3*, 405–413.
60. Nawalany, G.; Sokołowski, P. Building–Soil Thermal Interaction: A Case Study. *Energies* **2019**, *12*, 2922. [CrossRef]
61. Michalak, P. Selected Aspects of Indoor Climate in a Passive Office Building with a Thermally Activated Building System: A Case Study from Poland. *Energies* **2021**, *14*, 860. [CrossRef]
62. Thermishe Behaglikheit. Available online: http://www.bosy-online.de/thermische_Behaglichkeit.htm (accessed on 11 May 2022).
63. ASHRAE Handbook—Fundamentals. Available online: https://www.ashrae.org/technical-resources/ashrae-handbook (accessed on 11 May 2022).
64. Climate.OneBuilding.Org. Available online: https://climate.onebuilding.org/WMO_Region_6_Europe/POL_Poland/index.html (accessed on 11 May 2022).
65. Foster, J.; Sharpe, T.; Poston, A.; Morgan, C.; Musau, F. Scottish Passive House: Insights into Environmental Conditions in Monitored Passive Houses. *Sustainability* **2016**, *8*, 412. [CrossRef]

Article

Assessment of ANN Algorithms for the Concentration Prediction of Indoor Air Pollutants in Child Daycare Centers

Jeeheon Kim [1], Yongsug Hong [2], Namchul Seong [3],* and Daeung Danny Kim [4],*

[1] Eco-System Research Center, Gachon University, Seongnam 13120, Korea; jlaw80@naver.com
[2] Division of Human-Architectural Engineering, Daejin University, 1007 Hoguk-Ro, Pocheon 11159, Korea; yshong@daejin.ac.kr
[3] Department of Architectural Engineering Kangwon National University, Samcheok-si 25913, Korea
[4] Architectural Engineering Department, King Fahd University of Petroleum and Minerals (KFUPM), Dhahran 31261, Saudi Arabia
* Correspondence: inamchul@kangwon.ac.kr (N.S.); dkim@kfupm.edu.sa (D.D.K.)

Abstract: As the time spent by people indoors continues to significantly increase, much attention has been paid to indoor air quality. While many IAQ studies have been conducted through field measurements, the use of data-driven techniques such as machine learning has been increasingly used for the prediction of indoor air pollutants. For the present study, the concentrations of indoor air pollutants such as CO_2, $PM_{2.5}$, and VOCs in child daycare centers were predicted by using an artificial neural network model with three different training algorithms including Levenberg–Marquardt, Bayesian regularization, and Broyden–Fletcher–Goldfarb–Shanno quasi-Newton methods. For training and validation, data of indoor pollutants measured in child daycare facilities over a 1-month period were used. The results showed all the models produced a good performance for the prediction of indoor pollutants compared with the measured data. Among the models, the prediction by the LM model met the acceptable criteria of ASHRAE guideline 14 under all conditions. It was observed that the prediction performance decreased as the number of hidden layers increased. Moreover, the prediction performance was differed by the type of indoor pollutant. This was caused by patterns observed in the measured data. Considering the outcomes of the study, better prediction results can be obtained through the selection of suitable prediction models for time series data as well as the adjustment of training algorithms.

Keywords: indoor air pollutants; ANN model; training algorithm; child daycare center

Citation: Kim, J.; Hong, Y.; Seong, N.; Kim, D.D. Assessment of ANN Algorithms for the Concentration Prediction of Indoor Air Pollutants in Child Daycare Centers. *Energies* **2022**, *15*, 2654. https://doi.org/10.3390/en15072654

Academic Editors: Katarzyna Ratajczak and Łukasz Amanowicz

Received: 12 March 2022
Accepted: 1 April 2022
Published: 5 April 2022

Publisher's Note: MDPI stays neutral with regard to jurisdictional claims in published maps and institutional affiliations.

1. Introduction

People generally spend most of their time in buildings, with this time rapidly increasing due to the situation caused by the SARS-CoV-2 virus [1,2]. Thus, much attention has been paid to the improvement of indoor environmental quality including indoor air quality, thermal parameters, etc. [3–5]. Normally, the quality of indoor air is highly influenced by outdoor air pollution and other indoor sources [1,6]. Specifically, indoor sources originating from building materials, appliances, human activities, etc. have produced indoor air contaminants [7–9]. Regarding indoor pollutants, many studies have performed investigations aiming to reduce the concentration or prevent their occurrence [10–15]. While most studies have focused on indoor pollutants in residential and non-residential buildings, people have started to notice the importance of indoor pollutants in certain facilities such as child daycare centers.

Several studies have observed severe indoor pollutants in child daycare centers through field measurements. According to the study of Oh et al., high levels of $PM_{2.5}$ and PM_{10} were found in ten child daycare centers in South Korea, which were highly influenced by traffic conditions and vehicles through various openings [16]. In the case of child daycare centers in Paris, Roda et al. measured biological and chemical pollutants [17]. In their findings, some chemical pollutants were above the acceptable indoor levels in 28 child daycare centers. A similar result was observed in the findings of a study by Hwang et al. [18]. Both biological and chemical pollutants were measured in 25 child daycare centers. Specifically, the concentration of VOCs was the highest among the measurements in their data. In the field measurements performed by Madureira et al., severe concentrations of biological pollutants were found in nine child daycare centers [19]. Other studies observed high concentrations of indoor pollutants through their measurements in various child daycare centers [20–23].

To improve indoor air quality and prevent indoor pollution, the prediction of concentrations of various indoor pollutants affecting occupants' health is essential. Even though severe indoor air quality conditions in child daycare centers were reported, most IAQ studies have focused on indoor pollutants in residential or commercial buildings. In addition, most of these studies collected data on indoor pollutants through measurements, which is time-consuming and costly. Another technique for predicting indoor air pollutants is the use of simulations. In a study by Heibat et al., the researchers used a coupling method including CONTAM and WUFI for the prediction of CO_2, $PM_{2.5}$, and VOCs [24]. However, the accuracy of the simulation results can be highly dependent on the user's experience.

Recently, a prediction made by utilizing advanced computer applications was significantly recognized with the development of data-driven methods such as machine learning techniques [19,20]. Machine learning techniques have been used for extracting data patterns and quantifying the impacts of various design parameters [25]. The data-driven methods have been widely used in applications of engineering, medicine, and economics. Among various data-driven methods, artificial neural network (ANN) models, support vector machine regression (SVR), random forest, XGBoost, etc. have been applied for various purposes such as energy consumption predictions, thermal performance quantification, mechanical system diagnostics, and so on [26–30]. For the present study, the ANN model was chosen to predict indoor air quality in a child daycare center due to its high prediction accuracy [27]. According to several studies, the ANN model showed the best performance among the machine learning methods [31–33]. In general, different learning algorithms were used with a regular ANN model. However, the prediction results can become unstable caused by fluctuations in these training algorithms [34]. To provide more reliable prediction results, the performance of several training algorithms was tested. While the predictions made by using machine learning techniques have been widely employed in various fields, there were few studies available for indoor air pollutant prediction. In addition, the predictive performance by different training algorithms was rarely investigated. This study presents the difference in the predictive performance of indoor air pollutants by applying different training algorithms in the ANN model.

2. Machine Learning Applications for the Prediction of Indoor Air Quality

For the prediction of indoor air quality, Jeong et al. predicted indoor environmental parameters such as temperature, humidity, and CO_2 by using a machine learning technique [35]. By comparing the collected data, a highly correlated relationship between the data and prediction results was achieved. In addition, the IAQ management was conducted remotely by using IoT systems [36]. Through cloud data analysis, the comparison of measurement data with simulations was implemented to improve the mechanical exhaust systems. Li et al. used a machine learning method such as the random forest algorithm to predict $PM_{2.5}$ concentrations in residential buildings [37]. In their study, the random forest model showed excellent performance for the prediction of $PM_{2.5}$ concentration levels. Kallio et al. investigated the performance of four machine learning methods: Ridge regression, decision tree, random forest, and multilayer perceptron to predict indoor CO_2 concentrations [38]. The abovementioned machine learning models showed a better performance than statistical methods regarding indoor CO_2 concentration predictions. Another study by Taheri et al. performed comparisons of several machine learning algorithms including support vector machines, AdaBoost, random forest, gradient boosting, logistic regression, and multilayer perceptron [39]. In addition, Sassi et al. utilized a deep learning technique for data analysis and augmented reality (AR) to predict indoor air quality based on the monitoring by an IoT system [40].

Regarding the ANN model application, Saad et al. proposed an IAQ monitoring system to identify sources affecting IAQ levels. By utilizing ANN techniques, their study recognized the patterns of measured data and proved that the proposed system was able to measure indoor air quality levels. In addition, the sources affecting indoor air quality such as ambient air, presence of chemicals and fragrances, food and beverages, and human activity were classified successfully [41]. In the case of the study performed by Putra et al., ANN models were used to predict indoor air quality with the data, which were measured 8 h a day for several months. For this study, the authors utilized the Levenberg–Marquardt training method and proved that this training method produced good prediction results [42]. Moreover, Dai et al. constructed an ANN model by using indoor CO_2 concentration data sets in a residential building to predict indoor air quality with ventilation rates [43]. About 80% of the overall accuracy levels by the constructed ANN model were achieved and the authors proved that the indoor CO_2 concentration predicted by the ANN model was highly influenced by locations and outdoor air temperatures. According to the study of Egala et al., a practical approach was presented to train ANN models regarding the prediction of indoor CO_2 concentration [44]. By training the model with collected data for a month, computational errors were reduced and high predictive accuracy was achieved. To control HVAC (Heating, ventilation, and air conditioning) systems for improving indoor air quality, Tagliabue et al. used the ANN model [33]. Moreover, Amuthadevial implemented machine learning methods including nonlinear ANN models, statistical multi-level regression, neural purge, and deep learning short and long-term memory (DL-LSTM) to predict concentration levels of SO_2, CO, NOx, and O_3 [32]. Regarding indoor $PM_{2.5}$, PM_{10}, and NO_2 concentrations, Zhang et al. also used several machine learning techniques such as multiple linear regression (MLR), time series regression (TSR), and ANN models [31]. In their study, the ANN model showed the best performance among the machine learning methods.

3. Methodology

Figure 1 presents the research process for the present study. Due to high concentrations of CO_2, $PM_{2.5}$, and VOCs in child daycare centers, the measurement data of CO_2, $PM_{2.5}$, and volatile organic compounds (VOCs) provided by the Big Data Environment Platform were chosen and these were converted to input data for the ANN model. By using the input data, the concentrations of CO_2, $PM_{2.5}$, and VOCs were predicted with different training algorithms of the ANN model. The prediction results of different training algorithms were evaluated by CV (RMSE) (coefficient of variation of the root mean square error) and MBE (mean bias error). In addition, the suitability of each model was assessed by R^2 (coefficient of determination).

3.1. Collection of Training Data Set

The dataset used for the present study was composed of the measurement data provided by the Big Data Environment Platform [45]. For input data, major indoor pollutants such as CO_2, $PM_{2.5}$, and VOCs were measured at 5-min intervals in the child daycare facilities during the month of May 2021. The data consisted of 8929 sets for each pollutant of which 80% of these data were used for training and 20% for testing.

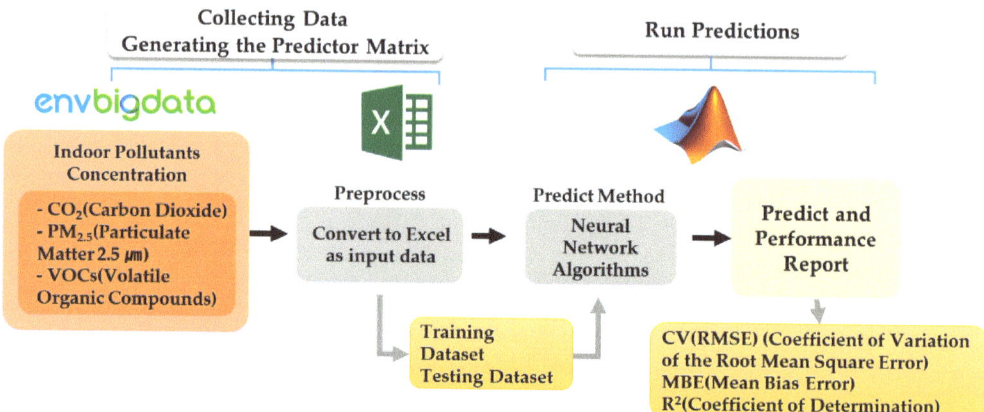

Figure 1. Schematic diagram of the process of predicting indoor air pollution concentrations.

3.2. Indoor Pollutant Concentration Prediction Model

The concentration of indoor pollutants was predicted by using the learning neural network models in the Neural Networks Toolbox of MATLAB [46]. A data multilayer neural network training shows excellent performance when this is optimized by using the slope of neural network performance against neural network weights and the Jacoby matrix of a neural network error. The slope and Jacoby matrix were calculated using the feed-forward back-propagation algorithms. As commonly used in ANN models, feed-forward networks can produce output data quickly avoiding delays [47]. In addition, the feed-forward back-propagation algorithms perform the iteration of updating weights and biases values of network parameters and back-propagate the error for training ANN models [34,48]. For the present study, three different feed-forward back-propagation algorithms were used, which were Levenberg–Marquardt (LM), Bayesian regularization (BR), and Broyden–Fletcher–Goldfarb–Shanno (BFGS) quasi-Newton (BFG). The performance for the prediction of indoor air pollution was evaluated.

The feed-forward neural network model is generally composed of an input layer, a hidden layer, and an output layer [49]. The hidden layer forms the structure of the ANN and the neuron exists in each layer. Since the number of neurons in the hidden layers mainly influences the calculation prediction and time, the number of neurons was fixed as 20. When the number of the hidden layers was changed to 1, 3, and 5, the predicted performance was compared (Figure 2). As one of the learning parameters, the number of epochs was 100. The total number of data were 8927, in which 80% and 20% of datasets were used for training and testing, respectively. Detailed conditions are summarized in Table 1.

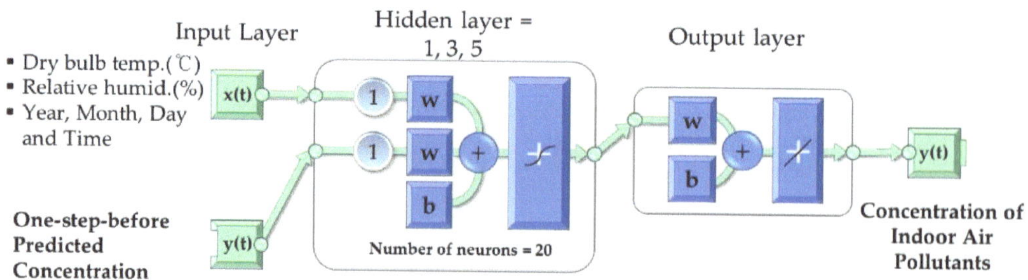

Figure 2. Schematic of the multilayer shallow neural network training algorithms for predicting indoor air pollution concentrations.

Table 1. Training parameters.

Parameter		Value
Number of hidden layers		1, 3, 5
Number of neurons		20
Epochs		100
Data	Training	7142 (80%)
	Testing	1785 (20%)

For input data, the measurement date and time, indoor thermal parameters such as temperature and humidity, and the prediction results of each pollutant fed back from the output layers were used. In the hidden layer, measurement data at the 5-min interval were received as an input signal from the input layer. The feed-forward neural network operations were performed through the internal neurons. The output layer predicted the indoor air pollutant concentrations after 5-min of the input signal point based on the hidden layer calculation result.

By using CV(RMSE) and MBE, the performance evaluation indicators of the predictive model were validated. CV(RMSE) refers to the degree of scattering of estimated values in consideration of variance, and MBE is an error analysis index that identifies errors by tracking how close estimates form clusters through data bias. The models will be declared to be calibrated if they are within the acceptable values of ASHRAE Guideline 14 (Table 2) [50]. The equation for obtaining CV(RMSE) and MBE are as follows.

$$\text{MBE} = \overset{n}{\underset{}{\textstyle\sum}}(y_i - \hat{y}_l) / [(n - p) \times \bar{y}] \cdot 100 \tag{1}$$

$$\text{Cv(RMSE)} = 100 \cdot \left[\sum (y_i - \overline{y_l})^2 / (n - p) \right]^{1/2} / \bar{y}, \tag{2}$$

where n is the number of data points, p is the number of parameters, y_i is the utility data used for calibration, $\hat{y}_i$ is the simulation predicted data, and $\bar{y}$ is the arithmetic mean of the sample of n observations. In addition, the suitability of the model was evaluated by using R^2.

Table 2. Acceptable Calibration Tolerances in building energy performance prediction.

Calibration Type	Index	ASHRAE Guideline 14 [50]
Monthly	MBE_monthly	±5%
	CvRMSE_monthly	15%
Hourly	MBE_hourly	±10%
	CvRMSE_hourly	30%

4. Results

For the present study, the concentrations of CO_2, $PM_{2.5}$, and VOCs in child daycare centers were predicted by using three different feed-forward back-propagation algorithms including LM, BR, and BGF. The performance for the prediction of indoor air pollution was evaluated by CV(RMSE) and MBE. Moreover, the suitability of each model was assessed by R^2.

4.1. CO$_2$

The prediction results of CO_2 are summarized in Figure 3. The indoor CO_2 concentration is greatly affected by human breathing. In addition, it can be seen that the concentration is altered by the number and activities of occupants. As shown in all graphs in Figure 3, the CO_2 concentration gradually increased to 1500–2000 ppm and then, decreased to 500 ppm. This trend regarding the CO_2 concentration repeated during the measurement period. In addition, a similar trend was observed in all prediction methods with three different training algorithms. Moreover, the values of training and testing predicted by the ANN model were close to the measurement data.

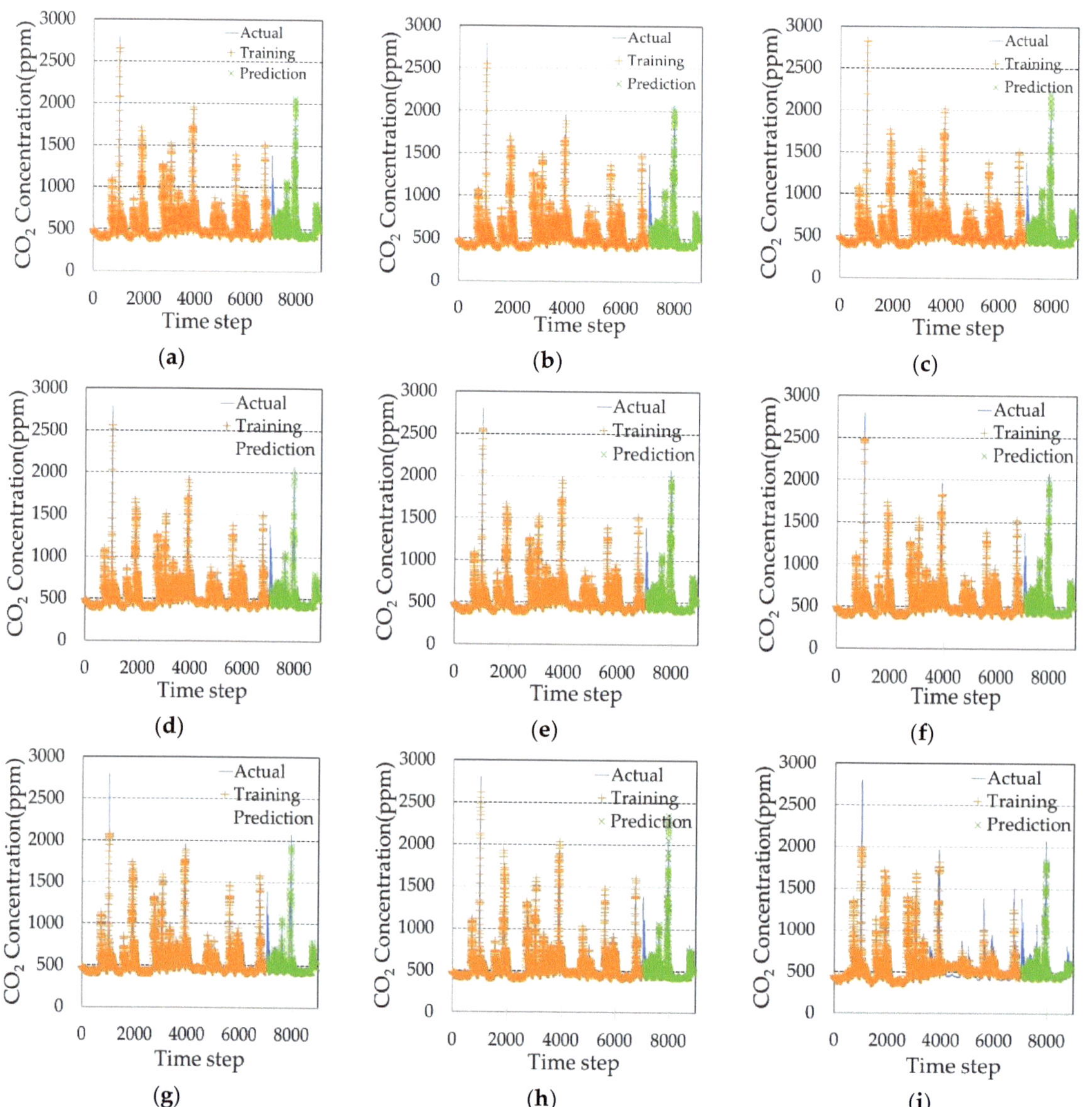

Figure 3. Prediction results of CO_2 concentration. (**a**) LM-Hidden layer-1; (**b**) LM-Hidden layer-3; (**c**) LM-Hidden layer-5; (**d**) BR-Hidden layer-1; (**e**) BR-Hidden layer-3; (**f**) BR-Hidden layer-5; (**g**) BFG-Hidden layer-1; (**h**) BFG-Hidden layer-3; (**i**) BFG-Hidden layer-5.

Table 3 presents CV(RMSE) and MBE of CO_2 prediction results. The CV(RMSE) of the LM model ranges from 4.04% to 4.47% and shows a low degree of dispersion under all conditions. The MBE ranges from 7.53% to 8.56%, which shows an excellent predictive performance due to bias reduction. For the BR model, CV(RMSE) and MBE show ranges of 4.06–4.21% and 7.53–8.06%, respectively. This also shows good predictive performance satisfying the acceptable range of the ASHRAE guideline 14. For both the LM model and BR model, the values of CV(RMSE) and MBE increased as the number of hidden layers increased. However, a slight increase in those values was observed. In the case of the BFG model, a somewhat larger increase in the CV(RMSE) and MBE was observed than that predicted by using the LM model and BR model. Moreover, the MBE for the BFG model with more than three hidden layers could not satisfy the acceptance criteria of the ASHRAE guidelines 14.

Table 3. CV(RMSE) and MBE of CO_2 Concentration Prediction Result.

Training Algorithm	Hidden Layers-1		Hidden Layers-3		Hidden Layers-5	
	CV(RMSE) (%)	MBE (%)	CV(RMSE) (%)	MBE (%)	CV(RMSE) (%)	MBE (%)
LM	4.04	7.53	4.06	8.03	4.47	8.56
BR	4.06	7.53	4.15	7.98	4.21	8.06
BFG	5.26	9.86	7.65	10.16	12.54	10.86

Figure 4 presents the R^2 of the prediction results. As shown, the high suitability values of 0.9999 and 0.9998 were observed for both the LM model and BR model due to a low degree of dispersion. In addition, a better prediction result was shown when the number of the hidden layer was one. Because of a large degree of dispersion generated by the BFG model, R^2 of the BFG model was large and the suitability was low. Specifically, R^2 measured 0.9393 when the number of the hidden layer was five.

4.2. $PM_{2.5}$

Figure 5 shows the prediction results of the indoor $PM_{2.5}$ concentration. For all models, $PM_{2.5}$ concentration is largely increased to about 160 µg/m^3 and a slight difference was observed in the training. In the prediction phase, the $PM_{2.5}$ concentration is maintained under 30–40 µg/m^3, which is close to the measurement data.

The CV(RMSE) and MBE of $PM_{2.5}$ concentration are shown in Table 4. For both LM and BR models, the range of CV(RMSE) and MBE are within the acceptable criteria of the ASHRAE guidelines 14 in all three hidden layers. Between the two models, the best performance of the prediction is observed when the BR model was used. However, in the case of the BFG model, the prediction performance decreased as the number of the hidden layers increased, even though the range of CV(RMSE) and MBE is within the acceptable criteria of the ASHRAE guidelines 14 when the hidden layer is one. Moreover, the CV(RMSE) is 27.90% when the number of the hidden layer is five for the BFG model.

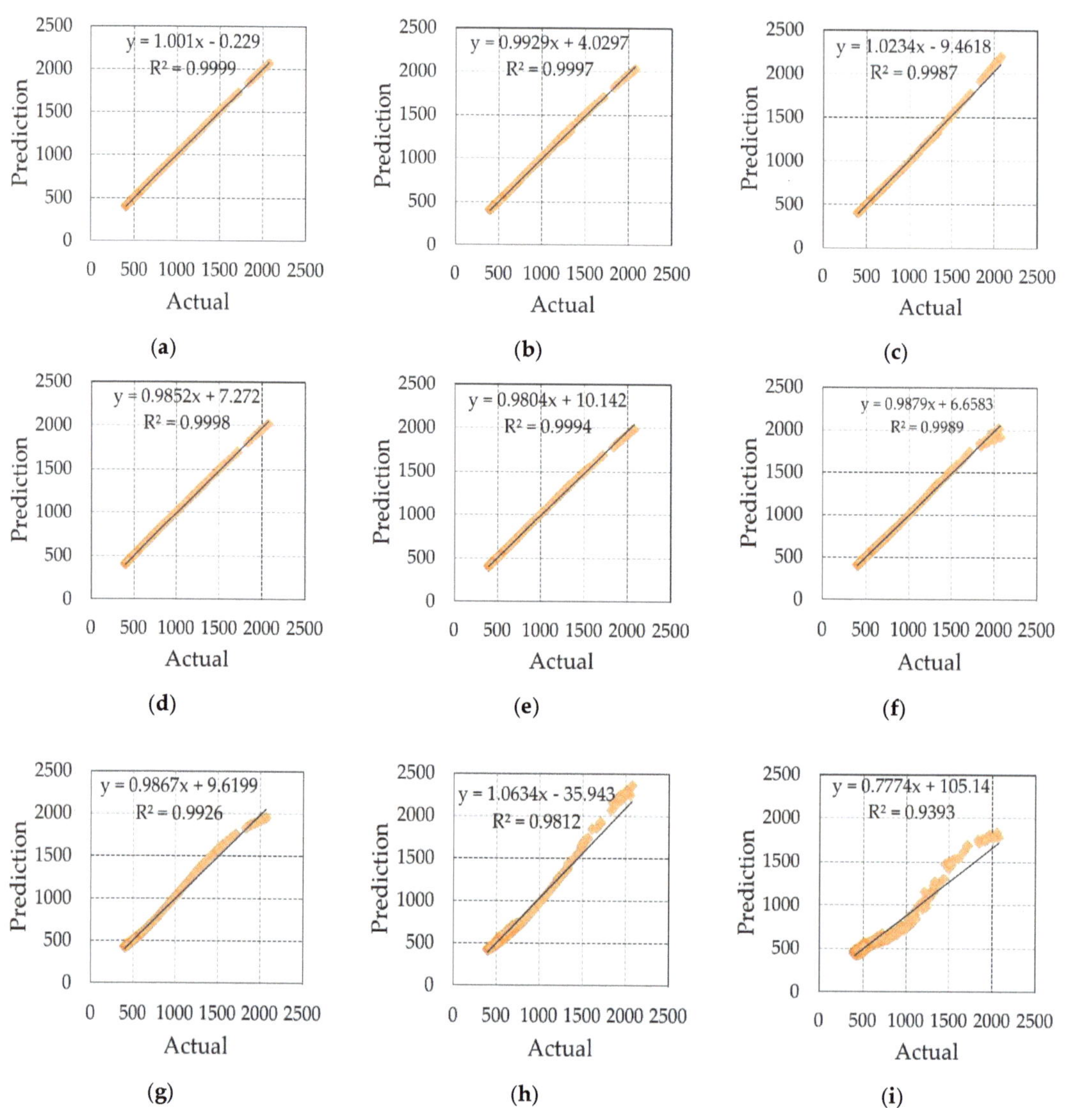

Figure 4. R^2 of prediction results of CO_2 concentration. (**a**) LM-Hidden layer-1; (**b**) LM-Hidden layer-3; (**c**) LM-Hidden layer-5; (**d**) BR-Hidden layer-1; (**e**) BR-Hidden layer-3; (**f**) BR-Hidden layer-5; (**g**) BFG-Hidden layer-1; (**h**) BFG-Hidden layer-3; (**i**) BFG-Hidden layer-5.

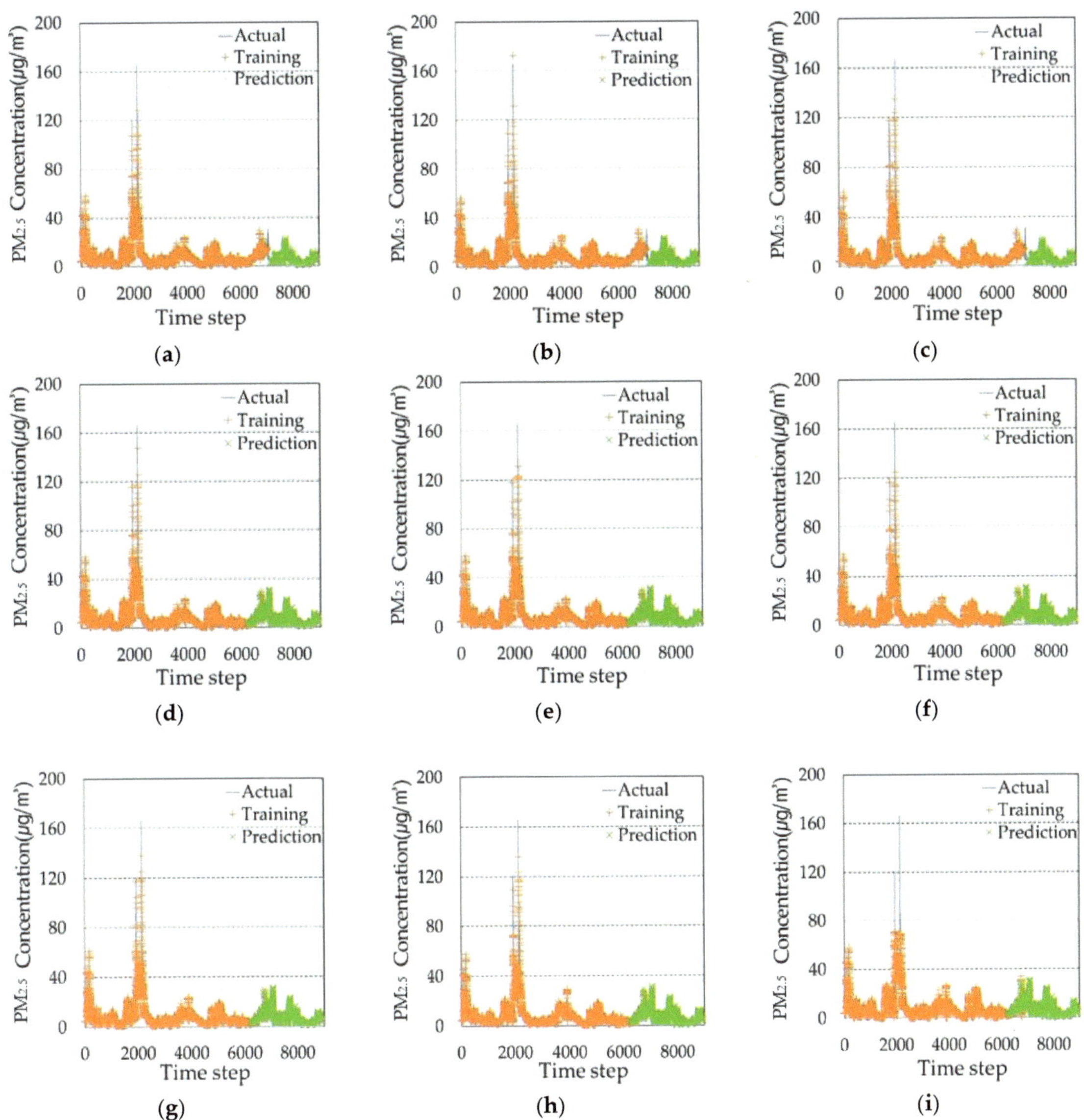

Figure 5. Prediction results of PM$_{2.5}$ concentration. (**a**) LM-Hidden layer-1; (**b**) LM-Hidden layer-3; (**c**) LM-Hidden layer-5; (**d**) BR-Hidden layer-1; (**e**) BR-Hidden layer-3; (**f**) BR-Hidden layer-5; (**g**) BFG-Hidden layer-1; (**h**) BFG-Hidden layer-3; (**i**) BFG-Hidden layer-5.

Table 4. CV(RMSE) and MBE of PM$_{2.5}$ concentration prediction result.

Training Algorithm	Hidden Layers-1		Hidden Layers-3		Hidden Layers-5	
	CV(RMSE) (%)	MBE (%)	CV(RMSE) (%)	MBE (%)	CV(RMSE) (%)	MBE (%)
LM	13.24	4.44	13.76	5.17	13.73	5.49
BR	13.17	3.62	13.27	3.91	13.31	4.25
BFG	13.79	5.66	18.60	6.52	27.90	6.69

For the R^2 of the PM$_{2.5}$ concentration prediction, high suitability for both LM and BR models is achieved (Figure 6). When the number of the hidden layers is one, the R^2 for the LM model and BR model are 0.9994 and 0.9998, respectively. In the case of the BFG model, the R^2 is 0.9984 when the number of the hidden layer is one, which shows high suitability. However, the R^2 is decreased to 0.9734 when the number of the hidden layers is increased.

4.3. VOCs

As shown in Figure 7, the indoor VOCs concentration prediction results are compared with the measurement data. In the training phase, the VOCs concentration is significantly increased to about 16,000 µg/m^3 and all the models show the difference in the VOCs concentration. The VOCs concentration for all the models is close to the measurement data in the prediction phase.

Table 5 summarizes the CV(RMSE) and MBE of the prediction results. The LM model shows the lowest CV(RMSE) among the models and the MBE of the LM model is only within the acceptable range of the ASHRAE guidelines 14. While the CV(RMSE) and MBE for the BR model can meet the acceptable criteria of the ASHRAE guidelines 14 when the number of the hidden layer was one, the CV(RMSE) and MBE cannot satisfy them when the number of the hidden layers is increased. A similar trend was observed for the BFG model. Specifically, the CV(RMSE) is increased to 26.52% when the number of the hidden layers is increased to 5.

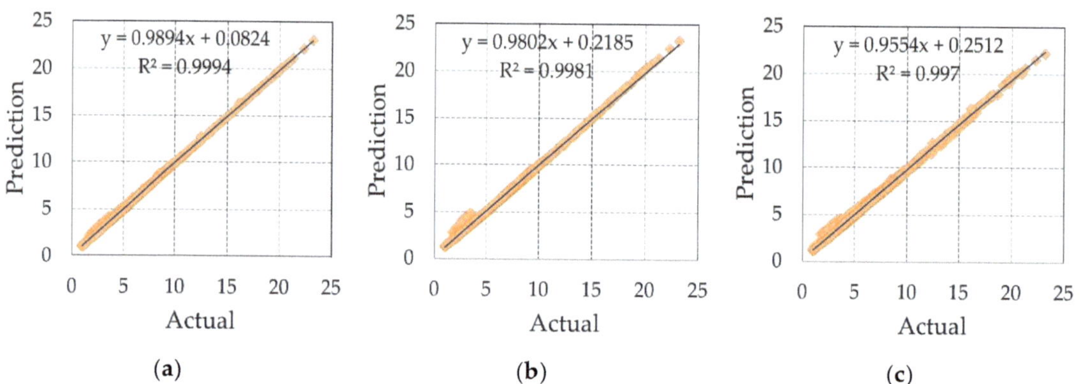

Figure 6. *Cont.*

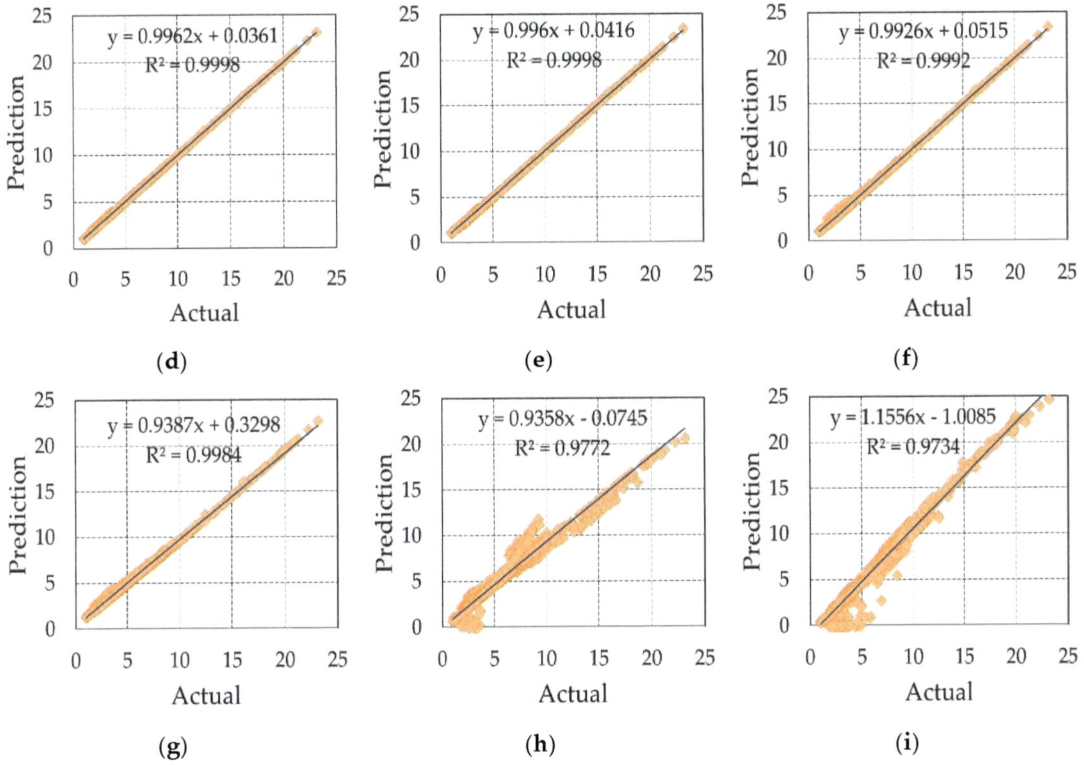

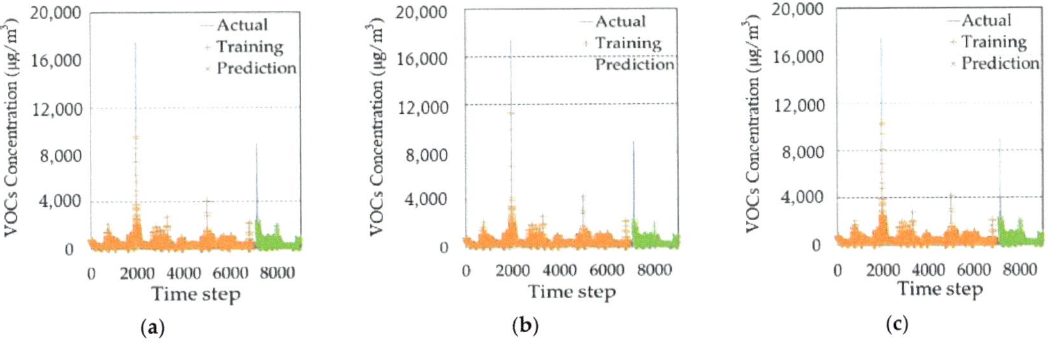

Figure 6. R^2 of prediction results of PM$_{2.5}$ concentration. (**a**) LM-Hidden layer-1; (**b**) LM-Hidden layer-3; (**c**) LM-Hidden layer-5; (**d**) BR-Hidden layer-1; (**e**) BR-Hidden layer-3; (**f**) BR-Hidden layer-5; (**g**) BFG-Hidden layer-1; (**h**) BFG-Hidden layer-3; (**i**) BFG-Hidden layer-5.

Figure 7. *Cont.*

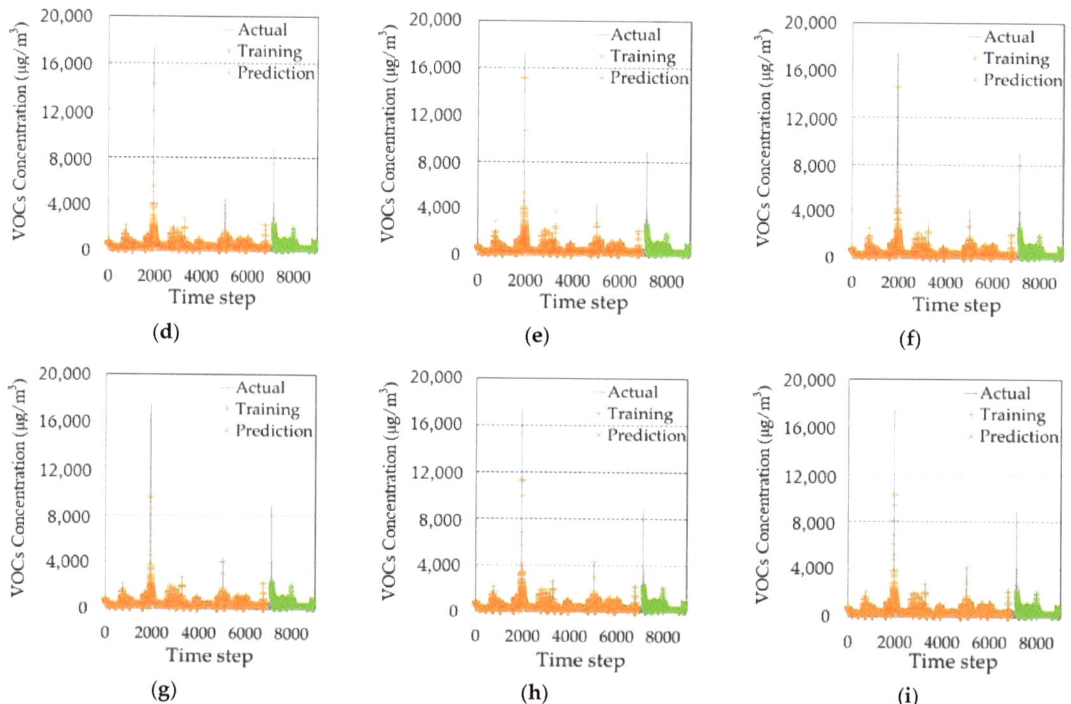

Figure 7. Prediction results of VOCs concentration. (**a**) LM–Hidden layer-1; (**b**) LM–Hidden layer-3; (**c**) LM–Hidden layer-5; (**d**) BR–Hidden layer-1; (**e**) BR–Hidden layer-3; (**f**) BR–Hidden layer-5; (**g**) BFG–Hidden layer-1; (**h**) BFG–Hidden layer-3; (**i**) BFG–Hidden layer-5.

Table 5. CV(RMSE) and MBE of VOCs concentration prediction result.

Training Algorithm	Hidden Layers-1		Hidden Layers-3		Hidden Layers-5	
	CV(RMSE) (%)	MBE (%)	CV(RMSE) (%)	MBE (%)	CV(RMSE) (%)	MBE (%)
LM	10.81	8.89	10.98	9.35	11.12	8.96
BR	11.06	9.10	14.11	10.19	15.65	10.37
BFG	14.57	10.68	14.58	10.79	26.52	11.25

Figure 8 presents the R^2 of the VOCs prediction results. When the number of the hidden layer is one, the LM model shows the highest suitability among the models. When the number of the hidden layer is increased to 5, the R^2 for the LM model is 0.9984. While the BR model shows a relatively lower R^2 (0.9998) than that of the LM model, it is highly suitable. When the number of the hidden layers is increased, the R^2 for the BR model is ranging from 0.9946 to 0.9966. In the case of the BFG model, the R^2 is 0.9974 when the number of the hidden layer is one. However, it is decreased to 0.9445 with the increase in the number of the hidden layers.

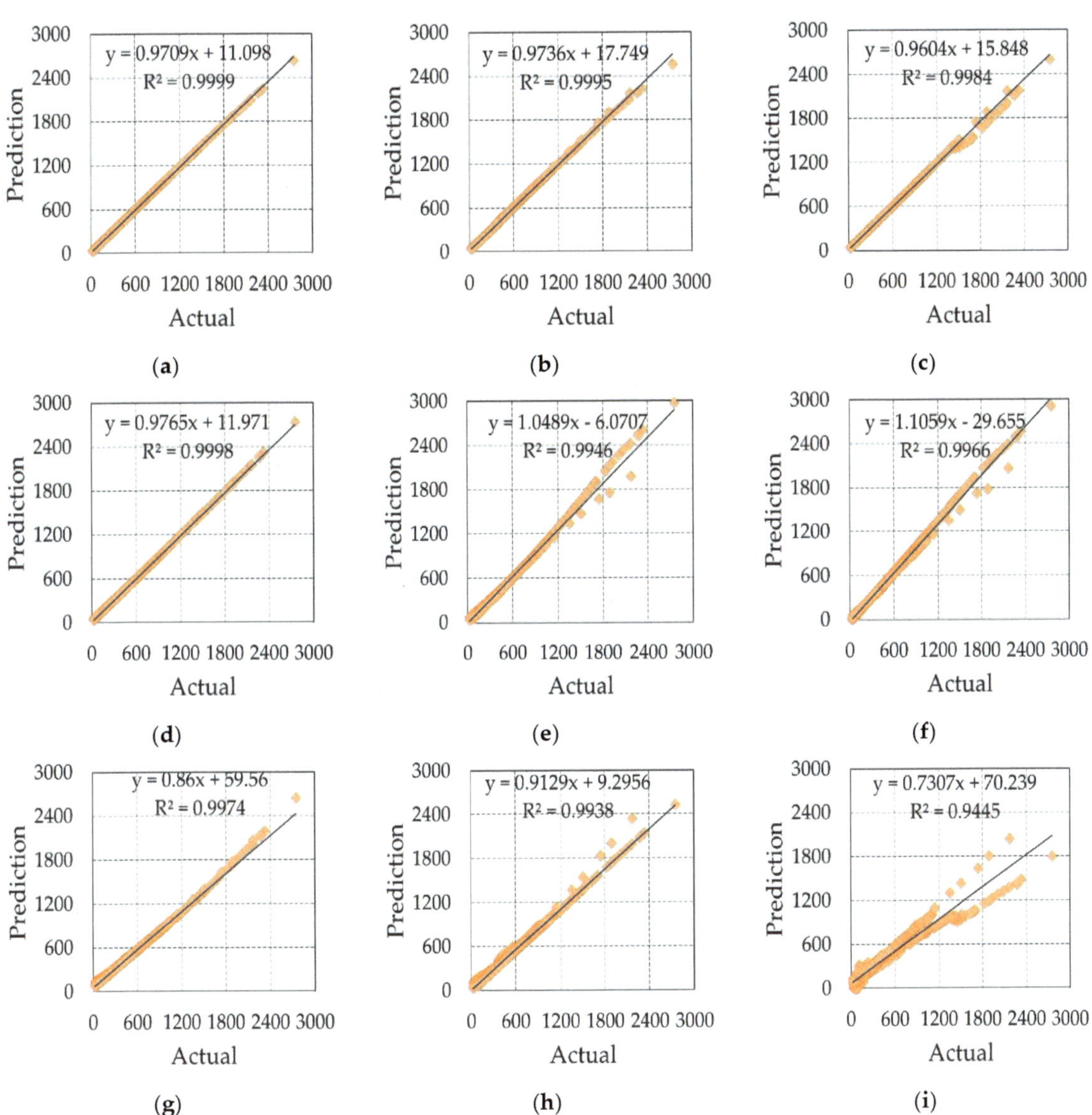

Figure 8. R^2 of prediction results of VOCs concentration. (**a**) LM-Hidden layer-1; (**b**) LM-Hidden layer-3; (**c**) LM-Hidden layer-5; (**d**) BR-Hidden layer-1; (**e**) BR-Hidden layer-3; (**f**) BR-Hidden layer-5; (**g**) BFG-Hidden layer-1; (**h**) BFG-Hidden layer-3; (**i**) BFG-Hidden layer-5.

5. Discussion

For the present study, the concentration of CO_2, $PM_{2.5}$, and VOCs was predicted using the feed-forward neural network model with three different back-propagation training algorithms, which were LM, BR, and BFG. Among the models, the LM model showed the best performance for the prediction of those indoor pollutants and the obtained results can meet the acceptable criteria of the ASHRAE guidelines 14. While the prediction for CO_2 and $PM_{2.5}$ by the BR model showed a good performance, the performance was decreased as the number of the hidden layers was increased in the case of VOCs concentration prediction.

The BFG model showed the lowest performance for the prediction of indoor pollutants and it was decreased largely as with the increase in the number of the hidden layers.

It was commonly observed that the prediction performance decreased as the number of the hidden layers increased. As shown, the number of the hidden layers in the ANN model can highly affect the prediction results. In addition, over-fitting data can be produced as the number of the hidden layers is increased [51]. Thus, the choice of the number of the hidden layers is one of the most important variables in the ANN model [52]. According to the study of Yeon et al., the number of hidden layers can be determined, when the CV(RMSE) is lowest [53]. In their study, the number of hidden layers was tested from one to three layers and one layer was selected based on the lowest CV(RMSE) value. This was similarly observed in the present study. Bui et al. also chose one hidden layer for identifying the impact of each building parameter on the energy performance [51]. In the case of the ANN model performed by Cho and Moon, they proved that the structured model with two hidden layers was suitable to predict the concentrations of CO_2, PM_{10}, and $PM_{2.5}$ in a school building [54]. As can be shown, the optimized number of hidden layers can be different based on the structured ANN model and it is carefully chosen by assessment of the prediction performance.

Moreover, the prediction performance differed based on the type of indoor pollutant. Based on the CV(RMSE), all models showed the best performance for the prediction of CO_2 concentration, while the poorest performance was shown for the prediction of $PM_{2.5}$ concentration. This can be caused by the difference in data patterns. In the case of the CO_2, the measured CO_2 concentration showed a regular pattern. For the measured concentration of $PM_{2.5}$ and VOCs, the data showed irregular patterns due to the outdoor pollutants through building openings. This can affect the prediction results in the training phase causing a low degree of dispersion. While these irregular data patterns caused low suitability, the high fitness of the R^2 (0.99) was achieved for all the models when the number of the hidden layer was one.

6. Conclusions

As the amount of time spent by people indoors is significantly increasing, much attention has been paid to the indoor air quality in buildings. Many studies have investigated indoor air pollutants to improve the IAQ. While most investigations were performed in residential or commercial buildings, a few studies have focused on the IAQ in child daycare centers in which, a high concentration of indoor air pollutants was reported. Moreover, investigations of indoor air pollutants, in general, were conducted through field measurements which are cost-intensive and time-consuming.

For the present study, the concentrations of indoor air pollutants such as CO_2, $PM_{2.5}$, and VOCs in child daycare centers were predicted by using a feed-forward neural network model with three different back-propagation training algorithms such as LM, BR, and BFG. For the training and validation, the data of the indoor pollutants measured in child daycare facilities for a month were used. The number of hidden layers in each model was set at one, three, and five, and other training parameters were the same for all models.

The results showed that all the models produced a good performance for the prediction of indoor pollutants compared with the measured data. Among the models, the prediction by the LM model met the acceptable criteria of the ASHRAE guideline 14 under all conditions. While the CO_2 and $PM_{2.5}$ concentrations predicted by the BR model satisfied the acceptance criteria of the ASHRAE guideline 14, the predictive performance decreased, when the number of the hidden layers increased. The BFG model showed the poorest performance for the prediction of indoor pollutants among the models under all conditions. Moreover, a large difference between the prediction and the measured data was observed by the BFG model when the number of the hidden layers increased.

While the predictions by using machine learning techniques have been widely used in various fields, there were few studies available for indoor air pollutant prediction. In addition, the predictive performance by different training algorithms was rarely investigated.

This study presented the difference in the predictive performance of indoor air pollutants by applying different training algorithms in the ANN model. Considering the outcomes of the study, better prediction results can be obtained through the proper selection of training algorithms for time series data. Furthermore, the outcome of the present study can be used as information for more sophisticated machine learning applications for improving indoor air quality and related predictions in child daycare centers. For further studies, other indoor pollutants such as CO and PM_{10}, etc. will be investigated for improving indoor air quality in child daycare centers.

Author Contributions: J.K. contributed to the project idea development and wrote a draft version; Y.H. prepared the data and contributed to the results; N.S. performed the data analysis; D.D.K. reviewed the final manuscript and contributed to the discussion. All authors have read and agreed to the published version of the manuscript.

Funding: This research was supported by Basic Science Research Program through the National Research Foundation of Korea(NRF) funded by the Ministry of Education (2021R1I1A1A01056761).

Institutional Review Board Statement: Not applicable.

Informed Consent Statement: Not applicable.

Data Availability Statement: Not applicable.

Conflicts of Interest: The authors declare no conflict of interest regarding the publication of this article.

References

1. Awada, M.; Becerik-Gerber, B.; Hoque, S.; O'Neill, Z.; Pedrielli, G.; Wen, J.; Wu, T. Ten questions concerning occupant health in buildings during normal operations and extreme events including the COVID-19 pandemic. *Build. Environ.* **2021**, *188*, 107480. [CrossRef]
2. Adam, M.G.; Tran, P.T.M.; Balasubramanian, R. Air quality changes in cities during the COVID-19 lockdown: A critical review. *Atmos. Res.* **2021**, *264*, 105823. [CrossRef] [PubMed]
3. Schibuola, L.; Tambani, C. High energy efficiency ventilation to limit COVID-19 contagion in school environments. *Energy Build.* **2021**, *240*, 110882. [CrossRef] [PubMed]
4. Sha, H.; Zhang, X.; Qi, D. Optimal control of high-rise building mechanical ventilation system for achieving low risk of COVID-19 transmission and ventilative cooling. *Sustain. Cities Soc.* **2021**, *74*, 103256. [CrossRef]
5. Gil-Baez, M.; Lizana, J.; Becerra Villanueva, J.A.; Molina-Huelva, M.; Serrano-Jimenez, A.; Chacartegui, R. Natural ventilation in classrooms for healthy schools in the COVID era in mediterranean climate. *Build. Environ.* **2021**, *206*, 108345. [CrossRef]
6. Szczepanik-Ścisło, N.; Ścisło, Ł. Air leakage modelling and its influence on the air quality inside a garage. *E3S Web Conf.* **2018**, *44*, 00172. [CrossRef]
7. Ye, W.; Won, D.; Zhang, X. A preliminary ventilation rate determination methods study for residential buildings and offices based on voc emission database. *Build. Environ.* **2014**, *79*, 168–180. [CrossRef]
8. Földváry, V.; Bekö, G.; Langer, S.; Arrhenius, K.; Petráš, D. Effect of energy renovation on indoor air quality in multifamily residential buildings in slovakia. *Build. Environ.* **2017**, *122*, 363–372. [CrossRef]
9. Kaunelienė, V.; Prasauskas, T.; Krugly, E.; Stasiulaitienė, I.; Čiužas, D.; Šeduikytė, L.; Martuzevičius, D. Indoor air quality in low energy residential buildings in lithuania. *Build. Environ.* **2016**, *108*, 63–72. [CrossRef]
10. Yang, S.; Yuk, H.; Yun, B.Y.; Kim, Y.U.; Wi, S.; Kim, S. Passive pm2.5 control plan of educational buildings by using airtight improvement technologies in south korea. *J. Hazard. Mater.* **2022**, *423*, 126990. [CrossRef]
11. Bralewska, K.; Rogula-Kozłowska, W.; Bralewski, A. Indoor air quality in sports center: Assessment of gaseous pollutants. *Build. Environ.* **2022**, *208*, 108589. [CrossRef]
12. Connolly, R.E.; Yu, Q.; Wang, Z.; Chen, Y.-H.; Liu, J.Z.; Collier-Oxandale, A.; Papapostolou, V.; Polidori, A.; Zhu, Y. Long-term evaluation of a low-cost air sensor network for monitoring indoor and outdoor air quality at the community scale. *Sci. Total Environ.* **2022**, *807*, 150797. [CrossRef]
13. Mikola, A.; Hamburg, A.; Kuusk, K.; Kalamees, T.; Voll, H.; Kurnitski, J. The impact of the technical requirements of the renovation grant on the ventilation and indoor air quality in apartment buildings. *Build. Environ.* **2022**, *210*, 108698. [CrossRef]
14. Kang, I.; McCreery, A.; Azimi, P.; Gramigna, A.; Baca, G.; Abromitis, K.; Wang, M.; Zeng, Y.; Scheu, R.; Crowder, T.; et al. Indoor air quality impacts of residential mechanical ventilation system retrofits in existing homes in chicago, il. *Sci. Total Environ.* **2022**, *804*, 150129. [CrossRef]
15. Al-Rawi, M.; Ikutegbe, C.A.; Auckaili, A.; Farid, M.M. Sustainable technologies to improve indoor air quality in a residential house—A case study in waikato, new zealand. *Energy Build.* **2021**, *250*, 111283. [CrossRef]
16. Oh, H.-J.; Nam, I.-S.; Yun, H.; Kim, J.; Yang, J.; Sohn, J.-R. Characterization of indoor air quality and efficiency of air purifier in childcare centers, korea. *Build. Environ.* **2014**, *82*, 203–214. [CrossRef]

17. Roda, C.; Barral, S.; Ravelomanantsoa, H.; Dusséaux, M.; Tribout, M.; Le Moullec, Y.; Momas, I. Assessment of indoor environment in paris child day care centers. *Environ. Res.* **2011**, *111*, 1010–1017. [CrossRef]
18. Hwang, S.H.; Seo, S.; Yoo, Y.; Kim, K.Y.; Choung, J.T.; Park, W.M. Indoor air quality of daycare centers in seoul, korea. *Build. Environ.* **2017**, *124*, 186–193. [CrossRef]
19. Madureira, J.; Paciência, I.; Rufo, J.C.; Pereira, C.; Teixeira, J.P.; de Oliveira Fernandes, E. Assessment and determinants of airborne bacterial and fungal concentrations in different indoor environments: Homes, child day-care centres, primary schools and elderly care centres. *Atmos. Environ.* **2015**, *109*, 139–146. [CrossRef]
20. St-Jean, M.; St-Amand, A.; Gilbert, N.L.; Soto, J.C.; Guay, M.; Davis, K.; Gyorkos, T.W. Indoor air quality in montréal area day-care centres, canada. *Environ. Res.* **2012**, *118*, 1–7. [CrossRef]
21. Harbizadeh, A.; Mirzaee, S.A.; Khosravi, A.D.; Shoushtari, F.S.; Goodarzi, H.; Alavi, N.; Ankali, K.A.; Rad, H.D.; Maleki, H.; Goudarzi, G. Indoor and outdoor airborne bacterial air quality in day-care centers (dccs) in greater ahvaz, iran. *Atmos. Environ.* **2019**, *216*, 116927. [CrossRef]
22. Zuraimi, M.S.; Tham, K.W. Indoor air quality and its determinants in tropical child care centers. *Atmos. Environ.* **2008**, *42*, 2225–2239. [CrossRef]
23. Araújo-Martins, J.; Carreiro Martins, P.; Viegas, J.; Aelenei, D.; Cano, M.M.; Teixeira, J.P.; Paixão, P.; Papoila, A.L.; Leiria-Pinto, P.; Pedro, C.; et al. Environment and health in children day care centres (envirh)—Study rationale and protocol. *Rev. Port. De Pneumol.* **2014**, *20*, 311–323. [CrossRef] [PubMed]
24. Heibati, S.; Maref, W.; Saber, H.H. Assessing the energy, indoor air quality, and moisture performance for a three-story building using an integrated model, part two: Integrating the indoor air quality, moisture, and thermal comfort. *Energies* **2021**, *14*, 4915. [CrossRef]
25. Olu-Ajayi, R.; Alaka, H.; Sulaimon, I.; Sunmola, F.; Ajayi, S. Machine learning for energy performance prediction at the design stage of buildings. *Energy Sustain. Dev.* **2022**, *66*, 12–25. [CrossRef]
26. Elbeltagi, E.; Wefki, H. Predicting energy consumption for residential buildings using ann through parametric modeling. *Energy Rep.* **2021**, *7*, 2534–2545. [CrossRef]
27. Olu-Ajayi, R.; Alaka, H.; Sulaimon, I.; Sunmola, F.; Ajayi, S. Building energy consumption prediction for residential buildings using deep learning and other machine learning techniques. *J. Build. Eng.* **2022**, *45*, 103406. [CrossRef]
28. Zhu, Y.F.; Yao, Y.; Huang, Y.; Chen, C.H.; Zhang, H.Y.; Huang, Z. Machine learning applications for assessment of dynamic progressive collapse of steel moment frames. *Structures* **2022**, *36*, 927–934. [CrossRef]
29. Chalapathy, R.; Khoa, N.L.D.; Sethuvenkatraman, S. Comparing multi-step ahead building cooling load prediction using shallow machine learning and deep learning models. *Sustain. Energy Grids Netw.* **2021**, *28*, 100543. [CrossRef]
30. Ding, Y.; Fan, L.; Liu, X. Analysis of feature matrix in machine learning algorithms to predict energy consumption of public buildings. *Energy Build.* **2021**, *249*, 111208. [CrossRef]
31. Zhang, H.; Srinivasan, R.; Yang, X. Simulation and analysis of indoor air quality in florida using time series regression (tsr) and artificial neural networks (ann) models. *Symmetry* **2021**, *13*, 952. [CrossRef]
32. Amuthadevi, C.; Ds, V.; Ramachandran, V. Development of air quality monitoring (aqm) models using different machine learning approaches. *J. Ambient Intell. Humaniz. Comput.* **2021**, 1–13. [CrossRef]
33. Tagliabue, L.C.; Re Cecconi, F.; Rinaldi, S.; Ciribini, A.L.C. Data driven indoor air quality prediction in educational facilities based on iot network. *Energy Build.* **2021**, *236*, 110782. [CrossRef]
34. Jang, H.-S.; Xing, S. A model to predict ammonia emission using a modified genetic artificial neural network: Analyzing cement mixed with fly ash from a coal-fired power plant. *Constr. Build. Mater.* **2020**, *230*, 117025. [CrossRef]
35. Jeong, J.-H.; Jo, J.-H.; Kim, S.-H.; Bang, S.-H.; Lee, S.-S. A study on machine learning model for predicting uncollected parameters in indoor environment evaluation. *J. Korea Inst. Inf. Electron. Commun. Technol.* **2021**, *14*, 413–420.
36. Scislo, L.; Szczepanik-Scislo, N. Air quality sensor data collection and analytics with iot for an apartment with mechanical ventilation. In Proceedings of the 2021 11th IEEE International Conference on Intelligent Data Acquisition and Advanced Computing Systems: Technology and Applications (IDAACS), Cracow, Poland, 22–25 September 2021; pp. 932–936.
37. Li, Z.; Tong, X.; Ho, J.M.W.; Kwok, T.C.Y.; Dong, G.; Ho, K.-F.; Yim, S.H.L. A practical framework for predicting residential indoor pm2.5 concentration using land-use regression and machine learning methods. *Chemosphere* **2021**, *265*, 129140. [CrossRef] [PubMed]
38. Kallio, J.; Tervonen, J.; Räsänen, P.; Mäkynen, R.; Koivusaari, J.; Peltola, J. Forecasting office indoor co2 concentration using machine learning with a one-year dataset. *Build. Environ.* **2021**, *187*, 107409. [CrossRef]
39. Taheri, S.; Razban, A. Learning-based co2 concentration prediction: Application to indoor air quality control using demand-controlled ventilation. *Build. Environ.* **2021**, *205*, 108164. [CrossRef]
40. Sassi, M.S.H.; Fourati, L.C. In Deep learning and augmented reality for iot-based air quality monitoring and prediction system. In Proceedings of the 2021 International Symposium on Networks, Computers and Communications (ISNCC), Dubai, United Arab Emirates, 31 October–2 November 2021; pp. 1–6.
41. Saad, S.M.; Andrew, A.M.; Shakaff, A.Y.M.; Saad, A.R.M.; Kamarudin, A.M.Y.; Zakaria, A. Classifying sources influencing indoor air quality (iaq) using artificial neural network (ann). *Sensors* **2015**, *15*, 11665–11684. [CrossRef]
42. Putra, J.C.P.; Safrilah; Ihsan, M. The prediction of indoor air quality in office room using artificial neural network. *AIP Conf. Proc.* **2018**, *1977*, 020040.

43. Dai, X.; Liu, J.; Zhang, X.; Chen, W. An artificial neural network model using outdoor environmental parameters and residential building characteristics for predicting the nighttime natural ventilation effect. *Build. Environ.* **2019**, *159*, 106139. [CrossRef]
44. Segala, G.; Doriguzzi Corin, R.; Peroni, C.; Gazzini, T.; Siracusa, D. A practical and adaptive approach to predicting indoor co2. *Appl. Sci.* **2021**, *11*, 10771. [CrossRef]
45. Big Data Environment Platform. Available online: https://www.Bigdata-environment.Kr/user/main.Do (accessed on 1 November 2021).
46. Matlab, The Mathworks, Inc. Available online: https://www.Mathworks.Com/?S_tid=gn_logo (accessed on 1 December 2021).
47. Golafshani, E.M.; Behnood, A. Application of soft computing methods for predicting the elastic modulus of recycled aggregate concrete. *J. Clean. Prod.* **2018**, *176*, 1163–1176. [CrossRef]
48. Koçer, M.; Öztürk, M.; Hakan Arslan, M. Determination of moment, shear and ductility capacities of spiral columns using an artificial neural network. *J. Build. Eng.* **2019**, *26*, 100878. [CrossRef]
49. Wang, R.; Lu, S.; Li, Q. Multi-criteria comprehensive study on predictive algorithm of hourly heating energy consumption for residential buildings. *Sustain. Cities Soc.* **2019**, *49*, 101623. [CrossRef]
50. American Society of Heating, Refrigerating and Air Conditioning Engineers. Ashrae Guideline 14-2002, Measurement of Energy and Demand Savings—Measurement of Energy, Demand and Water Savings. 2002. Available online: http://www.eeperformance.org/uploads/8/6/5/0/8650231/ashrae_guideline_14-2002_measurement_of_energy_and_demand_saving.pdf (accessed on 1 February 2021).
51. Bui, D.-K.; Nguyen, T.N.; Ngo, T.D.; Nguyen-Xuan, H. An artificial neural network (ann) expert system enhanced with the electromagnetism-based firefly algorithm (efa) for predicting the energy consumption in buildings. *Energy* **2020**, *190*, 116370. [CrossRef]
52. Lu, C.; Li, S.; Lu, Z. Building energy prediction using artificial neural networks: A literature survey. *Energy Build.* **2022**, *262*, 111718. [CrossRef]
53. Yeon, S.; Yu, B.; Seo, B.; Yoon, Y.; Lee, K.H. Ann based automatic slat angle control of venetian blind for minimized total load in an office building. *Sol. Energy* **2019**, *180*, 133–145. [CrossRef]
54. Cho, J.H.; Moon, J.W. Integrated artificial neural network prediction model of indoor environmental quality in a school building. *J. Clean. Prod.* **2022**, *344*, 131083. [CrossRef]

MDPI AG
Grosspeteranlage 5
4052 Basel
Switzerland
Tel.: +41 61 683 77 34

Energies Editorial Office
E-mail: energies@mdpi.com
www.mdpi.com/journal/energies